THE TRUTH OF
# THE ORIGIN OF
# THE UNIVERSE
## THE ORIGIN OF THE UNIVERSE - CREATION, SCIENCE FICTION OR SCIENCE MADNESS?

VOLUME-1

# THE TRUTH OF
# THE ORIGIN OF THE UNIVERSE

## THE ORIGIN OF THE UNIVERSE - CREATION, SCIENCE FICTION OR SCIENCE MADNESS?

**Dr. Sabrie Soloman**
*Ph.D., Sc.D., MBA, PE*

**KHANNA BOOK PUBLISHING CO. (P) LTD.**
Publisher of Engineering and Computer Books
4C/4344, Ansari Road, Darya Ganj, New Delhi-110002
**Phone** : 011-23244447-48    **Mobile:** +91-9910909320
**E-mail** : contact@khannabooks.com
**Website :** www.khannabooks.com

**ISBN:** 978-93-5538-278-8

**THE TRUTH OF THE ORIGIN
OF THE UNIVERSE (VOLUME-1)**
*by* **Dr. Sabrie Soloman**

**First Edition:** Dec 2024

*Published by:*
**Khanna Book Publishing Co. (P) Ltd.**
*CIN: U22110DL1998PTC095547*

*Visit us at:* www.khannabooks.com
*Write us at:* contact@khannabooks.com

To view complete list of books,
please scan the QR Code:

# DEDICATIONS

*William Thomson* – The voice of his words of wisdom echoed through the core of the inner chambers of my heart, engraving channels of living knowledge, which generated floods of insight to illuminate my life`s journey.

*William Thomson`s* pure and innocent soul has reached the gates of heavens to bring me knowledge of Eternity. His living loving example compelled me to follow even his shadow to touch Eternity, while my feet are still on Earth!

His knowledge imparted on my mind inspiring me to write the "**The Origin of The Universe**" book – distinct from most other science books, which composed by worldwide prominent scientists highlighting unjustified and unproven hypotheses. Through **William Thomson** the factual truth of the Universe is recognized and logically revealed to my modest intellect to compose "**The Truth of The Origin of The Universe**," Thus, I humbly dedicate this book to him.

*Joseph Rondinelli* – While the size of the universe may not be comprehended, so also the goodness of the heart of my friend Joseph Rondinelli may not be grasped. Joseph permitted me to touch the limitless boundaries of the goodness of his loving nature. His friendship with me permitted me to observe a living loving example to follow.

*Joseph Rondinelli* is bountifully living loving those whom he cherished. He brought heavens highest values to my humble living earth. While I was temporarily incapacitated, he was my feet, hands, and even eyes sitting for hours by my side, debating our intellectual capacities to understand life under the domain of our **Creator, The Origin of Our Universe**.

When my mind acted as an interrupted volcano spouting novel thoughts of **The Origin of The Universe**, he came to my dwelling to rescue me out of the dense-smoke of my scattered knowledge-suffocating my mind. He instantly changed my environment to a more tranquil atmosphere enabling me once again recollect my thoughts, helping me to **Create Order out of Chaos**. For this and much more I am forever, indebted.

# FOREWORD

Shortly before Frances R. Havergal, at age 36, wrote the hymn, "Lord, Speak to Me" she wrote in a letter, "… if I am to write any good, a great deal of living must go to a very little writing." Before putting pen to paper, a great deal of living has gone into what you are about to read.

When Dr. Soloman asked me to write a forward to his book on the origin of the universe, I was humbled and honored because this book is of unusual importance by an author with an exceptional life of service and greatly respected by all who know him and his work. I pray that those who read his 2,800 pages, 32 chapters will find its message clearly stated, giving clarity to a subject that has too long been delt with on a surface and non-scientific level.

Understanding the origin of the universe is key to understanding everything else and developing a truly Biblical worldview. A person's view on this subject is a window by which we view the world. Dr. Soloman gives us a verifiable frame to evaluate whether his findings are flawed or something that is understandably sensible to merit serious consideration.

When we tell others our view on evolution is based on scientific evidence rather than solely on the book of Genesis, we have a much stronger argument. Dr. Soloman builds the case that evolution, with his accompanying philosophy, is identified with a worldview at such a level that readers must sincerely struggle with how the theory could go against the evidence.

This book can train a whole new generation to confront and survive the worldview challenges they encounter all their life. Those of you who read this book in its entirety will be able to survive in a post-modern culture without being brainwashed by prejudice, faulty thinking, and preconceived assumptions.

Dr. Soloman has made exciting reading for the ordinary person, but it is thoroughly researched by scholarship to present a deep understanding of the subject. The situation we often find ourselves in is flawed thinking and lazy scholarship. To that end, I invite you to read Dr. Soloman's book which will prove to not only be enjoyable reading, but one of the most prolific and insightful tomes you will ever read on the origins of the universe.

**Dr. Peter W. Teague,**
President Emeritus
Lancaster Bible College | Capital Seminary & Graduate School

# PREFACE

## The Universe Had a Beginning

The scientific consent 100 years ago was that the universe was everlasting. This idea started to unravel with the inferences of Albert Einstein's Theory of Relativity back in 1916, where his equations pointed to an expanding universe. Yet he didn't like that conclusion and so supplemented a constant to his equation that invalidated the expansion. Later, he admitted it had been the major mathematical blunder in his life.

Then in 1929 astronomer Edwin Hubble acknowledged he observed galaxies expanding outward, which meant they had been much closer together in the past. Einstein, fascinated, wanted to see the indication for himself and in 1931 he went to the Mt. Wilson Observatory in Los Angeles, California. Einstein examined through the telescope, observed the evidence and then concluded, "I now see *the necessity* of a beginning." This began a change in the scientific boldness toward the cosmos.

Years after, in 1965, two U.S. scientists distinguished the leftovers of the original burst of energy of the creation occurrence characteristically so-called the "Big Bang." They both won a Nobel Prize in physics. One of them, Arno Penzias, later professed, "The best data we have [about the beginning of the universe] are exactly what I would have predicted had I nothing to go on but the first five books of Moses, the Psalms and the Bible as a whole" ("Clues to Universe Origin Expected," *The New York Times*, March 12, 1978, p. 1).

With the indication at hand, what was written in Genesis 1:1 truly surprised many scientists by its precision: "In the beginning, God created the heavens and the earth." Here it says the universe of matter and energy *appeared at a certain point in time* and was all created by an Ultimate Creator who existed before all of this occurred. It was an enormous proof of God's existence, with no real substitute enlightenment for a universe that, according to modern physics, *appeared out of nothing.*

## The Universe is Fine-Tuned for Life

Almost 50 years ago, in 1973, cosmologist Brandon Carter instituted that the independent constants or laws in physics have one exceedingly rare characteristic in common—*they are precisely the values needed to establish and sustain a universe capable of producing life.* This is another vast and virtually accepted proof for a universe that has been prudently designed.

Scientists have instituted some 30 constants or laws of physics that rule the universe. All are dissimilar to each other and yet are finely tuned to unbelievable proportions to make life possible. The indication points to "Someone" spending much too much time tuning all of these laws so they would work in unison.

Amazingly, the Bible exposed this truth long before any scientist revealed these facts. As Jeremiah 33:25 states, "But I, the Lord, have a covenant with day and night, and *I have made the laws* that *control* earth and sky" (*Good News Translation*).

## The Origin of Life – The Genetic Code

Opposing to what many have been led to trust, scientists have no genuine explanation for how life arose.

Even the well-known atheist and evolutionist Richard Dawkins acknowledged concerning the appearance of life, "*Nobody knows how it happened*" (*Climbing Mount Improbable,* 1996, p. 282). Moreover, one of the pioneers of the DNA code, the atheist Francis Crick, determined, "An honest man, armed with all the knowledge available to us now, could only state that in some sense, *the origin of life* appears at the moment to be *almost a miracle,* so many are the environments which would have had to have been gratified to get it going" (*Life Itself: Its Origin and Nature,* 1981, p. 88).

In the past 60 years, biologists have found that life started with a massive amount of accurate information *already embedded* in the cell. The human genome unaided is a molecule with almost 3 billion genetic letters, all precisely well-organized to give instructions to the cell. Furthermore, scientists have never instituted inorganic matter to generate a coded system of information and the machinery to interpret it. From the most primitive cells to human beings, all have the same elementary operating system of mind-boggling intricacy, with codes, transmitters and receivers all working together.

Furthermore, the origin of life aenigma has a "chicken-and-egg question"—which came first, the chicken or the egg? In this case, to get life to transpire, you need *both* the complete genetic code and the proteins—the machine parts—that read the code and build new proteins. Deprived of the code, you can't shape proteins. And without proteins, you can't process the code. So how could both have risen at the same time?

## Biological life Exquisitely Programmed – "Robotic Machines"

In order to comprehend what is fashioning inside a cell, a good illustration is picturing a large city swarming with life and movement.

Biochemist Michael Denton defines the cell this way: "To clutch the reality of life as it has been discovered by molecular biology, we must amplify a cell a billion times until it is twenty kilometers in diameter and looks like a giant airship large enough to cover a great city like London or New York. What we would then see would be an entity of unmatched intricacy and adaptive design . . .

"We would see around us, in every direction we looked, all sorts of robot-like machines. We would notice that the simplest of the functional components of the cell, the protein molecules, were astonishingly complex pieces of molecular machinery, each one consisting of about three thousand atoms arranged in highly organized 3-D spatial conformation.

"We would wonder even more as we watched the strangely purposeful activities of these weird molecular machines, particularly when we realized that, despite all our accumulated knowledge of physics and chemistry, the task of designing one such molecular machine—that is one single functional protein molecule—would be completely beyond our capacity at present" (*Evolution: A Theory in Crisis,* 1986, p. 329).

This is the reason biochemists have a hard time have faith in explaining that blind evolution can build such machinery—and obtain all the parts to work together from the start. Furthermore, to keep the human body operational, biologists calculate that "about *330 billion cells* are substituted *daily,* equivalent to about 1 percent of all our cells" (Mark Fischetti, "Our Bodies Replace Billions of Cells Every Day," *Scientific American,* April 1, 2021).

"We take life for granted," adds Douglas Ell, "because it is everywhere. Our planet is overrun by *biological machines.* There are at least 10 million different types (species) of machines; some estimate that tens of millions of other types (species) have not yet been discovered . . .

"Coordinated systems permit blue whales to dive thousands of feet below sea level without being crumpled and sing complex songs that travel across oceans. Other systems guide bees to do a dance that tells other bees where to find the best sources of pollen. There are systems for hiding, systems for fighting, systems for reproducing, systems for obtaining food, systems for communicating, and so on" (*Counting to God,* p. 110).

Such findings show that everything about life is programmed to the last detail and that virtually nothing has been left to chance. Does this superb design point to evolution or to God? The answer is clear.

## The Earliest Evidence of Life

Though Darwin titled his book *On the Origin of Species by Means of Natural Selection,* he was never able to substantiate that assumption. Many people adopt that the theory of evolution, with its countless mutations and natural selection as the means of that change, can account for the origin and development of all the living things on this planet. Yet this is sleight of hand, since evolution can account for *micro*evolution, or changes *within* the species (such as dogs of varying sizes, shapes and colors), but not *macro*evolution, or changes from one kind of creature to another. Natural selection can tell you something about the *survival* of the species, but nothing about the *arrival* of the species. It undoubtedly cannot trace the origin of the approximately *10 million species* on Earth. These are classified into some 33 main body types or phyla, such as sponges, worms, insects and mammals.

Darwin foretold that as more of the fossil record was revealed, it would display types of species gradually appearing, beginning with one or a few, and then multiplying from simple to more complex life forms. He wrote, "If numerous species . . . have really *started into life at once,* the fact would be *fatal* to the theory of evolution through natural selection" (*Origin of Species,* 1859, p. 305). Yet that is precisely what has been found—major body types appearing at what's considered the *beginning* of the fossil record rather than in deposits laid down later.

Scientists call this "the Cambrian Explosion," referring to major types of plants and animals suddenly appearing *fully formed* in that fossil layer. This is the *opposite* of what Darwin and evolutionists had claimed would be found—and they have no real explanation or answers. Of the 33 main body types, 23 of them (or 70%) appear at the recognized *beginning* stage of the fossil record.

The issue in question here, by analogy, would be as discovering together such diverse inventions as a washing machine, a refrigerator, a bicycle, a car and an airplane. While they do have some shared features, they have very discrete purposes and resolutions. Likewise, the major types of creatures found in the Cambrian layer, such as sponges, worms, trilobites and jawless fish, are very diverse, complex and appear suddenly, with no evidence of these main body types evolving from other creatures.

As paleontologist Niles Eldredge admitted: "If life had evolved into the wondrous profusion of creatures little by little, then there should be some fossiliferous record of those changes . . . But no one has found any evidence of such in-between creatures . . . All of the fossil evidence to date has failed to turn up any such missing links" (George Alexander, "Alternate Theory of Evolution Considered," *Los Angeles Times,* Nov. 19, 1978). Indeed, Darwin has been let down by the fossil record!

## Earth Has "Just Right" Conditions to Sustain Life

In 1966, Carl Sagan presented the famous TV documentary series *Cosmos*. He believed in order to have life you just needed two conditions—a right kind of star and a planet at the right distance. This supposition showed to be totally baseless.

Now, more than 50 years later, scientists have come to the comprehension that *more than 200 conditions* have to be "just right" for life to exist and thrive. The probabilities are about 1 in $10^{2,685,000}$. The probability of that happening comes out at about **1 in $10^{2,685,000}$**, or 10 followed by **2,685,000** zeros. For comparison, the Universe only has $10^{80}$ atoms. The infographic finishes by letting you know that the probability of you existing as you is pretty much zero.

As author Eric Metaxas explains: "Today there are more than 200 known parameters necessary for a planet to support life—*every single one of which must be perfectly met, or the whole thing falls apart.* Without a massive planet like Jupiter nearby, whose gravity will draw away asteroids, a thousand times as many would hit Earth's surface. The odds against life in the universe are simply astonishing" ("Science Increasingly Makes the Case for God," *The Wall Street Journal,* Dec. 25, 2014).

The Bible tells us: "The Lord is God. He made the skies and the earth. He put the earth in its place. He did not want the earth to be empty when he made it. He created it *to be lived on.* I am the Lord. There is no other God" (*Isaiah 45:18, Easy-to-Read Version*).

## The Universe Mathematically Designed

Amazingly, the universe has been discovered to be mathematically designed. It follows orderly laws that can be described in mathematical terms. Sir James Jeans, one of the great astronomers of the 20th century, remarked: "From the intrinsic evidence of his creation, the Great Architect of the Universe *now begins to appear as a pure mathematician . . .* The universe begins to look more like a great thought than like a great machine" (*The Mysterious Universe,* 1930, pp. 134, 137).

A substantial problem for evolutionists and atheists is this: *Evolution can't do math,* since it is founded on random variations and mutations, and math necessitates an intelligent agent who can first formulate a mathematical blueprint of laws before creating things so they will be logical. This is why the present cosmos can be traced back to mathematical rules.

As Einstein distinguished, "The most incomprehensible thing about the universe is that it is comprehensible." He meant that it could be understood in *mathematical* terms but that an explanation for that was outside math.

As far back as the early 1900s, scientists were discovering the laws that govern the subatomic realm, the tiny microcosm designated by quantum mechanics. It has very diverse rules than our macro world and appears to make room for such things as free will to arise.

Many scientists came to comprehend that *not all is determined by matter and energy.* Experiments illustrate that an observer can change a particle through means of observing it. The consequences are that we can determine the outcome of our lives by the choices we make.

It brings to mind what God said: "*Today I have given you a choice between life and death, success and disaster. I command you today to love the Lord your God. I command you to follow him and to obey his commands, laws, and rules. Then you will live . . .*" (*Deuteronomy 30:15-16, ERV*).

## Science Points to The Existence of God

Cutting-edge science is continually showing more intricacy and deeper design, not only in the cosmos, but also in all living things.

The biblical patriarch Job once challenged skeptics to look at the design of the creatures around them and notice they witness to a Supreme Designer and Creator. He stated: "Even birds and animals have much they could *teach* you; ask the creatures of earth and sea for their wisdom. All of them know that *the Lord's hand made them*" (*Job 12:7-9, GNT*).

Therefore, by exploring all the evidence and grasping where it leads, we hope you will believe in God, the Creator and the "Origin of the Universe." *K*eep believing in Him, and earnestly seek His will for your life!

**Sabrie Soloman**

# ACKNOWLEDGMENTS

The Author is extremely grateful for the opportunity to express his heartfelt acknowledgments to the individuals who have played a vital role in the successful publication of "The Truth of The Origin of The Universe" book by Khanna Book Publishing Company.

First and foremost, the author extends his deepest gratitude to the Director and Editor-in-Chief, Mr. Mukul Seth, for his unwavering dedication and expertise in shaping the book's narrative. His keen eye for detail and impeccable editing skills have significantly enriched the content and ensured its quality.

Also, the author expresses his sincere appreciation to the Editorial Team, comprising of Mr. Mukesh Sharma and Mr. Yogesh Kumar. Their valuable feedback, insightful suggestions, and meticulous attention to detail have greatly enhanced the overall coherence and readability of the manuscript.

A special mention goes out to the Marketing Team, led by Mr. Rakshit Khanna for his tireless efforts in promoting the book and reaching a wider audience. Their innovative marketing strategies and relentless pursuit of excellence have played a crucial role in generating interest and enthusiasm among readers.

Furthermore, I am truly grateful for the dedication and hard work of the Sales Team, consisting of Mr. Nand Lal, Mr. Surinder Kumar, and Mr. Roshan Lal. Their exceptional sales acumen and customer-centric approach have boosted sales and fostered positive relationships with clients and partners.

The author would also like to extend his heartfelt thanks to Creation Ministries International for its invaluable contributions to the research and development of the book. Their pioneering work in the field of creation science has laid the foundation for a deeper understanding of the origins of the universe.

**Sabrie Soloman**

# ABOUT THE AUTHOR

**Dr. Sabrie Soloman**, Ph.D., Sc.D., MBA, PE – He is the Chairman & CEO of American SensoRx, Inc., USA; Founder of Advanced Manufacturing Technology Post Graduate Studies at Columbia University, NY, USA; Professor of Advanced Technology at Columbia University. Dr. Soloman authored a numbers of technical books published and translated worldwide: Sensors Handbook (2 editions), Sensors and Control Systems in Manufacturing (2 editions); Affordable Automation; Introduction to Electromechanical Engineering; Modern Welding Technology; 3D Printing Technology; 3D Bioprinting Technology & Design to name a few. Dr. Soloman holds numerous Patents, Technical Awards, and several US Product Registrations. Dr. Soloman is considered an international authority on advanced manufacturing technology, robotics, biomedical engineering, pharmaceuticals, and automation in the microelectronic, automotive, beef, pork, poultry industries. He has been and continues to be instrumental in developing and implementing several industrial and modernization programs through the United Nations to Europe, Asia, and African Countries. He is the first to introduce and implement unmanned flexible synchronous/asynchronous manufacturing systems in the microelectronic and the meat industries, and the first to incorporate advanced vision technology in wide array of robot/micro-robot manipulators. Dr. Soloman was selected to deliver the US presidential closing address "Innovative Remote Sensors Technology," at the Universal Design Conference," New York, USA. Dr. Soloman was the President of the International Christian Union at New Castle-Upon-Tyne University. He debated the "Origin of the Universe" before numerous attendees presenting his case against intellectuals and proponents of Big Bang's Singularity in Great Britain.

# CONTENTS

## CHAPTER 2: THE EVOLUTION OF THE "BIG BANG" — 57–108



## CHAPTER 3: THE ORIGIN OF THE UNIVERSE — 109–162

## CHAPTER 4: THE HYPOTHESIS OF MULTIVERSE         163-208



## CHAPTER 5: THE MYSTERY OF GRAVITY         209-246

## CHAPTER 6: THE MYSTERY OF OUR UNIQUE EARTH     247–292

# INTRODUCTION

The search for the truth about the origin of the universe has been a fascinating journey filled with various theories, controversies, and discoveries. From the Eternal omnipotent-omnipresent Creator to the Big Bang theory and to the hypothesis of multiverse, from the mystery of gravity to the phenomenon of black matter and energy, there has been a myriad of ideas and speculations regarding how our universe came into existence. In this book, we will delve deep into some of these theories and concepts, examining their historical context, key figures, and their impact on our understanding of the universe.

One of the most widely accepted scientific theories, about the origin of the universe, absent of God, is the Big Bang theory. Proposed in the early 20th century, this theory suggests that the universe began as a singularity, a point of infinite density and temperature, and has been expanding ever since. While the Big Bang theory has gained widespread acceptance among mostly atheist scientists, there are still many unanswered questions and mysteries surrounding it. For example, what caused the Big Bang to occur? What was the nature of the universe before the Big Bang?

Another intriguing concept in the study of the universe is the hypothesis of multiverse. This theory suggests that our universe is just one of many parallel universes that exist simultaneously. Each universe may have its own set of physical laws and properties, leading to a vast array of possibilities and outcomes. The idea of a multiverse challenges our traditional understanding of the universe and raises profound questions about the nature of reality and existence.

In recent years, there has been a growing interest in the concept of intelligent design and molecular organisms are best explained by an intelligent designer rather than random processes. They point to the complex structures and systems found in nature, such as the molecular machinery of cells, as evidence of a higher intelligence at work. While intelligent design is a controversial topic, it has sparked important debates about the role of science and religion in understanding the universe.

In addition to scientific theories, there are also religious beliefs about the origin of the universe. Many religious traditions teach that the universe was created by a divine being, such as the God of Judeo-Christian faiths. According to these beliefs, the universe is finely tuned for life by its creator, who designed it with purpose and intention.

The mystery of gravity is another key aspect of the origin of the universe. Gravity is the force that holds planets, stars, and galaxies together, shaping the structure of the cosmos. While we have a good understanding of how gravity works on a macroscopic scale, there are still many unanswered questions about its fundamental nature and origin. Scientists are exploring new theories and models, such as string theory, in an effort to unify gravity with the other fundamental forces of the universe.

Moving beyond our own solar system, the black hole has captured the imagination of scientists and the public alike. Black holes are regions of spacetime where gravity is so strong that nothing, not even light, can escape. They are thought to play a crucial role in the formation and evolution of galaxies, and may hold clues to the ultimate fate of the universe. Studying black holes has led to many important discoveries and breakthroughs in our understanding of the cosmos.

Darwin's misconception of unguided evolution, the hypes and exaggerations about the Big Bang theory, and the concepts of space-time warp, time dilation, and understanding UFOs are all topics that have generated immense interest and speculation among scientists and the general public alike. The string theory, which seeks to unify the forces of nature and explain the fundamental particles of the universe, represents a cutting-edge area of research in theoretical physics.

Inaccuracies and substantial errors in carbon dating methods, the age of the Earth, the Earth's magnetic field, and 101 pieces of evidence for a young Earth are all contentious subjects that continue to fuel debates among scientists and scholars. The belief in God as the sole creator of life, the mind and power of the creator, the signature of the creator in DNA, and the idea that the universe is being held together by a divine force add spiritual and philosophical dimensions to our understanding of the cosmos. The substantial errors in carbon dating have raised questions about the reliability of this dating method. While carbon dating is a valuable tool for determining the age of organic materials, it is not without its limitations and potential sources of error.

The mystery and importance of the six-day creation, the eventual and predicted end of the universe, the potential control of AI over our future, the rise of the anti-christ and the false prophit, the church rapture, the tribulation and world war III, the second coming of Jesus Christ and His Millenium reign, and the ultimate fate of the cosmos are all thought-provoking subjects that invite further exploration and contemplation. As we continue to unravel the mysteries of the universe and uncover the truths behind its origins, we must approach this vast and complex topic with an open mind and a willingness to consider multiple perspectives and possibilities. The journey to understand the truth of the origin of the universe is a never-ending quest that challenges us to explore the deepest mysteries of existence, and the presence of the Creator in His universe.

# THE UNIVERSE

## THE VISIBLE UNIVERSE

### The Universe Before Time/Space/Matter

Infinity Existed Eternally prior to time, space and matter. The Personhood of Infinity resides in Eternity. Time is a created finite entity, which never fails to forward-travel solely in one-direction, However, the entity of time unceasingly gazes to its Infinite Eternal Architecture, realizing it is a mere insignificant quantum entity assembled before its Infinity-Creator. Nonetheless, the entity of time was enabled to be entwined with every existing point in the fabric entity of the space of the Universe. The Infinity Composer of Eternity apportioned time to measure the life span on planet Earth and its living beings including mankind.

***Fig.1.1:** The Universe Before Space, Time and Matter – In the Beginning God (The Eternal Infinity) Created Heavens and Earth. The Earth was Formless and Empty, and Darkness Covered the Deep Waters. And the Spirit of God (The Eternal Infinite Holy Spirit) was Hovering over the Surface of the Waters (emphasis added)*

**The universe:** It is overwhelmingly made up of things that cannot be seen. In fact, the stars, planets and galaxies that can be detected make up only 4 percent of the universe, according to astronomers. The other 96 percent is made up of substances that cannot be seen or easily comprehended.

**Dark matter:** It is a mysterious and invisible form of matter, constitutes 85% of the total mass of the universe. Despite its prevalence, its composition and origin remain unknown. Proposed theories include weakly interacting massive particles (WIMPs), axions, primordial black holes, and more.

**Dark Energy:** It's one of the universe's biggest mysteries: more remains unknown than known about dark energy. It affects the universe's expansion, so physicists are able to infer that dark energy makes up roughly 68% of the universe and it appears to be somehow tied to the vacuum of space.

**Black holes:** Black holes are the darkest things in our universe because they emit no light whatsoever in any wavelength. The reason there are no images of black holes themselves is because it is a fact of their physics that they cannot be seen.

**Where do black holes take you?** When matter falls into or comes closer than the event horizon of a black hole, it becomes isolated from the rest of space-time. It can never leave that region. For all practical purposes the matter has disappeared from the universe.

**The Great Attractor:** 150-250 million light years from our galaxy lies a dreaded space anomaly called "The Great Attractor" with a gravitational pull so powerful that it can pull entire galaxies towards itself, and collapse them into each other. What's scarier is that we still don't know what it is.

**Quasars:** Scientists have unlocked one of the biggest mysteries of quasars – the brightest, most powerful objects in the Universe – by discovering that they are ignited by galaxies colliding. First discovered 60 years ago, quasars can shine as brightly as a trillion stars packed into a volume the size of our Solar System.

**Quasar 3C273:** The hottest place in the universe could be quasar 3C273 with an estimated temperature of 10 trillion degrees Celsius. Although it is the hottest object in the system, the temperature of the Sun is quite low compared to some other objects.

**Life outside of The Earth:** So far, the only lifeforms found are those from Earth. No extraterrestrial intelligence other than humans exists or has ever existed within the Solar System.

**No life besides Earth:** While a galaxy may hold trillions of planets, our galaxy is so far the only known life-bearing world. Humans are too precious within our universe.

**Beyond the universe:** The trite answer is that both space and time were created at the point of creation. There is the creator encompass the universe. However, much of the universe exists beyond the observable universe, which is maybe about 90 billion light years across.

**The deadliest object in space:** A traveling star could be more dangerous than a traveling planet— or, more precisely, an object with a diameter of several tens of kilometers and a mass like a star, moving at a speed light.

**The space smell like:** Various astronauts have described it in similar yet varying ways: "burning metal," "a distinct odor of ozone, an acrid smell," "walnuts and brake pads," "gunpowder" and even "burnt almond cookie." Much like all wine connoisseurs smell something a bit different in the bottle.

**Twinkle -Twinkle Little Star:** The stars seem to twinkle in the night sky due to the effects of the Earth's atmosphere. When starlight enters the atmosphere, it is affected by winds in the atmosphere and areas with different temperatures and densities. This causes the light from the star to twinkle when seen from the ground.

**Why is the sky blue if space is black?** Sunlight reaches Earth's atmosphere and is scattered in all directions by all the gases and particles in the air. Blue light is scattered more than the other colors because it travels as shorter, smaller waves.

**A day in space:** The International Space Station (ISS) orbits the earth at speeds of approximately 17,000 miles per hour, which is considerably faster than the earth rotates on its axis. Therefore, the solar day on the ISS is considerably shorter than the earth day at just over 90 minutes in duration.

**Thunder in space:** There are lightning storms in space. These occur on other planets, such as Jupiter, where intense lightning storms have been observed. Lightning can also occur in the atmospheres of other planets and moons, such as Saturn's moon, Titan.

**Speed of light:** Light is the fastest thing in the universe. Nothing can go faster than speed of light 186,282 miles/s. That is the speed limit of the universe.

**Speed of dark:** Most of us already know that darkness is the absence of light, and that light travels at the fastest speed possible for a physical object. So, what does this mean? In short, it means that, the moment that light leaves, darkness returns. In this respect, darkness has the same speed as light.

**Stronger than black hole:** The most powerful supernova yet recorded (ASSASN-15lh) was 22 trillion times more explosive than a black hole during its final moments. It doesn't matter how small or how massive a black hole is, their closing fireworks are exactly the same. The only difference is how long it will take a black hole to explode.

**Where did God come from?** God didn't come from anything. God has just always been there. He has always existed. This is what the Bible means when it says that God is "from everlasting to everlasting" - Psalm 90:1 - Lord, you have been our dwelling place throughout all generations. Before the mountains were born or you brought forth the earth and the world, from everlasting to everlasting you are God. You turn men back to dust, saying, "Return to dust, O sons of men."

## DARK ENERGY- DARK MATTER

In the early 1990s, one thing was fairly certain about the expansion of the universe. It might have enough energy density to stop its expansion and recollapse, it might have so little energy density that it would never stop expanding, but gravity was certain to slow the expansion as time went on. Granted, the slowing had not been observed, but, theoretically, the universe had to slow. The universe is full of matter and the attractive force of gravity pulls all matter together. Then came 1998 and the Hubble Space Telescope (HST) observations of very distant supernovae that showed that, a long time ago, the universe was actually expanding more slowly than it is today. So the expansion of the universe has not been slowing due to gravity, as everyone thought, it has been accelerating. No one expected this, no one knew how to explain it. But something was causing it.

Eventually theorists came up with three sorts of explanations. Maybe it was a result of a long-discarded version of Einstein's theory of gravity, one that contained what was called a "cosmological constant." Maybe there was some strange kind of energy-fluid that filled space. Maybe there is something wrong with Einstein's theory of gravity and a new theory could include some kind of field that creates this cosmic acceleration. Theorists still don't know what the correct explanation is, but they have given the solution a name. It is called **dark energy**.

More is unknown than is known. We know how much dark energy there is because we know how it affects the universe's expansion. Other than that, it is a complete mystery. But it is an important mystery. It turns out that roughly 68%of the universe is **dark energy. Dark matter** makes up about 27%. The rest - everything on Earth, everything ever observed with all of our instruments, all normal matter - adds up to about 4% of the universe. Come to think of it, maybe it shouldn't be called "normal" matter at all, since it is such a small fraction of the universe. As a result scientists build $12 billion Large Hydron Collider Instrument, hoping to underdstand **Dark Matter/Dark Energy,** Figure 1.2.

Nonetheless, theorists were successful in naming the entity of the massless Dark Matter/Dark Energy "the God's Particle," Figure 1.2.

1. Since the Dark Matter/Dark Energy, which is in fact an unknown mysterious matter and unknown energy, they seemed as if their entities were not part of the Six-Day creation.
2. It would be logical to assume their entities were pre-existed before creation.
3. And since the only Personhood before the act of creation was and is and forever will be the Eternal Creator, who is called by name the "I AM."
4. And since the entire Universe is only less than 4%, while
5. the mysterious entity of the Dark Matter/Dark Energy is more than 96% and
6. The Personhood of Eternity-Creator stated that He **holds** everything together, Col. 1:1

*Fig.1.2: Large Hydron Collider*

It is logical to consider that the power of the Personhood of the Eternal "I AM" is indeed the Eternal Creator who is holding together the Universe and everything in it through His power. Also, the Personhood of Eternity sustains the Universe in a perfectly finely tuned order maintaining its harmonious existence.

Accordingly, Dark Matter/Dark Energy is a consistent apportioned segment of the Personhood Power of the Eternal-Infinite entity, which causes the universe to be held and sustained together in a perfect order.

Moreover, it is the description of the Eternal Infinite Personhood of the Redemer God who is in charge of executing the actual act of creation.

As stated in the scripture, Colossians chapter 1 verse 17; He existed before anything else, and He ***holds all creation together***. While Dark Matter/Dark Energy is a vital power to sustain the universe, it is still an insignificant energy as portion of the Infinite Power of the Eternal Personhood of God the Redeemer/Creator. The author believes that this Dark Matter/Dark Energy power is the uncaused-cause portion of the Infinite Power of The Eternal Redeemer and sustainer of the universe. He is the Eternal God —the Lord Jesus Christ.

## THE UNIVERSE – OBSERVABLE AND UNOBSERVABLE

As we understand that there are two diverse meanings for the universe. The foremost, is the observable universe. It is all that we are able to see or observe through our latest James Webb telescopes, Figure 1.3. More precisely, the observable universe is the region of space visible to us from Earth, Figure 1.4. The second meaning of the universe is the entire universe, which means everything that exists, everything that has existed, and everything that will exist.

***Fig.1.3***: *First Images from the James Webb Space Telescope–100 times as powerful as the Hubble- NASA. It Changes How We See the Universe. Credit: NASA GSFC/CIL*

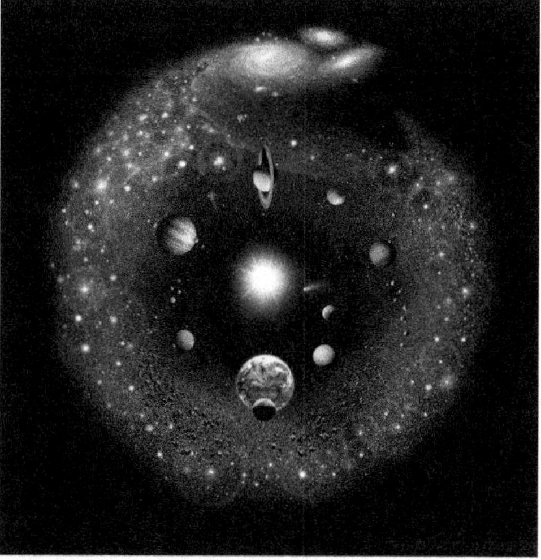

***Fig.1.4***: *The Observable Universe at a Location from Earth*

Many scientists believe the universe came to existence about 13.8 billion light-years ago. The light from the observable universe takes time to travel through space, irrespective of what direction we look. The observer on earth sees light from all direction, as if the observer is standing on a center of a circle. The light is been traveling to the observer distance about 13.8 billion light-years. Consequently, it is reasonable to consider that the observable universe must then be 2 times 13.8 equals 27.6 billion light years across. A light year is about 6,000,000,000,000 miles (6 trillion miles)

- Billion-Years is       Time
- Billion-Ligh-Years is     Distance
- A Light-Year       5.88 Trillion Years (Distance)

Nonetheless, the measurements taken were accurate only at the time they were measured. The space over time, has been expanding. Accordingly, the distant objects that provided light 13.8 billion light-years ago have since moved even farther away from us. Actually, today, the entire spectrum of distant objects is a slightly more than 46 billion light years away (dimension miles/kilometer) measured from an observer in the middle of the field of view, Figure 1.5. Multiply times 2, we get 93 billion light years, which is the diameter of the observable universe. A light year is equal to 5.88 trillion miles or 9.46 trillion kilometers per year, as light travels through interstellar space at 186,282 miles per second, 300,000 kilometers per second.

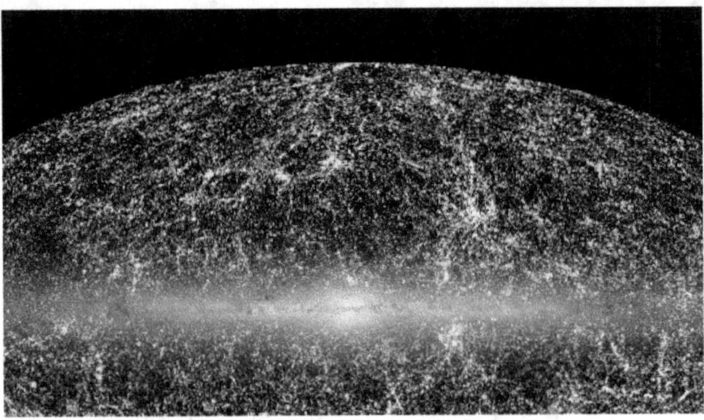

*Fig.1.5: The Distance of the observable Universe−93 billion Light Years*

In order to provide a feeling of the earth's size in the scale of the solar system, it is approximate less than a nano meter, which is equal to 1/1000,000,000 meter. It is approximately less than the size of a virus within the solar system. The size of the earth is incomprehensible with respect to the bewildering size of the universe.

## The Edge of the Universe

The observable universe is impressively huge, but then again, the entire universe, as far as we may comprehend, is an exceptionally larger. While the space cannot not be infinite, some scientists are loosely using the term infinite to describe the vastness of the entire universe. We, at this stage of our cosmic discovery may agree with some scientists to describe the universe as "edgeless." We have no such instrument to observe up to the edge of the universe.

## The Center of the Universe

With regard to the center of the universe; Since the observer is in fact, inside the vast domain the visible universe, the observer himself is the center of what is being observed. There the visible universe has a center, which is us on any point on Earth, Figure 1.4. We are at the center of the observable universe since the observable universe is just the domain of space visible from Earth. It is similar to an observation tower in the middle of a field. The tower can constantly observe a radius of its surrounding circle at any time. The circle of the visible universe is 13.8 light years. While the circle of the radius of the entire visible universe is slightly more than 46 light years.

Interestingly enough, to be more precise, each one of us represents the center of our own observable universe. However, we are certainly not at the center of the whole universe. Similar to an observing tower in the middle of a field present at the domain of the world. also, the tower is not at the center of the world. The tower is not the center of the world. It is the center of the field in the world that it can see, up to the horizon. Nonetheless, since you cannot see beyond the horizon, it does not mean there is nothing there.

When we look up to see the observable universe, we are looking up at the sky, we can only see voyaged light that spent approximately 13.8 billion light years as claimed by many scientists. The light emanated from distant stars and galaxies is now a radius of about 46 billion light years away. Anything farther galaxies or stars are beyond the horizon to be observed, Figure 1.4. However, each moment of time, we see new, older light traveled from slightly farther away, Figure 1.5.

Therefore, our sight of the cosmos is literally becoming larger all the time. If we continue watching the universe days, months, and years the light from more distant places will have time to reach us.

*Fig.1.6: Earth Event-Horizon as Observed from Space – Curtsy NASA*

Horizon is the voyage of humankind, seeing our place through what we observe, and how that sight has altered through time as we go outside the boundary of our horizon. With modern computation as the essential structure and dialectal of the universe, 'Horizon' proposes how its silent sound has been overheard and construed through the waves of time. Beginning with the scientific teachings of the modern perception of the universe, we see how our modern knowledge is molded by the Event Horizon of our universe, by what is visible and invisible to us, and how we still pursue responses from where we stand on how we perceive what is unobservable.

Amazingly, we are sitting at the center of our observable portion of the whole vast universe. Naturally, the question will arise; How big is the universe?

## How Big is The Universe

The observable universe is approximately 93 billion light years across. The whole universe is obviously much bigger, but cannot reach ***infinity***, as many scientists claim. ***Infinity*** is an attribute, which is exclusively reserved to describe the infinite-eternal superior being, which caused the universe to exist and continually expand in a highly organized order.

***Infinity*** is always singular and cannot be combined or be shared with another entity.

With respect to the edge of the universe; since the universe has an estimated quantity that may encompasses 2-3 trillion galaxies expanding away from each other at a specific speed, the universe accordingly has an undefined temporal edge. The edge cannot be clearly recognized until the finite earthly scientists discover a way to identify the true edge of the universe to consequently, render its center. Until then only speculative assumptions may be pondered. However, the visible universe has an edge at the horizon. It's 46 billion light years away in any direction, and the entire universe has a preliminary temporal edge.

With respect to a center of the universe; Again, the observable universe has a center, while the universe as a whole, is ambiguous to expound until a technique is revealed to find a center if there is one!

## Expanding The Universe

*Fig.1.7: The Universe Expanding Since the Existence of the Universe*

You might have heard, at some point, about the fact that our universe is constantly, everlastingly, expanding. This inherently means that either the space around us is literally stretching or getting further apart from each other. As the universe appeared, it started to expand its space outward forming our todays ever-expanding universe. The expansion did not cease and it has not come to a halt yet, Figure 1.7.

With respect to the universe becoming bigger; Indeed, the universe is becoming bigger. The space is expanding, which enables both the observable universe and the entire universe become bigger. Additionally, by the passing of time, we observe older and older light coming from farther and farther distance. Thus, our observable universe is becoming bigger in a similar manner.

That, in short, is our perception from the tower of an observer. Everyone on earth is at the center of the observable universe.

Everything a man can see and cannot see within the space is the universe. It includes every atom of solids, liquids and gases occupy the space is the universe. The universe, informally, includes everything there is. Mountains, oceans, moons, and hot nuclear spheres of hydrogen and a whole more. Nevertheless, while the universe is everything, it means that it embraces the things that we do not know, and that we do not know exist, or even the things that we accept as true exist, but have not thus far been seen or detected? E.g., would the future a par*t of the universe?*

## The Existence of the Invisible Universe

In the science of physics, we typically differentiate between these two philosophies of the universe as:

i. The observable universe, which is everything whose existence, up until now been able to confirm or observe, or could, in principle, observe if we pointed our telescopes at it, and

ii. The Universe, or the whole universe, which is everything that exists, has existed, or will exist, anytime, anywhere, notwithstanding of whether or not we are conscious of it yet or ever will be.

Undeniably, the main cause for this difference is so as we may dialog methodically about parts of the universe, we have not yet been capable to observe. The two fundamental reasons —we may not have been able to observe a portion of the universe yet either, it is futuristic, or it is very far away that light emitted from it has not had time to reach yet.

To clarify this matter, as it is claimed, the universe had a distinctive appearance 13.8 billion light-years ago. As light takes time to travel through space. Consequently, anywhere too far away from us is not yet visible, in the same manner as thunder from a distant lightning strike. It is not audible pending one waits certain time. If one were blind, the thunder would be the primary sign that the "storm's existence." Once it is heard, one may deduce that the storm must have existed.

## The Existence of Space/Time and Physical Laws

Therefore, within the observable universe, we know the existence of the followings:

i. Spacetime,  ii. Particles, and

iii. The physical laws, which describe how spacetime and particles interact

The physical laws gave rise to gravity, matter, forces and everything related. However, mathematics is being debated whether it should be considered as a part of the universe or it should be considered as a type of mathematical heavens exists outside the universe. The debate is still raging on among scholars, cosmologists, mathematicians, scientists and clergymen.

Interestingly enough, science fiction theorists, have successfully attempted to incorporate parallel universes, and multiverses, which proposed by science fiction research fanatics. They put forward several hypotheses that the universe is actually a small piece of a larger whole. That whole is in fact everything there is, Figure 1.8.

The first image of the observable universe from NASA's James Webb Space Telescope, Figure 1.6, the deepest and sharpest infrared image of the distant universe to date. Known as "Webb's First Deep Field," this image of galaxy cluster SMACS 0723 is teeming with features, as thousands of galaxies—counting the dimmest objects ever observed in the infrared—appeared in Webb's field of view for the first time. This portion of the observable universe covers an area of heaven above almost the size of a grain of sand detained at arm's length by somebody on earth.

*Fig.1.8: First Observable Image from James Webb Space Telescope: galaxy cluster–SMACS 0723–NASA*

## The Observable Universe – A Grain of Sand on a Beach

The region of heaven above earth or space that people may essentially observe with the assistance of technology, Figure 1.9. The observable universe, which can be supposed as a bubble with Earth at its center, is distinguished from the entireness of the universe, which is the entire cosmic system of matter and energy of which Earth, and therefore the anthropological race, is a part.

**Fig.1.9:** *Location of Earth with respect to the Its Milky-Way Galaxy*

As stated previously, the observable universe is nearly 93 billion light-years in diameter. This figure is resulting from numerous considerations. A light-year, is the distance light can travel in one Earth year, which is 9.46 trillion kilometers or 5.88 trillion miles. The projected Hypothetical age of the universe since it came to existence is 13.8 billion years. The light emanated by objects in space that people can see has been traveling in the direction of Earth for 13.8 billion years. That indicates that the observable universe is approximately 13.8 billion light-years in any direction from Earth and 27.6 billion light-years in diameter.

## CONSTANTLY EXPANDING UNIVERSE – HUBBLE'S LAW

Nevertheless, as stated by Hubble's law, space has been expanding since the universe came to existence. Accordingly, the observable universe, it remains to expand. The mathematical calculations of this enlargement demonstrate that objects, which emanated light 13.8 billion years ago, from a distance of 13.8 billion light-years, are presently farther away from Earth—46 billion light-years away, roughly. This illustrates that the observable universe is further than 46 billion light-years in all direction from Earth and nearly 93 billion light-years in diameter.

Specified the constant growth of the universe, the observable universe enlarges another light-year every Earth year. Simultaneously, light from objects that are always farther away endures to spread to Earth for the first time, indicating that individuals are able to see more and more of the universe with the transient of time. Although individuals may never be able to see the whole universe from Earth, only the comparatively small bubble of the observable universe, the sphere of observation is always expanding.

### The Cosmic Light Horizon

Observing the nightly sky with the aid of the most authoritative telescopes, one would be able to observe a firm distance in any direction. It is a type of *"cosmic horizon."* This is the observable universe, which is the slice of the universe that can be observed from Earth's location in spacetime (Milky-Way). The observable universe can be supposed as a spherical bubble with Earth at the center, one may not see outside the edge or horizon of the bubble. The cause is light travels at a limited speed, 186,282 mi/s. This edge at the vantage point of Earth may not observe beyond is called the *"cosmic light horizon"* or the *"particle horizon."*

As light travels at a finite speed of 186,282 mi/s, it takes time for light to travel to Earth. Also, the universe is *finite* in time and *space*, and has not always existed. Hence, as the universe existed for a *finite* period of time, when light travels at a finite speed, the light emitted by any matter, stars, or galaxies beyond a certain distance has not had sufficient time to make for Earth yet.

How do astronomers and scientists know the universe has existed for a *finite* amount of time?

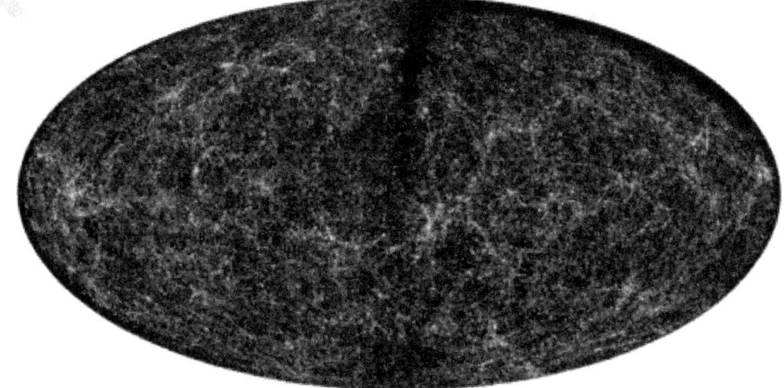

**Fig.1.10:** *This Image Illustrates the Observable Universe, Including All Visible Galaxies from Earth.–NASA*

Researchers, cosmologists and philosophers have come to the realization that the universe cannot be ageless or have existed infinitely through time, merely as light travels at a finite speed that can be measured. It may then be logical to conclude that if the universe has always existed, the night sky should be as *illuminated* as day, counting on each slice of the sky consuming a celestial object emanating light. Such light may have had sufficient time to spread to the Earth. This is known as *"Olber's Paradox,"* which states that the universe cannot be ageless. If it were ageless, all distances inside the universe would be visible owing to light from those distances may have finally enough time to reach Earth. Nonetheless, the night sky is still dark!

Instead, the universe is hypothetically only 13.8 billion years-old and dynamically expanding. Therefore, not all light in the universe has had sufficient time to reach the Earth's position in spacetime. Consequently, today the universe appears dark. This begs the questions:

i.   What is beyond the observable universe?

ii.  How large is the observable universe? and

iii. Does this mean the observable universe will consistently grow larger as light from great distances has more time to reach Earth?

Unfortunately, it is much more complicated than that.

## Observable Universe vs. Entire Universe

At this time, the greatest distant objects astronomers and scientists may observe are objects that have emitted light, which has been traveling toward Earth's present location within the universe for 13.8 billion years. Though, this does not lead us to believe astronomers may merely observe objects that are at most 13.8 billion light-years away (Distance). The universe is growing, the uttermost objects that scientists can see are in fact further than 13.8 billion light-years at this moment because those objects have been also expanding away from Earth's position since they first emitted the light that astronomers are currently observing from 13.8 billion years ago.

Nonetheless, how do astronomers, cosmologists and other scientists know the universe is expanding?

To comprehend this issue, the phenomenon recognized as *"redshift"* should essentially be comprehended.

## Red Shift Tells–Universe Expanding

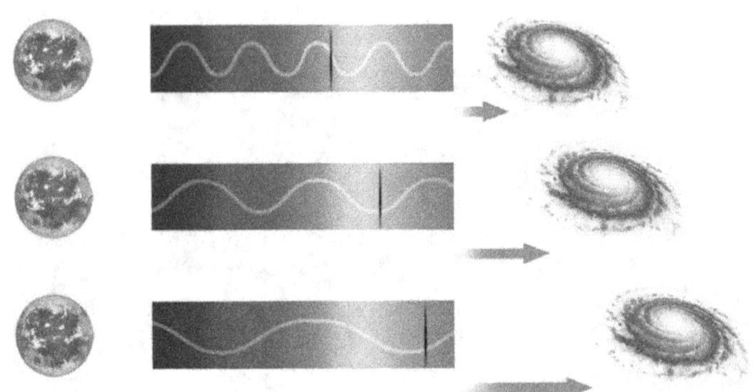

**Fig.1.11:** *Galaxies Moving away from Earth Indicated by the Red Shift Light Wave*

The galaxy moves away from Earth the light emitted from the galaxy, light wave length moves to the red domain, Figure 1.9. The **"Observable Universe"** is the slice of the universe that is observable from the viewpoint of Earth's position within the universe. The edge that separates the observable universe from the rest of the entire universe is called the *"cosmic light horizon,"* or *"particle horizon."* The cosmic light horizon is projected to be about 47 billion light-years from Earth. The light, which bends with the curvature of the space, will pass through this region that has not made it to Earth yet. Additionally, the light from objects beyond this region of space has been **cosmologically redshifted** to the point of being outside the visual light spectrum, which partially explains the darkness of the night sky.

## Beyond the Observable Universe

Beyond the observable universe is merely more universe.

How big is the observable universe compared to the whole universe?

The observable universe encompasses around 47 billion light-years (distance) in all directions from Earth, Figure 1.8. The observable universe is ~94 billion light-years in diameter, Figure 1.8. The entire universe on the other hand is possibly, and even likely, incomprehensible to phantom by human mind, founded upon the observable configuration of the universe from within the observable range of the entire universe. Consequently, it is conceivable that the entire universe is vastly larger than the observable universe.

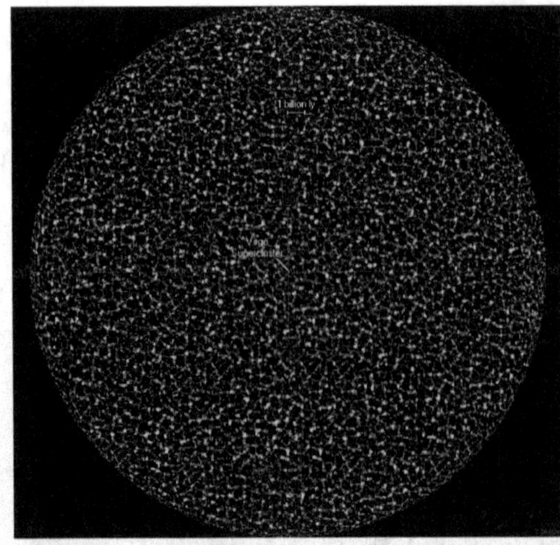

**Fig.1.12:** *The Map of the Observable Universe*

A map attempts to illustrate the total visible Universe, Figure 1.10. The galaxies in the universe incline to assemble into vast panes and superclusters of galaxies surrounding large voids providing the universe a cellular presence. As light in the universe solely travels at a fixed speed (186,282 mi/s), we may observe objects at the edge of the universe when it was up to 13.8 billion years ago.

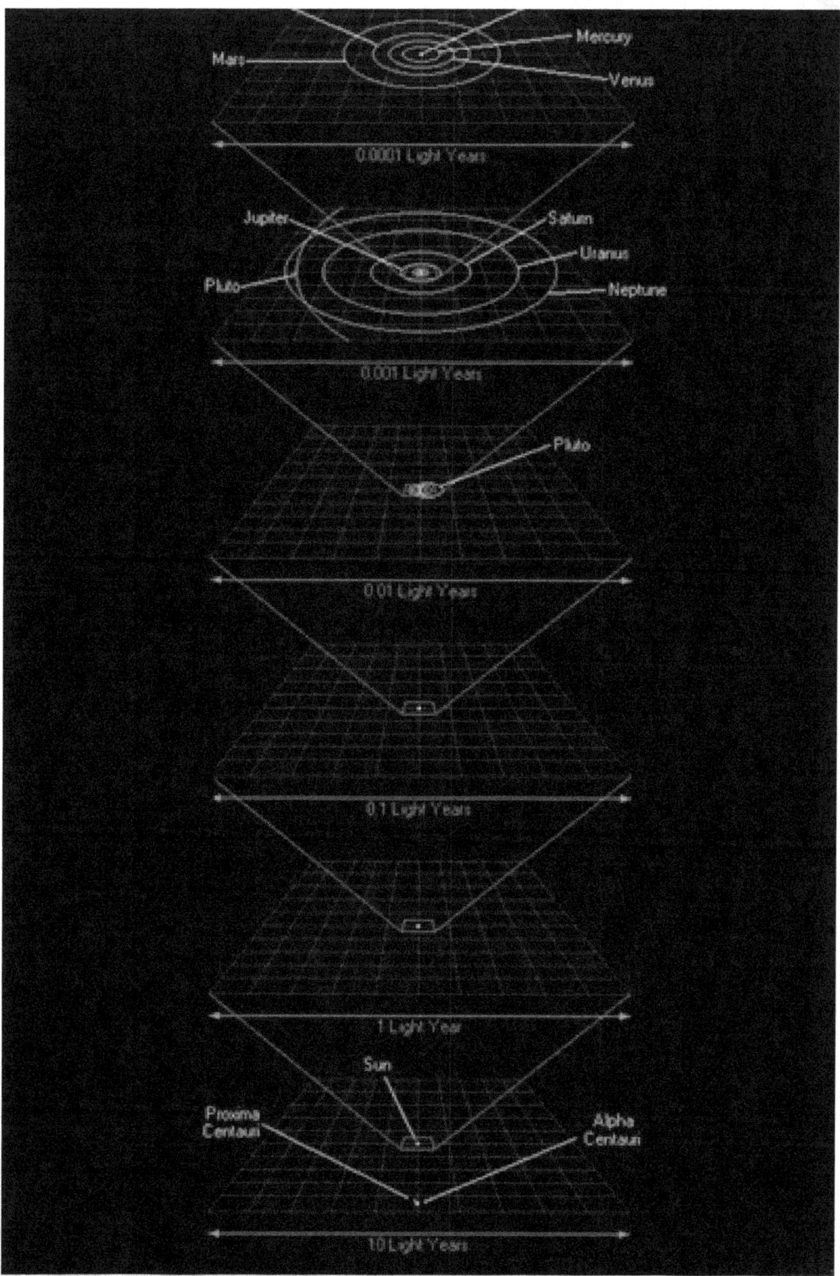


**Fig.1.13:** *The Visible Orbits Displayed According to their Distance from The Solar System*

## Objects between the Edge of the Solar System and the Nearest Star

There are certainly many asteroids beyond the orbit of Pluto. This is called the Kuiper belt and may contain tens of thousands of asteroids. The asteroids are likely objects formed from debris and dust from adjacent planets, originated from the solar system.

As shown in Figure 1.11, at a distance of one light year (5.88 trillion miles) it is believed that we are fenced by a vast cloud of ice-rich asteroids. Also, it is believed that there are possibly few hundred billion of these ice-rich asteroids dispersed over an enormous area. This cloud is named the "Oort cloud." This cloud is supposed to be the source of comets. Infrequently a few of these ice asteroids is pushed inwards towards the Sun. As they enter the solar system the ice begins to evaporate creating comets, Figure 1.12. Typically, the comets fly back to the solar system again. Infrequently, the comets are forced into short period orbits to remain within the solar system.

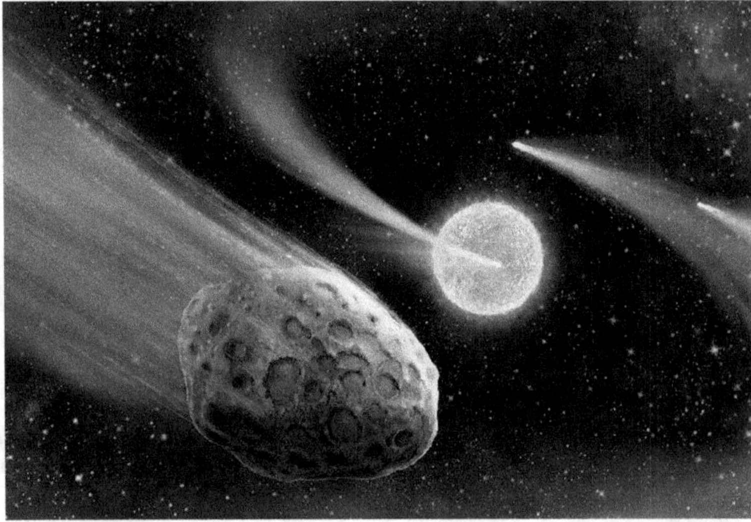

***Fig.1.14:*** *Scientists Detected Comets Outside Our Solar System*

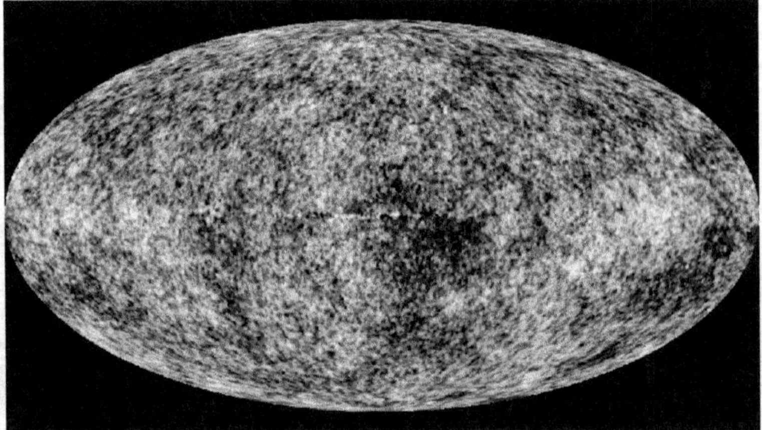

***Fig.1.15:*** *Temperature Variations Causes Super Clumped Together Clusters*

The temperature variations in Figure 1.13 describes a map of galaxies in the universe cluster together to form superclusters. The matters in the beginning when the universe appeared were cool clumps, represented by the darkest clumps in Figure 1.13. These clumps are exhibited in most galaxies. The hottest regions in the map represent low-density regions which eventually, became huge voids.

## HIGHLY ORGANIZED UNIVERSE – SINCE EXISTENCE

Once again, the question lurks – is there a center to the universe? – Again, there no center found yet, as there is no edge to be well defined, yet. The space in our finite universe is curved. Imagine you were traveling billions of light years (distance) in a straight line, you would be surprised to end up where you started due to the curve in the universe, Figure 1.16.

**Fig.1.16:** *Small Universe of Fewer Galaxies – Traveling the Universe in a Straight Line Ends up where you started*

Example: Imagine a very small universe of only fewer galaxies, you are traveling by a spaceship through this small universe. The spaceship moving among these stars cannot find the edge of this universe. Once the ship attempts to exits on one side of the universe, it will reemerge on the other side, Figure 1,14. The travelers inside the spaceship mistakenly perceive they are in an endless universe, with no boundary, and no center. However, the real fact is very different.

Where did the appearance of the universe happen?

Some scientists mistakenly created various hypotheses with a common assumption that an explosion occurred into the empty space. Naturally, no data or scientific experiments to substantiate such claim.

## Space-Time Created Simultaneously

Space and time were created simultaneously. They are inseparable twins. At the beginning of the universe, the space was then completely filled with all sort of created matters. The matters were highly organized. Eventually, the stars and galaxies appeared in the universe we see today.

Many scientists imagined that the universe may have begun from a simple point containing space, time and matter. However, these hypotheses could not be proven or even observed through the billion light years arrived to our Milky Way. However, it is a fact since the universe appeared, its galaxies and stars are constantly expanding away from each other. Figure 1.17. There is no center of the expansion.

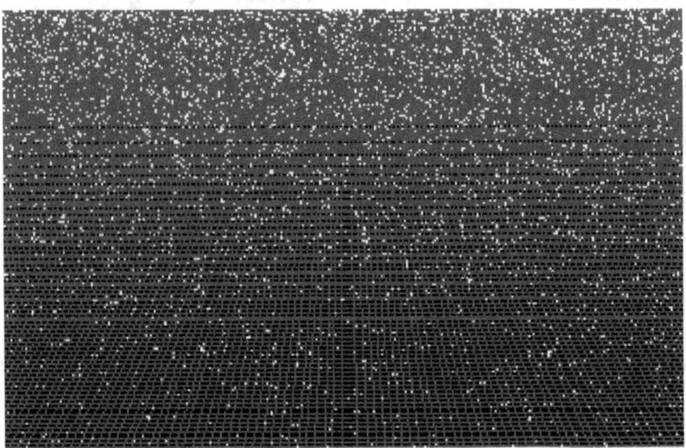

*Fig.1.17:* The Universe Galaxies are constantly expanding

The universe is expanding at all points, Figure 1.15. Observers in any adjacent galaxy will observe most of the other galaxies in the universe moving away from it. The entire universal galaxies and its myriads of stars must have appeared simultaneously, and instantly began to expand outwardly since the creation of time and space.

# DOES EARTH EXPAND WITH THE UNIVERSE OR SEPARATELY

The Earth is considered part of the solar system. The solar system is part of the Milky way galaxy. The solar system itself is under highly organized and miraculously tuned system for life and existence. The solar system itself does not expand or move out of its designate orbits. However, the Milky way galaxy is grouped with other galaxies, which are also held by gravities. Certain group of galaxies form a cluster of galaxies. These clusters of galaxies expand steadily. Gravity also holds galaxies together into groups and clusters. It is primarily the groups and clusters of galaxies are moving apart and expanding in the universe.

## What Does Exist Outside the Universe?

There is no outside of our universe. Furthermore, there are no knowledge or any information to faintly make any one believes that there is or there was another universe or universes somewhere else outside our universe.

## What Did Exist Before the Moment the Universe Appeared?

Before the appearance of the universe there was no time, no space and no matter. There was the *Infinity*, which is the infinite power, the infinite knowledge, and the infinite wisdom. The attributes of the *infinity* are its entity that permits time, space and matters to come to existence. They can be formed, sequenced and organized to create harmony and order. The entire universe, the visible and the invisible were already in the mind of the **Infinity**. This infinity has all the time in its entity past present and future. This infinite entity holds the time and space and matter in his infante existence. Time past, time present, and time future are as a straight line composed of many dots, drawn by Him the eternal. The initial point and the end point are well defined, and all the points in between to make the straight line are also exceptionally known to Him and well defined in the life of the universe.

## MULTI-UNIVERSES

Some theorists propose that our universe is a fragment of an infinity of other universes, called a multiverse, which are being continuously created, Figure 1.18. This is also a fanciful thought that hard to entertain, as no such evidence or scientific proof to this notation.

How did the Universe come to existence? This question will answer in detail in the following chapters.

**Fig.1.18:** *Multi Verse Hypothesis*

If the universe is 14 billion years old, how could galaxies have travelled more than 14 billion light years?

## The Age of the Universe

According to Einstein theory of relativity, the speed of light is absolute, 186,282 mi/s. This creates significant controversies, in particularly, among the scientific communities. Consequently, one may question the actual age of the universe.

a.  Either the age of the universe is inaccurate or

b.  The light-year age is inaccurate.

Scientists, devised a hypothesis stating that at the point of time and space came to existence the speed of light may have been numerous times greater than what had been proven by Einstein. Nonetheless, there are no data to substantiate such claim. On the contrary, many experiments were conducted utilizing an accelerometer namely (the Hydron Collider), exceed 186,282 mi/s. attempting to prove that the speed of light is absolute, which is nothing can go faster than the speed of light. Interestingly enough, the data collected have proven that the speed of light is indeed absolute and could not exceed 186.282 mi/s.

## Young Universe?

The logical conclusion one may derive is that the universe is a young universe as many scientists also claim. However, many scientists strongly believe that nothing can stop the universe expanding faster than speed of light. Contrary to the data collected at any local point within the universe, proving that nothing can also, travel faster than the speed of light.

**Fig.1.19**: *Gravitational Force of Objects stretch the Space*

Galaxies are comparable balls resting on an elastic membrane, which characterizes space. Once the elastic membrane is stretched, the balls move apart. However, the balls, which are close together will appear to move apart slowly. The balls, which are broadly separated will appear to move apart faster, Figure 1.16.

Imagine that someone is living in one of these balls. He will perceive that his ball is stationary, while the other nearby balls are moving away slowly and the distant balls are moving away faster.

Some scientists hypothetically, claim that the very far distant galaxies residing beyond the *horizon* can move away faster than the speed of light. However, the one residing at the visible local universe cannot observe them. Needless to say, the one observing his own part of the universe nothing is travelling faster than the speed of light. This concept also has not been proven worthy yet.

**Fig.1.20**: *The Distance Scale of the Universe*

Defining a distance in an expanding universe is a challenge not for the fainted heart. Imagine two galaxies are close to each other but moving away, Figure 1.17. Once the universe is only 1 billion years old, the first galaxy emanates a visible pulse of light. The second galaxy does not receive the pulse until the universe is 14 billion years old. At this time, the galaxies are parted by about 26 billion light years:

$$[26 = (14 - 1) \times 2];$$

the pulse of light has been travelling for 13 billion years. The observation the people receive in the second galaxy is an image of the first galaxy, as it was only 1 billion years old and when it was only about 2 billion light years away.

There are four dissimilar distance scales commonly found in cosmology. They are investigated below.

## The Size of the Universe

The observable universe seems to have a radius of 14 billion light years since the universe is about 14 billion years old. The light from more distant objects merely has not had time to reach us. Everybody in the universe will discover themselves at the center of their own observable universe. The precise scale of the universe is complex as the universe is expanding. Galaxies we observe near the edge of the visible universe emanated their light as they were much closer to us. Now, they are much further away.

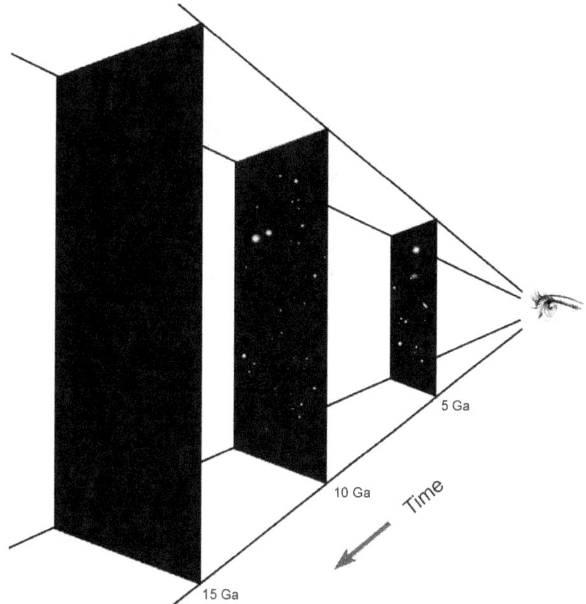


**Fig.1.21:** *Size and Age of the Universe*

The true size of the universe is obviously much larger than the noticeable universe. The geometry of the universe recommends that it may contain enormous incomprehensible size. Unless a superior creative power stops it, it will continue to expand, Figure 1.21. It apparent the observable universe is a minute speck in a much greater totality.

## THE HUBBLE DEEP FIELD

Hubble Space Telescope was pointed at a blank area of the sky in "Ursa Major," the for ten days in December 1995. A stunning picture was captured and shaped one of the greatest famous astronomy images of modern times - the Hubble Deep Field Image. A portion of this image is illustrated in Figure 1.19. Virtually every object in this image is a galaxy characteristically located 5 to 10 billion light years away. The galaxies in the image have exposed all configurations and colors. Some of these galaxies are young and blue, while others are old, red and dusty.

Two other similar images were captured by the same Hubble Space Telescope. They are called the "Hubble Deep Field South," produced in 1998. Also, the "Hubble Ultra Deep Field" *captured* in 2004.

*Fig.1.22: Hubble Deep Field Image, December 1995*

## A Slice of the Universe

it is conceivable to compose a slice of the universe through assembling distances collected from thousands of galaxies in a slight strip of the sky, Figure 1.23. The image was generated from the 2dF Galaxy Redshift Survey, which airs out into the universe to 3.5 billion light years. While not significant data were composed for galaxies beyond 3 billion light years, these kinds of image-plots illustrate the true formation of clustered the galaxies in the universe, even on the largest scales, which are about 52,000 galaxies plotted

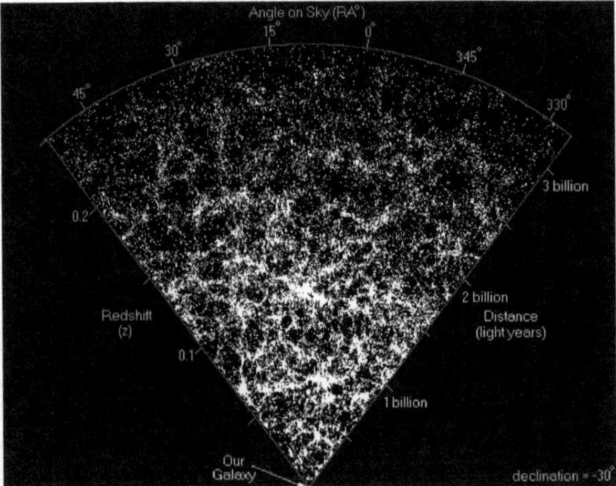


*Fig.1.23: Galaxies with Their Respective Red Shift and Time*

## THE MYSTERY OF THE VISIBLE UNIVERSE

When we were children, we have often gazed upon the stars and pondered at the greatness of the cosmos, fantasizing about visiting distant galaxies, stars, and planets.

That cosmic wanderlust is only augmented when comprehending that light takes a finite time, only 186,300 mi/s to travel to earth. We are observing these distant cosmic matters and amazed. Is it true that they existed thousands, millions, or even billions of years ago?

If the hypothesis of "Big Bang" is true, it will take much more than billions and billions of light-years to form. It may take even trillion light years to come close to validate the time scale of the "Big Bang" as the origin of the Universe.

If the Sole infinite power, who holds the time as a simple straight line in his entity, He, and He alone is the author of time, capable to make a blink, as a time of His desire, to cause the Universe to come to being —fully formed.

In my childish perception, every image captured by a telescope, and every glimpse at the night sky is a journey through space and back in my time.

Unquestionably, taking a cosmic journey like this is beyond my childish perception. Nonetheless, thanks to researchers and devoted scientists and cosmologists combined with years of development of Space Telescopes and collecting data along with space enthusiasts, I can enjoy the greatness of our universe and its maker!

Renowned great assembly scientists and astronomers have provided the means to observe a map of the observable universe to behold its awesome beauty.

## THE REDSHIFT OF THE OBSERVABLE UNIVERSE

Figure 1.24 – Illustrates the colorful dots sprinkled across the slice. Amazingly, each one of these dots is a galaxy positioned in its actual location. These galaxies have an extensive distinction in colors, varies from yellow to red to blue, Figure 1.24, is contributed to the convincing fact that our universe is expanding, shown by a tiny smidge of the observable universe.

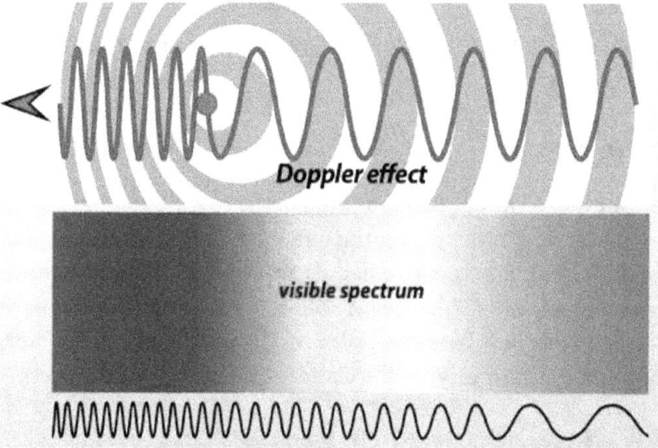


**Fig.1.24:** *Red Shift indicates object's wavelength is moving away*

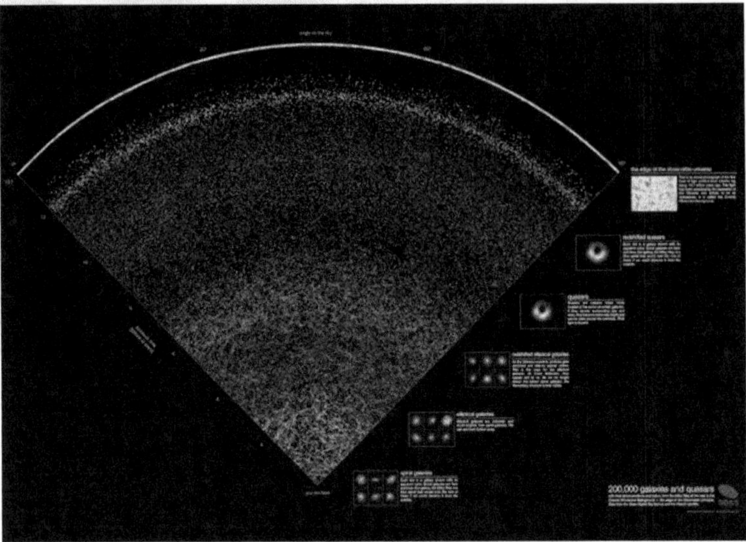

***Fig.1.25:*** *A slice of the Observable Universe Composed of 200,000 Galaxies and Quasars*

Simply put, as the universe expands, the wavelength of light is stretched traveling on the way to the red end of the electromagnetic spectrum. The further away a light source is, the more rapidly it is traveling away the more extreme redshift.

Accordingly, the red dots are far away galaxies, while the yellow dots signify galaxies at an intermediate distance. The blue-toned dots signify close galaxies. In actual fact, some could really be traveling in the direction of our Milky Way galaxy. Therefore, the wavelength of light they release is being compressed, making it seem "bluer."

The red and blue shift are comparable to the sound waves traveling on Earth. When a fire or a police siren, or a train whistle approaches us, soundwaves are compressed up. When the siren recedes, the soundwaves are stretched out. This is how an ambulance, fire engine, police car, or train whistle speeding past us appears to alter the tones, known as frequencies.

Observe the multicolored shell at the top of the observable slice-map, Figure 1.21, a hypothetical view of the existence of such scattered colors.

## BACK TO THE BEGINNING

The top of Figure 1.26 illustrates a hypothetical assumption of the first flash of radiation that traveled through the universe allegedly 13.8-billion-year history after the universe came to existence. This radiation is claimed to a multicolored shell of a fossilized light, contributed to an event in the universe's early history called the "*last scattering.*" Sometimes scientists call it "the surface of last scattering." However, cosmologists have lately call it *cosmic microwave background* (CMB). This claim of CMB will be investigated at a later chapter, as prominent scientists including the author dispute the claim.

### Faster Than Speed of Light – Fanciful Belief!

Contrary to the author's belief, many scientists claim that the early universe has rapidly expanded, even beyond the speed of light. After the initial expansion period, which is called the "Big Bang." Allegedly, the universe also cooled down, eventually reaching a temperature low enough to allow electrons to bond with protons. Again, no such solid data or experiments to validate such claims. It is pure speculation and hypothesis.

Scientists continue to image that when the universe cooled down after the first explosion of the "Big Bang" the first atoms were formed. Also, this formation created hydrogen, which is the lightest element in the universe. Hydrogen, in its simplest form is merely one proton orbited by a single electron happened about 380,000 years after the hypothesis of "Big Bang." This event is called the "***epoch of reconnection.***" It had a significant consequence beyond the creation of matter, or even the beyond "the all-knowing infinite creator Himself."

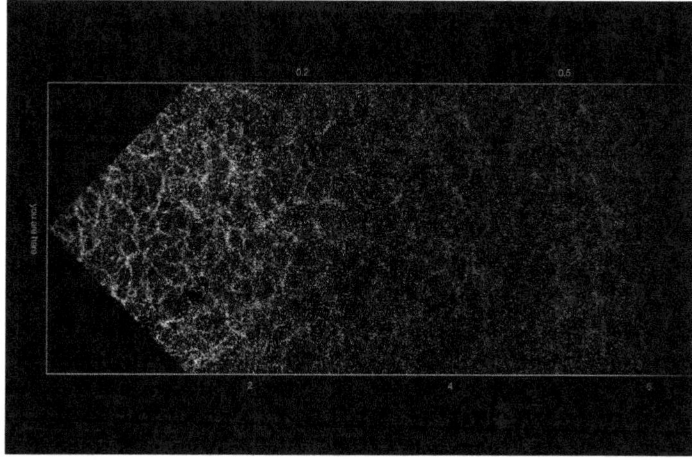

**Fig.1.26**: *The Edge of the Observable Universe –Image Credit: mapoftheuniverse.net*

Scientists pontificated that free electrons in the universe were limitlessly smattering photons, Photons are the essential particles of light. Consequently, photons were unable to travel, which make the universe in essence opaque. Also, it is claimed that "***the epoch of reconnection***" known as Cosmic Microwave Background (CMB) radiation has instantly changed the universe from being opaque to suddenly becoming transparent to light. This hypothesis will be discussed at a later chapter.

That first light travelling freely is hypothetically, the CMB. Also, it has its wavelength stretched by the expansion of the universe. This stretching caused the CMB to lose its energy. Consequently, redshift is displayed. At present, scientists claim that the CMB ubiquitously fills the universe with a uniform temperature of -2.73 Kelvin, which is equal to -464.584 °F.

Figure 1.26 represents the edge of the observable universe. Any light omitted outside this slice has not reached our Earth yet. The travel time is lengthier than the age of the universe.

Figure 1.26, does not illustrate any light marked on the map beyond the edge of the observable universe. There is no instrument on Earth known to us today that could perhaps "observe" it.

**Identical the earliest maps of early pioneers that presented the edge of the world, outside which lay ambiguities, this most contemporary map illustrates, for all purposes, the edge of the universe.**

Figure 1.27, presents the first primordial light is our galaxy, the Milky Way. The Milky Way doesn't actually hold any particularly special place in the universe, appearing as nothing more than a tiny insignificant speck at the base of the cosmic map.

Our solar system and planet lie within this speck, incomprehensibly small and unviewable, giving users a taste of how insignificantly small we are on a cosmic scale.

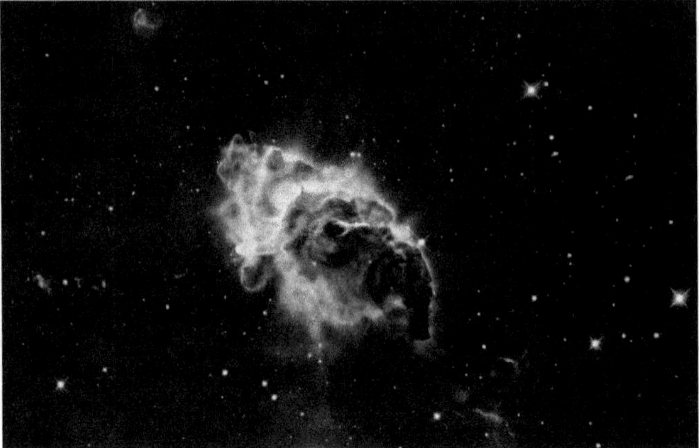

*Fig.1.27: First Primordial Light of Our Milky-Way Galaxy*

## The Entire Universe – The Visible and the Invisible

Thanks to the invention of powerful telescopes, humanity has expanded entree to the first-born light from the primary days of the universe. In spite of this is an implausible discovery, we may never be able to observe pass our observable universe. The accurate size of the entire universe is much enormous, and much greater than what we can ever observe.

Observable universe with respect to the entire universe is a mind-overwhelming contrast. No one could see past Earth's horizon. It does not mean there is nothing out there. The equivalent principle applies to the observable and the entire universe. The observable universe is roughly 46 billion light-years or 92 billion light-years across, however, the entire universe might be inconceivably larger. One light year is about 6,000,000,000,000 (6 trillion) Earth years.

As the technology significantly advanced over the years, astronomers were able to observe deeper and deeper into the past of our universe up until just a few moments after the hypothetical "Big Bang." As we observe back into the initial days of the universe, it has given us valuable understanding of the enormous size, Figure 1.28.

*Fig.1.28: Looking Back into the Initial Days of The Universe*

# NEW VIEW THE OBSERVABLE UNIVERSE

From our Earth perspective, Earth is positioned at the center of the universe, and to some extent, this is a fact. An observable universe is a spherical region of the entire universe. We cannot observe past the edge of the observable universe, although we know it is enormously larger. Technology means may advance further and further, to look deeper into space. Nonetheless, the technology has its physical limits created by the speed of light itself. Essentially, it seemed as if a line has been drawn beyond which, we would not be able to look any further.

Everything we observe and see with either our own eyes or with the aid of telescopes is a portion of the observable universe. One may see it because light has had sufficient time to reach the observer. If we look at the Sun, we are looking into the past. The light and the information that it carries, traveled 8 minutes and 20 seconds to us. If for some reason the Sun suddenly is "turned off," we would recognize it only after the last rays of sunlight reached us – 8 minutes and 20 seconds later.

## How Far is a Light-Year – Perspective

Here on Earth, we measure distance mostly in miles or kilometers. In some cases, we can also measure distance in time. In space, it is a completely different measuring ideology. The distances are so immense, that astronomers devised a different unit of measurement – light-year. The light-year is the distance that light travels in a vacuum of space in one Julian year or 365.25 days.

One light-year equals 9.46 trillion kilometers or 5.88 trillion miles. This number is basically too big for the human mind to grasp it. To equate that into a simpler understandable perspective a scientific term was coined describing the cosmic unit of length. It is called astronomical unit or "AU."

## The Astronomical Unit "AU"

The astronomical Unit "AU" is defined as the distance between *Earth* and *Sun*.

1 AU = 93 million miles, or 150 million km.

Robert Burnham noted that the number of astronomical units in one light-year and the number of inches in one mile are identical.

Turns out there are 63,000 astronomical units in one light-year, and 63,000 inches (160,000 cm) in one mile (1.6 km).

If we scale the astronomical unit at one inch, then the light-year on this scale represents one mile (1.6 km).

The nearest star to Earth, other than our own Sun, is "*Proxima Centauri*" which lies 4.2 light-years away.

If we scale down the Earth – Sun distance at one inch, then *Proxima Centauri* is located 4.2 miles (7 km) away from Earth.

Try to imagine 100, 1 000, 10 000, 100 000, 1 000 000 and more miles which on this scale represent light-years.

## The size of the Observable Infant Universe

Since we may have grasped how far a light year is, we may take a closer look at how big the observable universe is. We recognize that universe is nearly 13.8 billion years old. Similar to a ship in the middle of the ocean, the horizon at 5 kilometers (3,1 miles) only in every direction. The Earth's observers may aim their telescopes in any direction to peer 13.8 billion light-years away.

*Fig.1.29: The Detailed of The Infant Universe 13.8 billion Years Ago. Curtsy: NASA / WMAP SCIENCE TEAM*

Subsequently, light needs time to travel through the vacuum of space and reach the observer, we might consider that the radius of the observable universe is 13.8 billion light-years or 27.6 billion light-years across (in diameter). However, it, is much more than 27.6 billion light-years across. In fact, it is roughly 92 billion light-years across, because our universe is constantly expanding, and we need to take this inflation into account, Figure 1.25.

The Earth observers might observe a galaxy that lies 13.8 billion light-years away at the time of the hypothetical "Big Bang." The universe has continued to expand all over its lifetime. If expansion happened at a constant rate, that same galaxy is now 46 billion light-years away, making the diameter of the observable universe a sphere around 92 billion light-years across.

## The Mystery of The Entire Universe

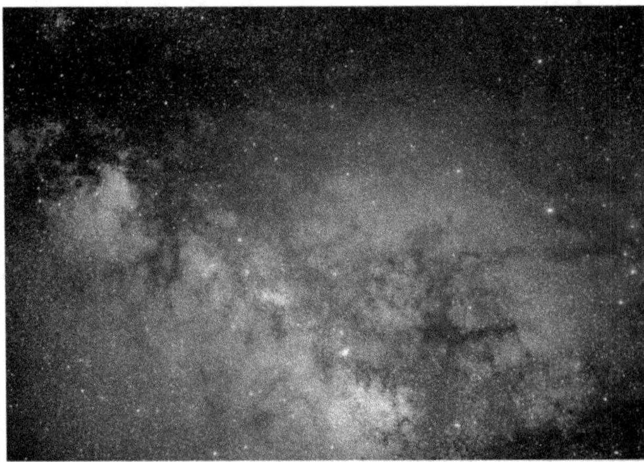

*Fig.1.30: The Mystery of the Size of the Entire Universe*

How big is the entire Universe is a great mystery. This mystery is one of the numerous mysteries of the universe. Since we are restricted to our bubble of the observable universe, there is no exact way to express how big the entire universe is. We cannot place a number on an overwhelming and enormous unknown, Figure 1.26.

Still, there are numbers of stimulating hypotheses, theories and calculations, which venture how big is the whole universe. If we undertake that the universe is finite, then by some calculations it could be as great as

7 trillion light-years (7,000,000,000,000 × 6,000,000,000,000 miles) or 250 times greater than the observable universe. That quantity was computed by a team of scientists led by Mihran Vardanyan at the University of Oxford in 2011.

If we look at other theories projected by a number of atheistic scientists, then the universe allegedly is infinitely large with no termination to it. In the infinite space, as they alleged, there is an infinite number of potentials for atoms to arrange themselves. That means if the universe truly is infinite, then somewhere out there is another planet Earth. Obviously, this is a preposterous Hypotheses. Imagine, there is infante number of you in different parts of infinite cosmos!

## 20,000 Stars Disappear Each Second

Scientists have computed that the universe is losing almost 20,000 stars each second. This is contributed to the huge expansion rate of the universe, which is 46 mi/s or 74 km/s. This expansion is shaped as a funnel, where the speed of expansion is smaller than that at the top of the funnel. As a result, the speed of expansion, Our entire expanding universe is making skies becoming emptier.

As a matter of fact, you may think of the expansion of the universe as a balloon that is being inflated bigger and bigger. This experiment is called "Expanding the Universe on a Balloon". If you would mark the galaxies on the surface of the balloon with a dot and kept inflating the balloon you would soon notice, that all the galaxies or dots in this experiment stay fixed in their position, but the distance between them is increasing.

We will still see stars in our very own Milky Way galaxy and for a long time we will see our local group of galaxies, but that is going to be it. The universe will get so big and will expand so fast, that it will be literally impossible to reach faraway galaxies unless we invent a way to travel faster than the speed of light (300,000 km/s, or 186,282 mi/s).

## Looking into The Past Consistently

Our observable universe has a radius of about 46 billion light-years even though allegedly it is only about 13.8 billion years old. Additionally, we are merely observing the past and view planets, stars, and very distant galaxies as they were a long time past.

Interestingly, we will look up to the skies to discover that the light from these distant stars voyaged numerous years to reach your eyes. It may well be that they are not even be there any longer. Envisage a distant star turning supernova, or even turning into a black hole, right at this moment Figure 1.31. Thanks to Hubble Telescope our bubble to expand and light from that blackhole has reached an observer on the Earth.

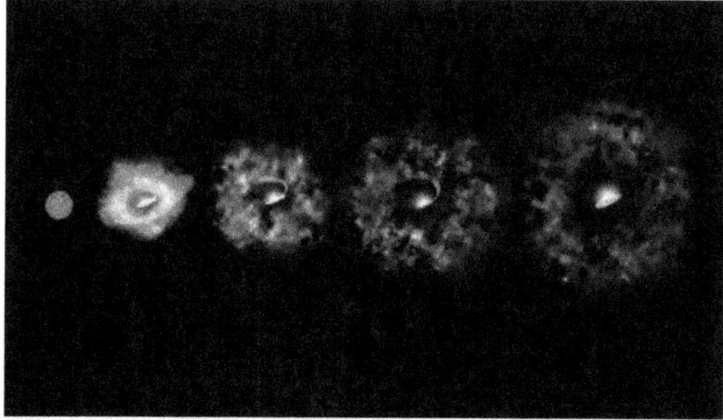

*Fig.1.31: A star Turned into a Black Hole Before Hubble's Very Eyes – Curtsy NASA*

But probably the most mind-blowing hypothesis, the claim that all of the matter and energy was condensed in the infinitely small point which blasted and contributed birth to time itself. No even Albert Einstein believed that – the foundations' fantasy of atoms in your body were already laid some 13.8 billion years ago.

## THE MARVEL OF OUR UNIVERSE

The Universe is a massive, on the face of it unending marvel of existence. Over the past two scores of years, we have acquired vast knowledge that the Universe stretches out beyond the billions of stars in our Milky Way, out across tens of billions of light years, containing close to 2-3 trillion galaxies all told, Figure 1.32.

**Fig.1.32:** *The First Cosmic Images from The James Webb Space Telescope–Unprecedented Views of Distant Galaxies, Bright Nebulae, and a Faraway Giant Gas Planet.–Curtsy: NASA*

And yet, that is just the **observable Universe**! There are virtuous reasons to have faith that the Universe remains on and on beyond the boundaries of what we can observe; how *far* does the universe go on? Endlessly? Or does it turn back itself at some scientists point?

To help us better understand this question, let's turn to something more familiar (and smaller) that we know how to measure the size of: the Earth.

## The Curvature of the Universe

From the highest point of a tall mountain, like Mauna Kea, you might hope to calculate the Earth's curvature, nonetheless, your labor would be in vain. From 14,000 feet up, the curvature of the Earth is completely indistinguishable from flat.

There are imageries out there someplace the Earth *seems* curved when you look out at the water, and certainly, they are not hard to find. Is that because of the Earth's curvature?

In fact, it is not. The view of curvature is contributed to the atmospheric distortion. If you were to attempt to calculate the circumference of the Earth from a photo like this, it will lead you to a smaller earth than *even the Moon*.

Moreover, the land surface of the Earth itself is not perfectly smooth. Many places on Earth are curved upwards, others downwards, Figure 1.33. A small visible region on Earth is unlikely to be a fair representation of the entire planet.

**Fig.1.33:** *Measuring the Curvature of Earth–Curtsy: James Elders*

There is a way that you would be able to measure, though, the shape and size of the planet Earth by take the appropriate measurements and use geometry. It is as simple by selecting three separate locations on Earth and drawing a triangle to connect those three points together, Figure 1.34.

On a flat surface of a triangle the three angles will always add up to 180°. Once you are on the surface of a sphere the **positive curvature** angles will add up to *more* than 180°. Knowing the distance between each of those three points will cause the total of the three angles is more than 180°. The three angles allow you to calculate the circumference of the Earth.

The farther away the three points are from one another, the less important the mountains' height, valleys and oceans, The converse would have been true if the Earth were shaped with **negative curvature**, like a saddle, as shown below, Figure 1.35.

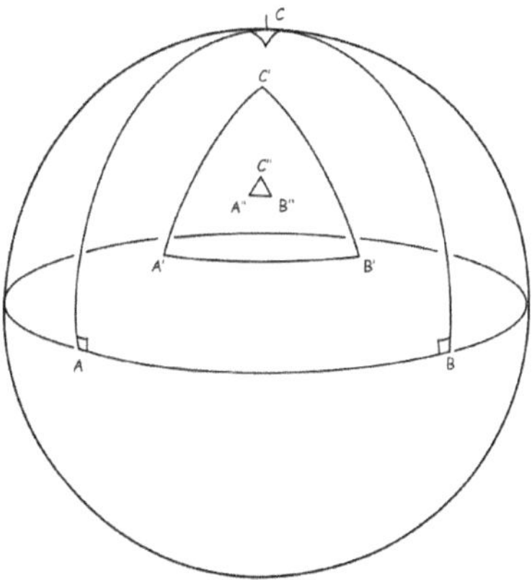

**Fig.1.34:** *Selecting Three Distant Points on Earth Surface*

**Fig.1.35:** *Negative Earth Curvature*

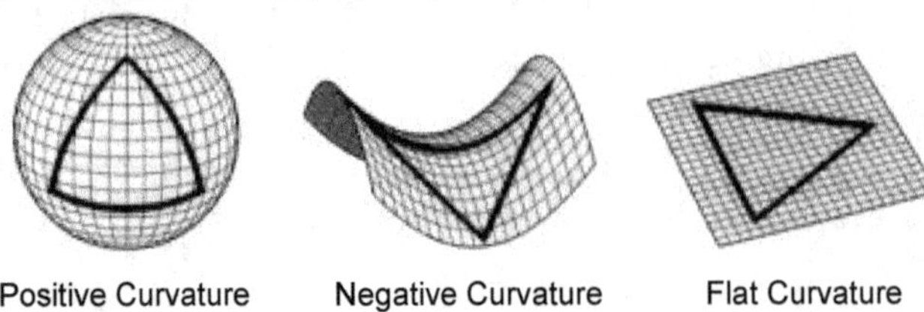



*Fig.1.36: Three Types of Rarth Curvature*

A surface of negative curvature has any three points form a triangle whose three angles sum to *less than* 180°, Figure 1.32, When knowing the distances and measurements of all three angles allows you to calculate the radius of curvature, Figure 1.37.

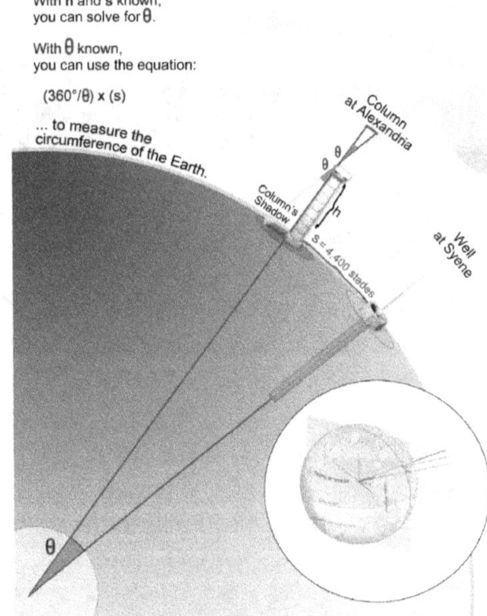



*Fig.1.37: Using Alexandria Column and Its Shadow, The Curvature of Earth was Calculated*

It would not be until the 20th Century that we were actually able to achieve altitudes capable of measuring the curvature of the Earth from space.

By 1948, we were creating mosaics of the Earth by stitching together multiple images of the Earth from space, and there could no longer be any doubt as to its circumference, Figure 1.38.

*Fig.1.38: Space Images Stitched together to Calculate the Earth Curvature. Curtsy Johns Hopkins / U.S. Navy*

The universe curvature is more difficult. However, it is still just a geometric construct, albeit a slightly more complicated one,

According to the theory of Einstein's General Relativity, the amount that the space of our Universe is curved is directly proportional to the amount of matter and energy that contained in the space, Figure 1.39.

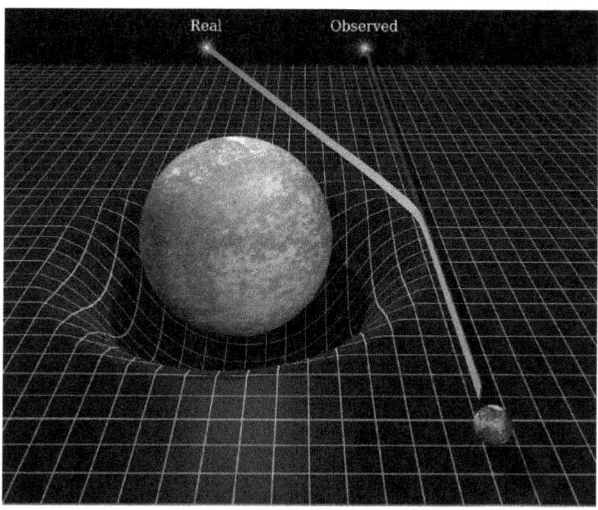


**Fig.1.39:** *Space is Curved Proportional to Mass object – Curtsy–Dave Jarvis Gallery*

The dense heavy masses as the Sun causes large amounts of curvature in a small space, which is significant enough to bend starlight by an amount significant enough to be observed by 1919's technology. Also, the Earth, curves the spacetime around itself.

Would the entire Universe ever close back in on itself, and if so, how big it is. In principle, just as any three points on a surface can aid us to calculate that surface's curvature, we can accomplish the exact same procedure with the Universe. Take any three points that are far enough apart, measure the distances between those points and the relative angles between them as well, and we will be able to figure out not only how the space is curved, but also what the radius of curvature is!

There are three possible cases.

i.  one Curve is where the Universe is positively curved, like a higher-dimensional sphere, Figure 1.40 A

ii.  one is where the Universe is totally flat, like a higher-dimensional grid, Figure 1.40 B, and

iii.  one where the Universe is negatively curved, like a higher-dimensional saddle, Figure 1.40 C.

In the context of general relativity, it's the energy density — the amount of matter and all other forms of energy — that determine this curvature.

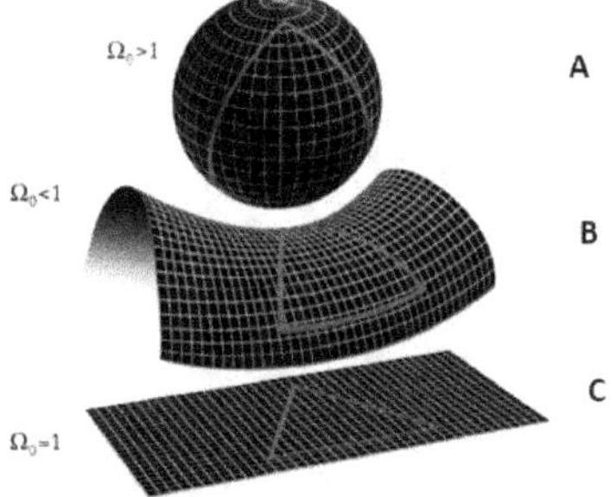


**Fig.1.40:** *Possible Curvature of the Universe*

As no one has a man-made object far away to express the necessary distances to measure curvature. However, we have light signals from the time when the Universe was just 380,000 years old, which tells us the shape of the Universe 46 billion light years away (Distance).

It is claimed that the fluctuations in the cosmic microwave background, allegedly is the — leftover glow from the big bang, may provide a window allowing us to see how our Universe is curved.

The first robust measurements of this came from the "Boomerang experiment," in the late 1990s, where scientists first agreed, instead of having significant positive, negative curvature, or flat as the Universe was indistinguishable from flat. The procedure is somewhat similar to finding Earth curvature. Nonetheless, by walking outside your location on Earth attempting to measure the curvature of the Earth within the horizon of 3 miles or 5 km. You would find that the Earth is *consistent* with being flat. However, it could also be positively or negatively curved on a larger scale of measuring.

The Universe is similar if we are able to make that measurement more precise by employing the PLANCK satellite, we measure the temperature fluctuations over the sky measuring an area using a resolution of 1/10 of a degree, Figure 1.41.

Astonishingly, the findings revealed the **Universe is consistently flat**. If the Universe does curve back and close on itself, its radius of curvature would have been at least 150 times as large. Thie finding disputes any notation made by Hawking that the Universe turns back on itself,

We are probably observing less than 0.0001% of the volume of the entire Universe. Once we consider "dark matter" and "dark energy" in addition to the expansion of the Universe, we realize that we will never observe more of the Universe we observe at present, or at any time in the future history of mankind. Accordingly, all that we currently observe —the billions of stars in each galaxy all of the galaxies, all lighting up our Universe.

*Fig.1.41: Planck Telescope Measuring Temperature Variation. Planck is a European Space Agency mission with significant participation from NASA. It was launched into space in May 2009, and now orbits the second Lagrange point of our Earth-sun system, about 1.5 million km (930,000 miles) away. Curtsy—NASA Planck mission.*

## The beginning of the Universe Hypothesis

Many Atheistic scientists, and some theistic scientists also argue that roughly 13.75 billion years ago, our universe came into existence. soon subsequently, allegedly, primordial light began shooting across the cosmos and scattering throughout the early universe. At this interval, the universe itself was also expanding. Also, atheistic scientists claim that the inflation of the universe decelerated immediately after the first initial burst. However, since then, the rate of expansion has been steadily cumulative due to the influence of dark energy/ dark matter.

Basically, since its beginning, the cosmos has been growing at an ever-increasing rate. Cosmologists approximate that the oldest photons that we can observe have voyaged a distance of 45-47 billion light-years since the alleged "Big Bang." That means that our observable universe is some 93 billion light-years wide. These 93 some billion light-years encompass all of the galaxies, quasars, stars, planets, nebulae, black holes, and everything else that we could possibly observe. However, the *observable universe* only encompasses the light that has had sufficient time to reach us. A much more universe occurs beyond what we are able to observe.

The observable universe is claimed to be 93 billion light-years across, while it is only 13.8 billion years old. Light hasn't had enough time to travel that far. Ultimately, understanding this facet of physics is the key to understanding what lies beyond the edge of the observable universe and whether we could ever get there.

To break this down, according to special relativity, objects that are close together cannot move faster than the speed of light with respect to one another. However, scientists falsely hypnotized that there is no such law for objects that are enormously distant from one another as the space between them is, itself, expanding.

In summary, it is not that objects are traveling faster than the speed of light, but that the space between objects is expanding, causing them to fly away from each other at amazing speeds. This claim has no imperial proof nor observational evidence. Einstein, disclaimed assumption due the absolute nature of the speed of light, 186,282 mi/s.

Ultimately, this false assumption that space expanded faster than the speed of light. It means that we could only reach the edge of the observable universe if we develop a method of transport that permits us to either:

1. Travel faster than the speed of light (something which most physicists think is impossible)
2. Transcend spacetime, by utilizing wormholes or warp drive, which most physicists also think is impossible.

According to the theory of cosmic inflation, the entire universe's size is at least $10^{23}$ times larger than the size of the observable universe. We, therefore, missing much too much universe.

## OUTSIDE THE OBSERVABLE UNIVERSE

Regrettably, since we cannot observe it or measure it, we cannot distinguish what lies beyond the limits of the observable universe. Though, we have several theories regarding the existence in the countless unknowns.

### The Unknown

Notwithstanding the strangeness of the unknowns of the universe, astronomers think space outside of the observable universe could be an infinite expanse of what we observe in the cosmos around us. It is distributed much the same as it is in the observable universe.

Another alternative of beyond the observable universe, is infinity. This idea does not sound plausible. It means that, someplace out there, there is alternative person who is undistinguishable to you in every conceivable way, and there is also a you who is only *slightly* different from you in every possible way. This notion seems inconceivable. Nonetheless, the infinity is inconceivable.

### Dark Flow

Another theory deals with something called "dark flow." In 2008, astronomers revealed something very bizarre and unanticipated—galactic clusters were all flooding in the same track at immense speed of more than two million miles per hour. One possible cause of this massive structures outside the observable universe is applying gravitational influence. As for the structures themselves, they could be anything enormous consists of matter and energy on scales we can hardly imagine. It could be even bizarre warps in space-time that are funneling gravitational forces from other universes. Simply, we do not know the character of these massive objects. Recent analyses claimed to discredit the "dark flow" model Figure 1.40.

## A Universe of Universes (Multi-Verse)

Another alternative involves a universe of universes. Some scientists have considered that our entire universe could exist in a small "bubble" in the midst of a vast array of other bubbles. Theorists call this a "multiverse." Fascinatingly, the idea declares that these universes can come into contact with one another through gravity, which can flow between these parallel universes, and when they connect, another Big Bang as the one that made our universe may occur.

## 10 Times More Galaxies Discovered

Astronomers came to the astonishing conclusion that there are at least 10 times more galaxies in the observable universe than previously thought. The consequences have clear suggestions for galaxy formation, and also aids to shed light on an ancient astronomical inconsistency, why is the sky dark at night?

In analyzing the data, a team led by Christopher Conselice of the University of Nottingham, U.K., claimed that 10 times as many galaxies were crowded into a specified volume of space in the initial universe than found today. Most of these galaxies were comparatively small and faint, with masses similar to those of the satellite galaxies surrounding the Milky Way.

As they amalgamated to form larger galaxies the population density of galaxies in space decreased, which explains that galaxies are not evenly disseminated throughout the universe's history. The research team stated that there is strong evidence that a substantial galaxy evolution has taken place throughout the universe's history, which intensely condensed the number of galaxies through unions between them — therefore, plummeting their total number. This provides us a confirmation of the so-called top-down development of structure in the universe.

One of the most important questions in astronomy is how many galaxies the universe contains. The landmark "Hubble Deep Field," telescope, in the mid-1990s, provided the first true vision into the universe's galaxy population. Following sensitive observations such as "Hubble's Ultra Deep Field" exposed a myriad of faint galaxies. This guided to an approximation that the observable universe contained about. The new research demonstrated that the true estimate is at least 2000 billion galaxies.

"Conselice" and his team grasped this deduction using deep-space images from Hubble and the previously published data from other research scientists. They thoroughly rehabilitated the images into 3-D, in order to make precise measurements of the number of galaxies at diverse epochs in the universe's history.

Furthermore, they utilized new mathematical models, which permitted them to conclude the existence of galaxies that the present generation of telescopes cannot observe. This guided to the astonishing deduction that in order for the numbers of galaxies we now see and their masses to add up, there must be an additional 90 percent of galaxies in the observable universe that are too faint and too far away to be seen with present-day telescopes.

These countless small faint galaxies from the early universe combined over time into the larger galaxies we can now observe.

"It perplexes the mind that over 90 percent of the galaxies in the universe have yet to be investigated. The James Webb Space Telescope is provided data to study these ultra-faint galaxies.

The lessening number of galaxies as time advances, also pays to the solution for "Olbers' paradox," first formulated in the early 1800s by German astronomer Heinrich Wilhelm Olbers; Why is the sky dark at night if the universe contains an infinity of stars? The team came to the supposition that there essentially is such an plenty of galaxies —that in every patch in the sky contains part of a galaxy. Though, starlight from the galaxies is unseen to the human eye and most modern telescopes due to the other known features that decrease visible and ultraviolet light in the universe. Those features are the reddening of light due to the expansion of space, the universe's dynamic nature, and the absorption of light by intergalactic dust and gas. All combined keeps the night sky dark to our vision.

When we look out into the Universe, the stuff we can observe must be close enough for light to have reached us since the Universe began. The universe is about 14 billion years old, so at first glance it is easy to think that we cannot see things more than 14 billion light years away.

The Universe is expanding, the most distant visible things are much further away than that. In fact, the photons in the cosmic microwave background have travelled a cool 45 billion light years to get here. That makes the visible universe some 90 billion light years across.

While that is big, the universe is certainly much bigger. However, how much bigger the inover could be.

## Mass, Size, and Density of the Universe

The mass, size, and density of the universe encompass very big numbers. Scientists write such numbers using a special code that keeps them short. When they write a number like 6e7 then the number after the "e" indicates by how many places the decimal point must be shifted to the right. For instance, 6e7 is equal to a 6 with 7 zeros behind it, or 60,000,000, and 6.1e7 is equal to 61,000,000. If the number after the "e" is negative, then the decimal point is shifted to the left. For instance, 6e-3 is equal to a 6 with 3 zeros in front of it and the decimal point after the first zero, or 0.006, and 6.1e-3 is equal to 0.0061.

The mass density of visible matter, i.e., galaxies in the Universe is estimated at:

3e-28 kg/m$^3$ (3e-31 times the mass density of water.

The radius of the visible Universe is estimated at:

1.7e26 m — 18 thousand million light years, plus or minus 20 percent.

This yields a total mass of the visible matter of about:

6e51 kg (1.3e52 lb), which is equivalent to

the weight of 4e78 hydrogen atoms.

Since nine out of ten atoms and ions in the Universe are in the form of hydrogen, this is a sensible approximation for the number of atoms in the Universe, founded on the visible galaxies only. Maybe a correction factor of the order of 2 has to be inserted to account for the warping of space on very large scales.

Though, there is substantial ambiguity about the mass density of all matter, visible and invisible, and energy, through.

**Einstein's E = mc$^2$ equation.**

As one studies, the movement of matter in and around galaxies, then it seems that up to about 10 times more mass is dragging at the matter, through its gravity, than is described for in the visible stars. This is the "missing-mass" difficulties. If this factor of ten holds through the Universe, then the total mass in the Universe would be about 6e52 kg. If the missing mass were mostly in the form of hydrogen atoms, which is not at all clear, then the number of atoms would be about 4e79.

A present popular theory of the development of the Universe, the so-called Inflation Theory, forecasts that the mass density of the Universe should be close to the so-called critical density that separates an open universe that always grows from a closed universe that ultimately breakdowns again. This critical mass density is presently equal to 6e-27 kg/m$^3$. If the Universe is at the critical density, then the total mass of the Universe is closer to 1e53 kg, and the number of atoms, assuming that most of the mass is in the form of hydrogen atoms, about 6e79.

It seems, then, that the number of atoms in the Universe is at least about 4e78, but perhaps as many as 6e79. It is suggest to consider 1e79 as a sensible approximation. That is, 10 000 000 000 000 000 000 000 000 000 000 000 000 000 000 000 000 000 000 000 000 000 000 000 000 000 atoms.

Obviously, in addition to material particles there are also much of photons and neutrinos flying around the Universe. It is projected that there are about 1e9 times as many photons and neutrinos as atoms in the Universe.

Clearly, we may not directly measure the size of the universe but cosmologists have various models that suggest how big it ought to be. E.g., one consideration is that if the universe expanded at the speed of light during inflation, then it ought to be $10^{23}$ times bigger than the visible universe.

## No Periphery to the Finite Universe

**Fig.1.42:** *The Finite Universe – Curtsy–NASA/JPL*

As far as we know, there is no edge to the universe. Space spreads out indefinitely in all directions. Moreover, galaxies fill all of the space through-out the entire indefinite universe. This deduction is reached by logically combining two observations, Figure 139.

First, the portion of the universe that we can observe is uniform and flat on the cosmic scale. The uniformity of the universe indicates that galaxy groups are spread out more consistently on the cosmic scale. The flatness of the universe indicates that the configuration of spacetime is not curved or warped on the cosmic scale. This indicates that the universe does not wrap around and connect to itself like the surface of a sphere, which would lead to a finite universe.

The flatness of the universe is actually a result of the uniformity of the universe, subsequently focused groups of mass cause spacetime to be curved. Moons, planets, stars, and galaxies are examples of focused groups of mass, and therefore they do indeed warp spacetime in the area around them. Though, these objects are so small compared to the cosmic scale, that the spacetime warping which they cause are negligible on the cosmic scale.

If one averages over all of the moons, planets, stars, and galaxies in the universe in order to obtain a large-scale figure for the mass distribution of the universe, you find it to be constant.

The second observation is that our angle of the universe is not special or different. Since the part of the universe that we can see is flat and uniform, and since our angle of the universe is not unusual, all parts of the universe must be flat and uniform. The only method for the universe to be flat and uniform accurately *everywhere* it is still finite universe. It expands constantly beyond any instrument to grasp its ends.

As time marches on, more and more points in space have had time for their light to reach us. Hence, our observable universe is continually increasing in size. You may think then that after an eternity of time, the entire universe will be observable to humans. There is, however, a difficulty that avoids this. The universe itself is still expanding. Although the present expansion of the universe is not as fast as during the hypothetical "Big Bang," it is just as real and important. Consequently, of the expansion of the universe, all galaxy groups are becoming constantly farther away from each other. Many galaxies are so far away from the earth that the expansion of the universe causes them to recede from the earth at a speed allegedly faster than light. Although special relativity averts two local objects from ever traveling faster than the speed of light with respect to each other, it does not allegedly avoid two distant objects from traveling away from each other faster than the speed of light thus the expansion of the universe.

Meanwhile these distant galaxies are receding away from earth at a hypothetical speed faster than light, the light from these galaxies will never reach us, no matter how long we wait. Consequently, these galaxies will continuously be outside of our observable universe. the real universe is also growing. The edge of the observable universe cannot sustain with the expansion of the universe so that numerous galaxies are eternally beyond our observation.

## DECLINING THE BIG BANG

### Background Heat

Space seems to be immersed in a low-level background heat which is supposed to be the "after-glow" or "smoking gun" of the original "big bang" blast. This is known as the "cosmic microwave background radiation" (CMBR). It has been computed precisely, and in excessive detail, and is said to have a temperature constant with what would be predictable from the "big bang."

Theistic scientists, though, point out that this background heat is essentially a chief *problem* for big bang theory. This is contributed to its temperature is effectively the same across the universe and this would *not* be predictable from a big bang. A orthodox "explosion" would leave behind an irregular pattern of heat, not the extremely even pattern actually detected.

Evolutionists, obviously, are well conscious of this difficulty, entitled the "horizon problem," and some claim to have an answer. Allegedly, very soon after the early "bang," and for only a fleeting period, the universe expanded at a much higher rate—actually faster than the speed light—and this allowed the background heat to be smoothed out. This very speedy expansion is known as "inflation."

Though, rendering to Paul Steinhardt, Albert Einstein Professor in Science at Princeton University, "inflation is very flexible … [and] can be adjusted to give any result … any outcome is possible." Hence, he says, "it is not possible to find evidence to support or refute inflation". Theistic scientists would reach agreement, and argue that it is really no more than a story so-called upon to explicate away facts that point to the big bang theory being erroneous.

### Abundance of Light Elements

Big bang theory is supposed to have correctly foretold the amounts of light elements that we essentially find in the universe. Hydrogen would be predictable to be the most common, followed by helium, deuterium and lithium—which is what is observed. Predominantly, it is said, the amount of helium (25%) is consistent with computations founded on big bang theory. Historically, nevertheless, this claim has been very contentious. As pointed out by Professors Burbidge and Hoyle,

*"We have now reached the stage where it is argued that the existence of helium … is taken, together with the microwave background radiation, as primary evidence in favor of the … big bang … . However, this argument is only powerful if there is no other way to explain the helium abundance and the microwave background radiation."*

They then contended that this helium was formed by the burning, or fusion of hydrogen in stars, rather than the big bang, and that the light emitted by the stars and absorbed by dust clouds produced the cosmic microwave background radiation.

According to Professor Burbidge, neither the observed wealth of helium, nor the level of the cosmic microwave background, were actually forecasts of big bang theory. Scientists had already computed the wealth of helium and the theory was adjusted so that it would provide the "right result." Denoting to the parameter, the number in the big bang theory, which would result in it "forecasting" the amount of helium in the universe, he wrote:

*"It is chosen to make things come out right … ."*

This is the reason the big bang theory cannot be argued to *explain* microwave background or to explain a cosmic helium value close to 0.25 [i.e., 25%]. … if you actually trust in a big bang, you can choose parameters which will make observation and theory agree, but the quarrel is not based on basic theory" (emphasis original).

Evolutionary cosmologists occasionally claim that big bang theory can forecast the amount of the light elements utilizing information, data gained from satellite measurements. This, though, cannot be so because their theory depends on the existence of "dark matter," a form of matter that cannot be observed. This "dark matter," which some secular cosmologists do not believe exists, is also required to clarify how galaxies and stars form by natural processes.

## Toleration the Big Bang

While many theistic leaders have not merely tolerated the "big bang" idea, nonetheless. incorporated it unreservedly. At last, faithful theistic scientists can confidently use science to prove there is a true infinite creator of the universe.

Nevertheless, the value of yielding to the lure of secular satisfactoriness, at least in physics and astronomy, has been heavy. Adopting the "big bang" into theistic thought is as riding the wooden horse within the walls of Troy. The previous section spread in this book may highlight the idea of the Big Bang:

1. The big bang forces acceptance of an arrangement of events completely mismatched with the Bible, e.g., earth after sun instead of earth before sun

2. The big bang's billions of years of astronomical development are not only founded on naturalistic assumptions, they are conflicting to the words of Jesus Himself, who said people were there from the beginning, not towards the end of an interminably long 'creation' process.

3. The leisurely evolution of the stars, then solar system and planets, as well as earth, in big bang ensures that "big bang theists" are perpetually pulled into compliant "geological evolution," which is millions of years for the earth's fossil-bearing rocks to be laid down. Consequently, theists end up repudiating the "global Flood," and accepting death, carnage and disease, as seen in the fossils, before Adam. This eliminates the "Fall" and the "Curse" on creation from any consequence on the real world, in addition to eliminating *the* biblical answer Theists have continuously had to the problem of suffering and evil, God made a perfect world, ruined by sin.

4. Wedding one's doctrine to today's science means that one is assuredly to be widowed tomorrow.

## Reasons Big Bang Theory Busted

In actual fact, the signs are robust that precisely that is occurring, and that those who have "sank their teeth" in the "big bang" for its allegedly irrefutable science have "caught wind." A bombshell "Open Letter to the Scientific Community" by 33 leading scientists has been published on world-wide, ("Cosmology statement,") and in *New Scientist* (Lerner, E., Bucking the big bang, *New Scientist* 182(2448)20, 22 May 2004). An article on www.rense.com titled 'Big bang theory busted by 33 top scientists' (27 May 2004) says, "Our ideas about the history of the universe are dominated by 'big bang theory.' But its dominance rests more on funding decisions than on the scientific method, according to Eric Lerner, mathematician Michael Ibison of Earthtech.org, and dozens of other scientists from around the world."

The open letter includes statements such as:

a.  The big bang today relies on a growing number of hypothetical entities, things that we have never observed—inflation, dark matter and dark energy are the most prominent examples. Without them, there would be a fatal contradiction between the observations made by astronomers and the predictions of the big bang theory.'

b.  'But the big bang theory can't survive without these fudge factors. Without the hypothetical inflation field, the big bang does not predict the smooth, isotropic cosmic background radiation that is observed, because there would be no way for parts of the universe that are now more than a few degrees away in the sky to come to the same temperature and thus emit the same amount of microwave radiation. … Inflation requires a density 20 times larger than that implied by big bang nucleosynthesis, the theory's explanation of the origin of the light elements.' [This refers to the horizon problem, and supports what we say in Light-travel time: a problem for the big bang.]

c.  'In no other field of physics would this continual recourse to new hypothetical objects be accepted as a way of bridging the gap between theory and observation. It would, at the least, raise serious questions about the validity of the underlying theory [emphasis in original].'

d.  'What is more, the big bang theory can boast of no quantitative predictions that have subsequently been validated by observation. The successes claimed by the theory's supporters consist of its ability to retrospectively fit observations with a steadily increasing array of adjustable parameters, just as the old Earth-*centered cosmology of Ptolemy needed layer upon layer of epicycles.*'

The nonconformists say that there are other clarifications of cosmology that do make some effective predictions. These other models do not have all the responses to oppositions, nevertheless, they say, *"That is scarcely surprising, as their development has been severely hampered by a complete lack of funding. Indeed, such questions and alternatives cannot even now be freely discussed and examined."*

## SILENCING YOUNG SCIENTISTS

Those who desire Atheists to agree to take the "big bang" as a "science fact" indicate its near-universal acceptance by the scientific community. Though, the 33 dissidents describe a situation acquainted to many creationist scientists: "An open discussion of ideas is missing in most mainstream conferences … uncertainty and opposition are not endured, and young scientists learn to keep on silent if they have something negative to say about the standard "big bang model." Those who distrust the big bang fear that saying so will cost them their funding."

Evolutionist and historian of science, "Evelleen Richards," has observed that it is tough even for rival *evolutionary* theories to grant a hearing when challenging the ruling paradigm—This should provide some awareness of the difficulties Theistic creationists face.

In fact, the prominent secular scientists' state: *"Even observations are now interpreted through this biased filter, judged right or wrong depending on whether or not they support the big bang."* Accordingly, disagreeing data on red shifts, lithium and helium abundances, and galaxy distribution, among other topics, are ignored or scorned.

Science is a delightful human instrument, but then, it desires to be comprehended, not worshipped. It is imperfect, moving, and is strictly limited as to what it can and cannot govern. Devoted honest Theistic scientists have often clarified, in the place of a scientific idea, the "big-bang" idea is more a rigid religious one—built on the religion of humanism. As these "big-bang" adversaries point out:

*"Giving support only to projects within the big bang framework undermines a fundamental element of the scientific method—the constant testing of theory against observation. Such a restriction makes unbiased discussion and research impossible."*

Furthermore, contrary to the naïve assertions of many who should know better, it is not in any sense a matter of "looking into a telescope and "seeing"? the "big bang" billions of years ago." As continuously, observations are interpreted and filtered through worldview lenses. Those who advanced the "big bang" were directed by secular worldview filters just as much as those who are now crying that the "emperor has no clothes."

They desired a universe that created itself; their adversaries want an eternal, uncreated universe. From a Theistic viewpoint, both are in open rebelliousness of their Creator's account of what actually occurred.

With Darwinism on the run, the Adversary of souls is looking for seducing believers into espousal a more subtle, yet far fatal method of escaping the authority of the Scripture. With progressive creationism "big-bangers" raging through the evangelical community, think they are on a winning side.

The big bang, has become a ruling paradigm, supported by fallacious logic and ignoring many scientific problems.

# APPENDIX

## An Atlas of the Universe–A Glossary

### What is considered a light-year?

Light-year is **the distance light travels in one year**. Light zips through interstellar space at 186,000 miles (300,000 kilometers) per second and 5.88 trillion miles (9.46 trillion kilometers) per year.

### Abell catalogue (A, ACO)

A catalogue of 2712 *rich clusters of galaxies* produced by George Abell in 1958 from careful examination of the Palomar Sky Survey plates. It was extended in 1989 by George Abell, Harold Corwin and Ron Olowin to include an extra 1364 rich clusters in the far southern hemisphere not covered by the original Palomar Sky Survey. It contains most of the richest clusters of galaxies within 3 billion *light years*.

### Abell cluster

See *Rich cluster of galaxies, Abell catalogue.*

### Absolute magnitude

The *apparent magnitude* an object would have if placed at a distance of exactly 10 parsecs (=32.6 *light years*). A *supergiant star* might have an absolute magnitude of -8 whereas a dim *red dwarf* might have an absolute magnitude of +16. The Sun has an absolute magnitude of +4.8–about half way between the two extremes.

## Angular Size

The apparent size of an object expressed as an angle. It is measured in degrees, minutes and seconds.

## Apparent magnitude

The system used to give the brightness of stars in the sky. Brighter stars have lower numbers and dimmer stars have higher numbers. The dimmest objects visible with giant telescopes have a magnitude of +30. A good portable telescope might see down to magnitude +15. Binoculars can see down to magnitude +9 and the faintest naked eye stars have a magnitude of +6. Very bright objects have a negative magnitude, the brightest star has a magnitude of -1.4, the full Moon has a magnitude of -12.7 and the noon Sun has a magnitude of -26.8.

Arcminute (')

A measure for small angles. 1 arcminute = 1/60 degree.

Arcsecond (")

A measure for very small angles. 1 arcsecond = 1/60 arcminute = 1/3600 degree.

Association

See *OB association*.

## Astronomical Unit (AU)

The average distance between the Earth and the Sun. It is equal to 149 597 871 km. It is often used for distances in a solar system or for distances between companion stars.

## Barnard catalogue

A catalogue of 349 dark nebulae north of declination -35° produced by E Barnard in 1927.

## Barred Galaxy

A galaxy with a bright central bar of stars.

## Bayer name

The combination of a Greek letter and the name of a constellation (Alpha Centauri, Epsilon Orionis etc.) used to identify bright stars. The system was first used by Johann Bayer in 1603. Brighter stars in a constellation usually have a letter near the beginning of the alphabet, and dimmer stars usually have a letter nearer the end of the alphabet. A few faint stars were given lower-case Roman letters from a to z or upper-case Roman letters from A to Q (p Eridani, N Velorum etc.) See also *Flams teed number*.

## Billion

1 billion = 1 000 000 000.

## Black hole

The ultimate cosmic plughole formed when a high mass supergiant star explodes in a supernova explosion at the end of its life creating a super-dense point in space where nothing can escape the gravitational pull. A star probably has to have a mass of more than 40 solar masses to create a black hole which typically have a mass of about 3 solar masses. Black holes can be detected by the disrupting effects they have on neighboring stars. The centers of most galaxies including our own are believed to have super-massive black holes which have sucked in thousands of stars.

## Blue Apparent Magnitude

The apparent magnitude of a star or galaxy when viewed through a blue filter. This magnitude system is commonly used for nearby galaxies because, historically, pictures of galaxies were photographed with photographic plates sensitive to blue light.

## Bonner Durchmusterung catalogue (BD, CD, CP)

A star catalogue of 325 037 northern stars produced between 1859 and 1862 with a supplement of 134 833 southern stars produced in 1886. Later came two more large southern star catalogues: the Cordoba Durchmusterung (613 959 stars) produced between 1892 and 1932, and the Cape Photographic Durchmusterung (454 877 stars) produced between 1895 and 1900.

## Bright Star catalogue (HR, BS, Yale)

A star catalogue of nearly all the stars brighter than *magnitude* +6.5 published by the Yale University Observatory. The original version was published in 1908 as the Harvard Revised Photometry Catalogue and it contains 9096 stars. The current version is the fifth edition.

## Brown dwarf star

A failed star that was too small at its birth for nuclear reactions to occur in its core. They may be very common, but because they only glow very dimly, they are very hard to detect. Brown dwarfs are **not** brown, they begin their lives by glowing a dull red and then fade. Brown dwarfs are more massive than planets, and range in mass from 10 to 80 times the mass of Jupiter (or 0.01 to 0.08 times the mass of the Sun). There are two main types of brown dwarfs–the hotter ones (1500K to 2500K) are type L; the cooler ones (below 1500K) are type T. The hottest brown dwarfs are sometimes classified as very cool type M *red dwarf stars*.

## Cluster of galaxies

A concentration of *galaxies* bound together by gravity. The term 'cluster' usually refers to a large collection of many tens or hundreds of *galaxies* and the term 'group' is used for a much smaller grouping.

## Constellation

Random patterns of stars in the night sky produced by the chance alignment of stars of different luminosities and distances. There are 88 constellations–48 were listed by the ancient Greeks, and another 40 were added after 1590.

## Dark matter

The visible stars and nebulae make up only a small fraction of all the matter in the universe. The rest is in a form that is not easy to detect, but clearly exists because of the effect it has on the motion of stars in *galaxies* and the motion of galaxies in clusters. Dark matter probably consists of various types of subatomic particles.

## Declination (Dec)

See *Equatorial coordinates*.

## Dwarf galaxy

A small *galaxy* usually containing any number of stars between a million and several billion. There is no official size below which a galaxy is designated a dwarf but any *galaxy* with a diameter below 30 000 *light years* can be considered a dwarf.

### Dwarf star

A normal *main sequence star* like the Sun that is burning hydrogen in nuclear reactions in its core. The brightest dwarf stars can be much larger than the Sun. See also *Giant star*, *Supergiant star*.

### Elliptical galaxy

A galaxy with a spherical or oval shape. They are classified as type E and range in shape from type E0–circular, to type E7–very elliptical.

### Equatorial coordinates (RA, Dec)

The most common coordinate system used by astronomers. It is the equivalent of the Earth's latitude and longitude projected onto the sky except that longitude is called Right Ascension and latitude is called Declination. For historical reasons, Right Ascension is not measured in degrees but in 'hours'–24 hours being equivalent to 360 degrees. Another complication is that this coordinate system is very slowly moving with time–star positions for 1950 are slightly different to those for the year 2000 for example.

### ESO catalogue (ESO)

The ESO/Uppsala Catalogue of Galaxies. An extension to the UGC catalogue listing 18 422 bright *galaxies* south of declination -20° published in 1982. Note that the combined UGC and ESO catalogues do not cover the declination zone between -2.5° and -20°.

### Flamsteed number

The combination of a number and the name of a constellation (61 Cygni, 36 Ophiuchi etc.) used to identify naked-eye stars. The numbers were applied to John Flamsteed's star catalogue published in 1725. Not all naked-eye stars have a Flamsteed number and most stars in the far southern hemisphere do not have one. See also *Bayer name*.

### Galactic coordinates (l , b)

A coordinate system based on the plane of the Galaxy, it is centered on the Sun with the zero point of longitude and latitude pointing directly at the galactic center. The symbols used for galactic coordinates are l (longitude) and b (latitude). The zero point of Galactic latitude and longitude is at RA=17h45m37s Dec=-28°56'10", and the Galactic north pole is at RA=12h51m26s Dec=+27°07'42", (epoch 2000 coordinates).

### Galactocentric coordinates

The same coordinate system as *galactic coordinates* except that this system is centered on the center of the Galaxy. The small uncertainty in the distance to the galactic center prevents this system from being widely used.

### Galaxy

A vast concentration of millions or billions of stars. There are four main types of galaxies: *Elliptical Galaxies*, *Lenticular galaxies*, *Spiral Galaxies* and *Irregular galaxies*. Our galaxy contains 200 billion stars, but the largest galaxies contain many trillions. See also *Dwarf Galaxy*.

### Giant star

A star the size of the Sun will end its life after several billion years by expanding greatly because of changing energy balances at the core of the star. The surface temperature drops and the star becomes redder, this lasts several million years before the star throws off its outer layers and becomes a *white dwarf*. See also *Dwarf star*, *Supergiant star*.

### Giclas catalogue (G)

The usual name given to the Lowell Proper Motion Survey catalogue produced in the 1970's containing 11 747 stars with a high proper motion.

### Gliese catalogue (Gl, Wo, GJ)

The usual name given to three catalogues of nearby stars compiled by W Gliese (and later by H Jahreiß) in 1957, 1969 and 1993. The third catalogue which listed 3803 stars within 25 *parsecs* was only released in a preliminary form.

### Globular star cluster

A spherical cluster of many thousands of stars. Globular clusters usually have a size of 50 to 150 *light years* and are found scattered in a spherical halo surrounding *galaxies*.

### Group of galaxies

A concentration of *galaxies* bound together by gravity. The term 'group' usually refers to a small collection of a few tens of galaxies and the term 'cluster' is used for a much larger grouping.

### Harvard Revised catalogue (HR)

See *Bright Star catalogue*.

### Henry-Draper catalogue (HD)

A catalogue of the *spectral classes* of 272 150 stars produced between 1918 and 1924.

### Hipparcos catalogue (HIP)

A catalogue of 118 218 stars surveyed by the Hipparcos satellite launched in 1989. It collected the *parallaxes* of these stars providing accurate distances to tens of thousands of stars within 1000 light years.

### Hubble constant (H)

The value which describes the rate at which the universe is currently expanding. The Hubble constant helps to determine the size and age of the universe, and is also used to convert the *redshift* of a *galaxy* into a distance estimate. The Hubble constant is not known precisely but it is somewhere in the range of 60 to 80 km/s/Mpc.

### IC catalogue (IC)

The Index Catalogue. A supplement to the *NGC catalogue* listing an extra 5386 *star clusters*, *nebulae* and *galaxies*. It is actually a combination of two catalogues, the first published in 1895 and the second in 1908. See also NGC catalogue.

### Irregular galaxy

A *galaxy* with a very irregular shape and no obvious elliptical or spiral structure. They are classified as type Irr. Irregular galaxies with a crude spiral-like structure are often classified as type Sm (or type SBm if they also have a central bar).

### Large Galaxy

See Galaxy.

## Lenticular galaxy

A *galaxy* that is shaped like a lens. They are of an intermediate type between an *elliptical galaxy* and a *spiral galaxy*. They are classified as type S0 (or type SB0 if they also have a central bar).

## Light year (ly)

The distance light travels in a year. It is equal to 0.3066 *parsecs*. It is 9461 billion km or 63 240 *astronomical units*.

## Luyten, Willem

An astronomer who from the 1930's to the 1980's found thousands of nearby stars in several huge surveys including the Bruce Proper Motion Survey in 1944 and the Luyten-Palomar survey conducted in the 1970's. He produced many proper motion catalogues including the Luyten catalogue (L), the Luyten-Palomar catalogue (LP), the Luyten Two-Tenths Arcsecond catalogue (LTT), the Luyten Four-Tenths Arcsecond catalogue (LFT), and the Luyten Half-Second catalogue (LHS).

## Magnitude

See *Apparent magnitude, Absolute magnitude.*

## Main sequence star

A normal star like the Sun that is burning hydrogen in nuclear reactions to produce its energy. Other types of stars include: *giant stars, supergiant stars* and *white dwarfs.*

## MCG catalogue (MCG)

The Morphological Catalogue of Galaxies. A catalogue of 31 000 *galaxies* north of declination -45° produced between 1962 and 1974.

## Minute

See *Arcminute.*

## Nebula

A cloud of interstellar gas and dust. Bright nebulae glow with light emitted by the gas of which they are composed (emission nebulae) or by reflected starlight (reflection nebulae) or both. Dark nebulae consist of clouds of gas and dust that are not illuminated. *Planetary nebulae* are shells of gas ejected by stars.

## Neutron star

The core of a supergiant star which has collapsed during a supernova explosion so much that it consists entirely of neutrons. Most stars between 8 and 60 solar masses end their lives like this usually producing a neutron star with a mass of about 1.4 solar masses. Neutron stars are only 10 kilometers across and have an incredible density–a teaspoon of neutron star material would have a mass of hundreds of millions of tons. See also Black hole.

## NGC catalogue (NGC)

The New General Catalogue. A catalogue of 7840 of the brightest *star clusters, nebulae* and *galaxies* published by J Dreyer in 1888. See also *IC catalogue.*

### OB association

A loose group of tens or hundreds of very bright stars scattered over several hundred *light years* of space. The stars in an OB association were formed in the same star-forming region and are slowly moving apart. They are usually found in the spiral arms of galaxies.

### Open star cluster

A cluster of stars usually containing several hundred members packed into a region usually less than 20 *light years* in size. They are normally found near regions of star formation in the spiral arms of *galaxies*.

### Orange dwarf star

Stars with a luminosity in between the Sun-like *yellow stars* and *red dwarfs*. They are classified as type K stars.

### Parallax

The tiny periodic shift of the apparent positions of nearby stars due to the changing position of the Earth as it orbits the Sun. The nearer the star is, the larger the shift. The distance to stars in parsecs is simply 1/parallax, (or in light years it is 3.2616/parallax) where the parallax is in arcseconds.

### Parsec (pc)

The distance a star has to be to have a parallax shift of 1 arcsecond. (No star is actually this close). 1 parsec = 3.2616 light years.

### PGC catalogue (PGC)

The Catalogue of Principal Galaxies. A catalogue of 73 197 of the brightest galaxies published in 1989. Since 1989, galaxies have continued to receive PGC numbers so that there are now over 1 million galaxies with PGC numbers.

### Planetary nebula

An expanding envelope of gas surrounding a hot white dwarf. It is formed at the end of a giant star's life when the core contracts ejecting the outer atmosphere of the star creating both the white dwarf and the nebula. The intense radiation from the central white dwarf makes the nebula glow. Planetary nebulae disperse within 50 000 years. They are called planetary nebulae because to early astronomers they looked a bit like planets.

### Proper Motion

The slow steady shift of the apparent positions of nearby stars over many years because of their independent motion within the Galaxy. Even the nearest and fastest stars require centuries to move a degree or more.

### Quasar

A *galaxy* with an extremely luminous nucleus outshining the parent *galaxy* by several hundred times. They lie billions of light years away and were a feature of the early universe. The energy source of a quasar is probably matter falling into a supermassive black hole.

### Radial Velocity (RV)

The speed of an object in the direction towards or away from the observer. In an expanding universe a galaxy with a larger radial velocity generally lies further from the observer than one with a smaller radial velocity.

### Red dwarf star

The smallest and dimmest stars. About 80% of all stars are red dwarfs although none are visible from Earth with the naked eye. Because they shine with less than 1% of the Sun's output they live a long time–the smallest are likely to last trillions of years. They are classified as type M stars.

### Recessional Velocity

The radial velocity of a galaxy caused by the expansion of the universe.

### Redshift (z)

An increase in the wavelength of light caused either by the source of the light moving away from the observer or by the expansion of the universe. In an expanding universe galaxies with large redshifts lie at greater distances than galaxies with small redshifts. Redshifts can also be produced by light climbing out of a strong gravitational field such as a black hole.

### Rich cluster of galaxies

A cluster of galaxies containing hundreds of large galaxies. The nearest rich cluster is the Virgo cluster. The richest clusters can contain more than a thousand large galaxies. Most of the richest clusters in the nearby universe are listed in the Abell catalogue.

### Right Ascension (RA)

See Equatorial coordinates.

### Ross, Frank

An astronomer who searched for high proper motion stars between 1925 and 1939, producing a list of 1070 nearby stars.

### Second

See Arcsecond.

### Solar system

A star together with the planets, moons, asteroids, comets and dust which orbit it.

### Spectral Classification

The system used to classify stars. Stars fall into seven main categories: O, B, A, F, G, K and M ranging from hot blue-white stars to cooler red stars. A number (0 to 9) is usually added to denote a sub-class. A roman numeral is sometimes added to denote the size of the star. Supergiant stars are class I, *Giant stars* are class III, and ordinary main sequence stars are class V. (Types II and IV are in between types). The Sun is type G2V, whereas Arcturus–an orange *giant star* - is type K2III.

### Spiral galaxy

A galaxy with spiral arms. There are two main types, those with central bar–SB, and those without–S (or SA). Spiral galaxies are also subdivided into types a, b, c (and sometimes d), depending on how tightly wound the spiral arms are.

### Star classification

See Spectral classification.

### Star cluster

See Open star cluster, Globular cluster.

### Stellar association

See OB association.

### Supercluster

A large concentration of hundreds or thousands of groups of galaxies. Superclusters range in size from 100 million to 500 million light years and are usually embedded in large sheets and walls of *galaxies* surrounding large voids in which very few galaxies exist. Superclusters formed in the early universe when matter clumped together under the influence of gravity.

Super galactic coordinates (L , B)

A coordinate system based on the approximate plane of the Virgo *Supercluster*. The super galactic plane passes through the Sun and the middle of the Virgo cluster. Several nearby superclusters also lie close to this plane. The symbols used for super galactic coordinates are L (longitude) and B (latitude). The zero point of Super galactic latitude and longitude is at RA=02h49m14s Dec=+59°31'42", and the Super galactic north pole is at RA=18h55m01s Dec =+15°42'32", (epoch 2000 coordinates).

### Supergiant star

A star bigger than about 10 solar masses will become a supergiant star at the end of its life as hydrogen burning ceases causing the stars outer layers to expand. Supergiant stars are the largest and brightest of all stars and they usually end up exploding in a supernova explosion and creating a neutron star or a black hole. See also Dwarf star, Giant star.

### Supernova

A catastrophic stellar explosion that can briefly outshine an entire *galaxy* of billions of stars. It can occur when a supergiant star exhausts all its nuclear fuel causing the core of the star to collapse releasing a vast amount of energy which blasts away the outer parts of the star and leaves behind a *neutron star* or in extreme cases a black hole.

### Supernova remnant

The remains of a star visible as an expanding *nebula* of gas that have been ejected at high speed by a supernova explosion.

### Temperature

In astronomy, temperature is measured with the Kelvin scale (symbol K) which is equal to °C + 273°. Thus, a midday Earth temperature of 20°C is equal to 293K and the Sun's surface temperature of 5500°C is about 5770K.

### Trillion

1 trillion = 1 000 000 000 000.

### UGC catalogue (UGC)

The Uppsala General Catalogue of Galaxies. A catalogue of 12 939 bright galaxies north of declination -2.5° published in 1973. See also ESO Catalogue.

### Variable star

A star that varies in brightness. There are many types, some stars can change in brightness in a matter of minutes whereas others change slowly over many months. The first 334 variable stars discovered in a constellation are given a one or two letter code such as R Scuti or UV Ceti. Other variable stars are designated V335, V336, etc. Proxima Centauri for example is known to variable star astronomers as V645 Centauri.

### Visual magnitude

Another name for Apparent magnitude.

### White dwarf star

The dying remnant of a giant star that has blown away its outer layers to reveal an intensely hot core. Only the youngest white dwarfs are actually white, over billions of years they slowly cool and change color to yellow or orange (although the coolest white dwarfs are bluer than expected because the compressed atmosphere of hydrogen filters out the red light). They eventually become a dead black dwarf, although because the universe is less than 15 billion years old none have yet cooled this much. They are classified as type D stars.

### Wolf, Max

An astronomer who searched for high proper motion stars between 1919 and 1931, producing a list of 1566 nearby stars.

### Yale catalogue

See Bright Star catalogue.

### Yellow dwarf star

Any small yellow star like the Sun. They are classified as type G stars.

# THE EVOLUTION OF THE "BIG BANG"

## THE HYPOTHESIS OF BIG BANG

### What Is a Theory?

In science, a theory is an effort to explain a specific feature of the universe. Theories cannot be proven, nonetheless, they can be invalidated. If observations and tests substantiate a theory, it becomes robust and typically more scientists will agree to take it. If the evidence opposes the theory, scientists must either abandon the theory or revise it in light of the novel evidence.

### What is Big Bang Theory

Humans have gazed at the stars and wondered how the universe developed into what it is today. It has been the topic of clergymen, philosophical, astronomers and scientific discussion and deliberation. People who have toiled to reveal the secrecies of the universe's development include such renowned scientists as Sir Isaac Newton, Albert Einstein, Edwin Hubble and Stephen Hawking.  One of the most well-known and widely accepted models among the academic community, for the universe's development is the Big Bang theory.

Although the Big Bang theory is famous, it's also widely misunderstood, and rejected by prominent scientists. A common misperception about the theory describes the origin of the universe. That's not quite accurate. The big bang is an attempt to explain how the universe developed from a very tiny volume of zero, and infinitely dense state into what it is today. It does not attempt to explain what originated the creation of the universe. Also, the theory does not address what came prior the Big Bang or even what lies outside the universe.

Additional misconception is that the big bang may not be a kind of explosion. That is not the truth either. The Big bang is an attempt to describe the expansion of the universe stated by a single point that contained time, space and matter as well as energy. While some versions of the theory refer to an incredibly rapid expansion. The expansion, which is claimed by a number of scientists and cosmologists that the expansion was possibly faster than the speed of light. Einstein had described the speed of light to be an absolute value that does not exceed 186,282 mi/s.

Tallying up the Big Bang theory is a challenge. It encompasses concepts that contradict the manner we distinguish the world. The earliest periods of the Big Bang concentrate on an instant in which all the distinct forces of the universe were part of a unified force, came out of a zero volume, which is referred to by "Singularity." The well-known laws of science begin to break down, the further back you look. Eventually, you may not be able make any methodical theories about what is happening, as thus far, scientists and science have not made logical evidence of the theory, yet. They claimed that science does not apply in this "Big Bang."

## SIMPLE MEANING OF THE BIG BANG THEORY

The Big Bang theory defines the development of the universe from the instant immediately after it came into existence up to nowadays. Big Bang is one of numbers of scientific models that tries to clarify the reason the universe is constructed method it is. The theory styles more than a few forecasts, several of which have been nor proven through observational data, deprived of any scientific data. As a result, while it is the most popular and accepted theory among some scientific and the public, it has not satisfied prominent scientists such as Albert Einstein, and even Hawking himself. It has been specifically widely accepted by hyper active journalists.

## TIME BEFORE TIME

The most important concept to get across a logical individual, is when conversing about the Big Bang expansion, Figure 2.1. Numerous people consider that the big bang is about an instant in which all the matter and energy in the universe was focused in a zero volume of a point, (Point of Singularity). At that instant, a time before time, exploded, spouting matter across space. Again, the "Bing Bang" instant spouted space before space. The universe came to existence. In fact, in the mind of some cosmologists, scientists, journalists and the public claimed that the Big Bang clarifies the expansion of space itself, which means everything confined within space is scattering apart from everything else.

Nowadays, when we gaze at the night sky, we see galaxies parted by what appears to be enormous expanses of vacant space. It is claimed by the believers of the Big Bang that at the earliest instants of the big bang, all of the matter, energy and space we could see was compacted to an area of zero volume and hypothetical infinite density. Cosmologists name this a "Singularity."

What was the universe like at the commencement of the Big Bang? Rendering to the Bing Bang model, it was enormously compressed and very hot. There was so much too much energy in the universe during those first rare instant that matter as we identify it could not formulate. But then again, the universe expanded fast, which indicates it developed less condensed and cooled down. As matter expanded, it started to arrange itself to radiation, which commenced to lose energy. However, in lone a few seconds, the universe shaped out of a singularity that overextended across space.

*Fig.2.1: The Universe Hypothetically Sprang into Existence 13.7 billion Years Ago – Curtsy - LIBRARY via Getty Images*

One consequence of the big bang was the establishment of the four basic forces in the universe. These forces are:

1. Electromagnetism
2. Strong nuclear force
3. Weak nuclear force
4. Gravity

At the commencement of the big bang, these forces were all share of an integrated force. It was only soon after the big bang started to permit the forces to be disconnected into present state today. These forces that were on one instant part of a unified entity, which is another mystery of the Big Bang even to scholarly scientists. Countless physicists and cosmologists are still intensively occupied on establishing the "Grand Unified Theory," which may enlighten how the four forces were once cohesive and how they relay to one another.

## The Origination of the Big Bang Theory

Hubble hypothesized that the universe inflates as time passes. It means that billions of years passed, the universe would have been very small, but denser. If we go back abundantly far, the universe would breakdown into an area with infinite density, holding all the matter, energy, space and time of the universe. The big bang theory originated according to reversed engineering.

Numerous individuals had a significant difficulty to yield their mind to this theory. Among them was the celebrated physicist, Albert Einstein. Einstein pledged to the certainty that the universe was static. A static universe does not vary. It has permanently been and continually will be the same. Einstein wished his theory of general relativity would provide him a profound thoughtful of the building of the universe.

Upon achieving of his theory, Einstein was astonished to realize that according to his computation, the universe would have to be expanding or contracting. This finding opposed his credence that the universe was static. Einstein examined further for a conceivable enlightenment. He suggested a cosmological constant -- a number, which contained within his general theory of relativity, described away the seeming requirement for the universe to expand or contract.

As enlightened with Hubble's discoveries, Einstein acknowledged that he was in error. The universe did appear to be expanding, and Einstein's particular theory reinforced the supposition. The theory and observations provided upsurge to a few forecasts, several of which have since been experiential.

One of those forecasts is that the universe is both homogeneous and isotropic. Fundamentally, it illustrates the universe appears the same no matter the perception of the observer. On a limited level, this prediction appears inaccurate. In spite of everything, not every star has a solar system of planets like ours. Not every galaxy appears the same. But on a macroscopic level that span millions of light years, the dispersal of matter in the universe is statistically homogeneous. Even if we were across the universe, our observations of the structure of the universe would appear the same as those here on Earth.

Another prediction was that the universe would have been deeply hot during the initial stages of the big bang. The radiation from this span of time would have been phenomenally huge, and there would have to be some indication of this radiation left over. In the meantime, the universe must be homogeneous and isotropic, the evidence should be evenly disseminated through the universe. Scientists revealed evidence of this radiation as early as the 1940s, though, at the time they did not know what they had found. It was not until the 1960s when two distinct teams of scientists revealed what we now call the cosmic microwave background radiation (CMB). The CMB allegedly is the remnants of the intense energy emitted by the primordial fireball in the big bang. It was claimed once intensely hot, but now has cooled to a chilly 2.725 degrees Kelvin (-270.4 degrees Celsius or -454.8 degrees Fahrenheit).

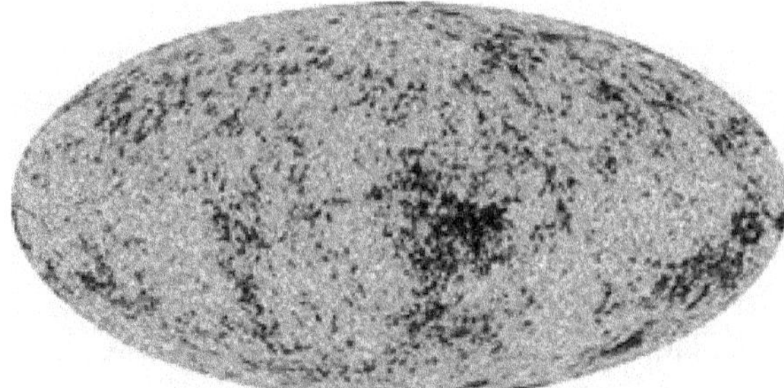

*Fig.2.2:* Cosmic Microwave Background Radiation – Wilkinson Microwave Anisotropy Probe – Curtsy NASA

These observations aided the big bang theory as the major model for the evolution of the universe.

*Fig.2.3:* Distant Galaxy NASA's Spitzer Space Telescope. Curtsy NASA

## The Initial Instant of the Existence of the Universe

Due to the restrictions of the laws of science, we cannot provide any deductions about the instant the universe came into existence. As an alternative, we may look at the period directly succeeding the existence of the universe. As for now, the earliest instant scientists argue happens at

$$t = 1 \times 10^{-43} \text{ seconds} = (1/10^{43}) \text{ seconds}$$

Where, "t" is the time after the universe send light to Earth.

Consider the number 1.0, move the decimal place to the left 43 times.

Cambridge University considers the assumption of the earliest instants as "quantum cosmology." It is the earliest instant of the big bang, where the universe was a single point that classical physics could be instituted. In its place, quantum physics were in view. Quantum physics cope with physics on a subatomic scale.

Much of the conduct of particles on the quantum scale appears bizarre to us, since the particles appear to disobey our comprehension of proven traditional physics. Scientists attempt to discover the link between quantum and classical physics.

At $$t = 1 \times 10^{-43} \text{ seconds,}$$

the universe claimed to be unbelievably small, dense and hot. This homogenous area of the universe crossed a region of only $1 \times 10^{-33}$ centimeters, which is $3.9 \times 10^{-34}$ inches. Nowadays, that equivalent stretch of space extents billions of light years. Throughout this stage, big bang theorists trust, matter and energy were enteined. The four primary forces of the universe were also a cohesive force. The temperature of this universe was

$$1 \times 1032 \text{ degrees Kelvin,} = 1 \times 1032 \text{ degrees Celsius,} = 1.8 \times 1032 \text{ degrees Fahrenheit.}$$

As miniscule fractions of a second passed, the universe expanded speedily. Cosmologists denote to the universe's expansion as "inflation." The universe doubled in size several times in less than one second.

As the universe expanded, it cooled. At around

$$t = 1 \times 10^{-35} \text{ seconds,}$$

matter and energy decoupled.

Cosmologists call this "baryogenesis" - baryonic matter is the type of matter we may see.

In contrast, we may not see "dark matter," however, we recognize it is present by the way it affects energy and other matter. All through baryogenesis, the universe jam-packed with a nearly equal amount of "matter and anti-matter." There was more matter than anti-matter, therefore, while most particles and anti-particles annihilated each other, some particles endured.

These particles would later amalgamate to form all the matter in the universe.

A period of particle cosmology trailed the quantum age. This period starts at

$$t = 1 \times 10^{-11} \text{ seconds.}$$

This is a phase that scientists can simulate in lab conditions with the "particle accelerators." That may illustrate that we have some experimental data on what the universe may have been like at this time. The unified force broke down into components. The forces of electromagnetism and feeble nuclear force split off. "Photons" became more than matter particles, but the universe was too dense for light to shine within it.

Following, came the period of normal cosmology, which begins

0.01 second after the start of the big bang.

From this moment on, scientists assume they have a good understanding on how the universe progressed. The universe sustained to enlarge and cool, and the subatomic particles shaped during baryogenesis began to bond together. They shaped neutrons and protons. By the time a full second had passed, these particles could form the nuclei of light elements like hydrogen, in the state of its isotope, deuterium, helium and lithium. This process is known as nucleosynthesis. However, the universe was still very dense and hot for electrons to link these nuclei and formulate steady atoms.

Scientists believe this was the most engaged second in the existence of the universe.

## The Next 13.8 billion Years

Much too Much occurred in that opening second of the big bang. However, that is merely the beginning. After 100 seconds, the universe's temperature cooled to 1 billion degrees Kelvin, 1 billion degrees Celsius, 1.8 billion degrees Fahrenheit. Subatomic particles sustained to amalgamate. The distribution of elements was almost 75% hydrogen nuclei and 24% helium nuclei, the other 1% consisted of other light elements as lithium.

The temperature of the universe was still highly elevated for electrons to bond with nuclei. Instead, electrons bumped with other subatomic particles named positrons, making more photons. Nonetheless, the universe was too compressed to permit light to shine inside of it.

The universe sustained to enlarge and cool. After about 56,000 years, claimed many scientists, the universe had cooled to

9,000 degrees Kelvin = 8,726 degrees Celsius = 15,740 degrees Fahrenheit.

At this time, the density of the matter distribution in the universe harmonized the density of radiation. After another 324,000 years, the universe had sufficiently expanded to cool down to 3,000 degrees Kelvin = 2,727 degrees Celsius, = 4,940 degrees Fahrenheit.

Lastly, protons and electrons could amalgamate to formulate neutral hydrogen atoms.

It was at this time,

380,000 years after the initial event, when the universe became transparent.

Light could shine all over the universe. The radiation that humans would later recognize as cosmic microwave background radiation sealed into place. Scientists claim that as we examine the CMB today, we may extrapolate a hypothetical image of what universe looked like then.

For the next

100 million years or so, the universe continued to expand and cool.

Small gravitational variations triggered particles of matter to cluster together. Gravity triggered gases in the universe to collapse into tight pockets. As gases shrunk, they become denser and hotter.

Some 100 to 200 million years after the initial creation of the universe, stars molded from these compartments of gas.

Stars began to cluster together to form galaxies. In the end, some stars became supernova. As the stars burst, they expelled matter across the universe. This matter encompassed all the heavyweight elements we find in nature, everything up to uranium. Galaxies in sequence shaped their own clusters.

Our own solar system was shaped around 4.6 billion years ago.

Nowadays, the temperature of the universe is

= 2.725 degrees Kelvin = -270 degrees Celsius, = -455 degrees Fahrenheit,

which is only 2 degrees above absolute zero. The homogenous section of the universe we can hypothesize about reaches

$1 \times 10^{29}$ centimeters across = $6.21 \times 10^{23}$ miles.

This is greater than the distance we are able to physically see utilizing our most innovative astronomical instruments.

The Big Bang Model Heighted Substantial Credibility Problems

Once scientists propagated their Big Bang theory, many prominent scientists have probed and disapproved the model. Describing below samples of the most common disapprovals of the Big Bang theory.

It vehemently violates the first law of thermodynamics, which articulates that you cannot create or destroy matter or energy.

Criticizers claim that the big bang theory proposes the universe started out of nothing. Advocates of the big bang theory tell us that such disapproval is unjustified for two reasons.

i.     The big bang does not discourse the creation of the universe, but rather the evolution of it.

ii.    Since the laws of science break down as you approach the creation of the universe, there is no cause to have faith in applying the first law of thermodynamics.

Myriads of critics say that the construction of stars and galaxies violates the law of entropy, which advocates that systems of change become less organized over time. However, if we interpreted the early universe as totally homogeneous and isotropic, then the present universe demonstrates signs of obeying the law of entropy.

Myriads of astrophysicists and cosmologists contend that scientists have misinterpreted proof as the redshift of celestial forms and the cosmic microwave background radiation. They quote the absence of exotic cosmic bodies that should have been the creation of the big bang according to the theory.

*Fig.2.4: The Planck Satellite Collected Data Attempting Refining Big Bang Theories. Curtsy - Eric Estrade/AFP/GETTY IMAGES*

The early inflationary phase of the big bang seems to violate the law that nothing can travel faster than the speed of light. Advocates have a few diverse replies to this disapproval. One is that at the beginning of the big bang, the "theory of relativity" did not apply. Consequently, there was no problem with traveling faster than the speed of light. Additional connected reply is that space itself can swell faster than the speed of light, as space falls outside the domain of the theory of gravity.

## OTHER MODELS DESCRIBING THE BEGINNING OF THE UNIVERSE

There are numbers of different models trying to clarify the construction of the universe. Nonetheless, none of them have as extensive acceptance as the Big Bang model.

## The Steady-State Model

The steady state model of the universe proposes the universe continuously had and will continuously have the same density. The theory resolves the seeming evidence that the universe is expanding by suggesting that the universe makes matter at a rate equivalent to the universe's rate of expansion.

## The Ekpyrotic Model

The ekpyrotic model proposes our universe is the consequence of an impact of two 3D – worlds on a veiled fourth dimension. It does not conflict with the Big Bang theory entirely. Once a particular amount of time it lines up with the proceedings defined in the Big Bang theory.

## The Big Bounce Model

The big bounce theory proposes our universe is one of a succession of universes that first expand, then contract again. The cycle reiterations after numerous billion years.

## Plasma Cosmology

Plasma cosmology tries to define the universe in terms of the electrodynamic characteristics of the universe. Plasma is an ionized gas, which illustrates it is a gas with free wandering electrons that can conduct electricity.

There are numerous other models as well. Is it possible that one of these models, or other models may appear one day to substitute the big bang model as the current recognized model of the universe. As time permits and our capability to investigate the universe upsurges, we will able to make more accurate models of how the universe developed.

Learn more about the truth of the big bang and its corresponding topics.

## Dark Matter – Dark Energy

*Fig.2.5: Maps of Dark Matter – Curtsy NASA*

Astronomers tinted the dark matter concentrations in the giant galaxy cluster Abell 1689 color blue, Figure 2.5. They reckoned out the position of those focusses by utilizing gravitational lensing

Cosmologists are still investigating some means to interpret the origin and the destiny of the universe. The stargazing scientists have positioned their investigation in front of their telescopes for a long time attempting to explain one of astronomy's most perplexing mysteries of the universe. Figure 2.5. The mystery is known as dark matter.

At present, scientists do not recognize the full comprehension of "Dark Matter," they deiced to call it Dark Matter or Black Matter as in the phrase of "Black Box," referring to the unknow contents."

Scientists know too little, for example, that dark matter behaves differently than "normal" matter, such as galaxies, stars, planets, asteroids and all of the living and nonliving entities on Earth. Astronomers classify all of such substance as "baryonic matter." They recognize that its fundamental unit is the atom, which itself is encompasses smaller subatomic particles, such as protons, neutrons and electrons.

Unlike baryonic matter, "dark matter" neither emanates nor absorbs light or any other forms of electromagnetic energy.  Astronomers realize it exists since unknown entity in the universe is applying momentous gravitational forces on things we can see and easily recognize as a cause. When they quantity the effects of this gravity, scientists approximated that "dark matter" adds up to 23% of the universe. Baryonic matter accounts for just 4.6%. Moreover, another cosmic mystery known as "Dark Energy" makes up the rest – a overwhelming 72%.

## The Character Entities of Dark Matter

While the author has declared his findings regarding Dark Matter/Dark Energy in the previous chapter, he is still devoting further study in a later chapter in this book:

i.     What is dark matter?

ii.    Where is its origin?

iii.   Where is it today?

iv.    How could we sense it while we may not see it?

v.     And do we gain in solving its mystery-puzzle?

Some scientists have attempted to hypothesize dark matter as the mystery of hardening the standard model of particle physics. It is fundamentally possible to alter our perception and comprehension of the world around us.

## EVIDENCE OF DARK MATTER

The study of galaxies is an interesting subject, which captivated astronomers for centuries.  Also, the study of our solar system revealed its position among massive body stars. Followed by then discovery that myriads of other galaxies existed beyond the "Milky Way." By Edwin Hubble in 1920s, along with other scientists were cataloging thousands of "island galaxy clusters," documenting information regarding the size, rotations and distances of galaxies from Earth.

Astronomers attempted to measure the mass of a galaxy. Nonetheless, scientists have to devise means to calculate the size and the weight of a galaxy.  One indirect technique was discovered to measure the galaxy weight through the light intensity.  The more luminous a galaxy emits, the more mass it possesses.  Another technique was also used by calculating the rotation of a galaxy, then tracking the stars within in the galaxy moving around its center. The variations in the rotational velocity should provide a reliable indication region of varying gravity and therefore mass.

When astronomers measured the rotations of spiral galaxies in the 1950s/60s, they stumbled into a perplexing discovery. They predicted the observed stars near a galaxy's center, where the visible matter is more concentrated, move faster than stars at the edge. Amazingly, they observed instead that stars at the edge of a galaxy have the identical rotational velocity as stars near the center.

Astronomers observed this first with the "Milky Way," Astronomer, "Vera Rubin" in the 1970s, had confirmed the same phenomenon as she made a detailed quantitative measurements of stars in several other galaxies, including the Andromeda (M31), which the closest to our Milky Way.

The consequence of all of these outcomes declared two options:

i.   To some degree there was fundamentally something wrong with our comprehension of gravity and rotation, which appeared improbable given that Newton's laws that endured numerous tests for centuries. Or,

ii.  More probable, galaxies and galactic clusters must encompass an unseen form of matter – "dark matter" – accountable for the observed gravitational properties. Consequently, astronomers concentrated their consideration on "dark matter."

## Dark Matter the New Discoveries

*Fig.2.6: Einstein Double Ring – Hubble Gravitational field – Elliptical Galaxy Warping the Light of Two Galaxies. Curtsy NASA*

Astronomers continued to discover perplexing information as they examined the widespread galaxies of the universe. A few valiant stargazers focused their consideration to galactic clusters – counted as few as 50 and as numerous as thousands of galaxies bound together by gravity.  Astronomers have hoped to find pools of hot gas that had previously gone unnoticed. This may explain the mass being attributed to "dark matter."  When they turned the Chandra X-ray Observatory telescopes, toward these clusters, they found massive clouds of superheated gas.  However, these clusters were not sufficient to provide justification for the discrepancies in mass. The measurement of hot gas pressure in galactic clusters has illustrated that there must have been about five to six times as much dark matter as all the observed stars and gas.  Else, there would not be sufficient gravity in the cluster to prevent the hot gas from escaping.

Galactic clusters have given us other hints about "dark matter."  Deriving from Albert Einstein's general theory of relativity, astronomers have illustrated that clusters and superclusters can distort space-time with their enormous mass. Light rays emitting from a distant object behind a cluster pass through the distorted space-time, causes the rays to bend and converge as they move toward an observer. Consequently, the cluster behaves as a large gravitational lens, similar to an optical lens.

The distorted image of the distant object may appear in three conceivable behaviors depending on the shape of the lens:

i.   Ring – image seems as an incomplete or whole circle of light recognized as an "Einstein ring," Figure 2.6. This occurs when distant object, lensed galaxy and observer/telescope are flawlessly aligned.

ii.  Oblong or elliptical – image becomes split into four images and appears as a cross known as an "Einstein cross." Cluster – image seems as a series of banana-shaped arcs and arc-lets.

When measuring the angle of bending, astronomers can calculate the mass of the gravitational lens, the greater the bend, the more massive the lens. Employing this technique, astronomers have established that galactic clusters, certainly have high masses surpassing those measured by luminous matter. Consequently, they have given extra indication of "dark matter."

## Mapping Dark Matter

**Fig.2.7:** *A Composite Image Merging Galaxy Cluster "Abell 520" Superimposed - Cluster's Concentration of Starlight "Orange," Hot Gas "green" and Dark Matter "Blue." – Curtsy  NASA*

As astronomers gathered clues about the existence – and staggering amount – of dark matter, they turned to the computer to create models of how the strange stuff might be organized. They made educated guesses about how much baryonic and dark matter might exist in the universe, then let the computer draw a map based on the information. The simulations showed dark matter as a weblike material interwoven with regular visible matter. In some places, the dark matter coalesced into lumps. In other places, it stretched out to form long, stringy filaments upon which galaxies appear entangled, like insects caught in spider silk. According to the computer, dark matter could be everywhere, binding the universe together like some sort of invisible connective tissue.

Since then, astronomers have worked diligently to create a similar dark matter map based on direct observation. And they've been using one of the same tools – gravitational lensing – that helped prove the existence of dark matter in the first place. By studying the light-bending effects of galaxy clusters and combining the data with optical measurements, they have been able to "see" the invisible material and have begun to assemble accurate maps.

In some cases, astronomers are mapping single clusters. For example, in 2011, two teams used data from Chandra's X-ray Observatory and other instruments such as the Hubble Space Telescope to map the distribution of dark matter in a galaxy cluster known as Abell 383, which is located about 2.3 billion light-years from Earth. Both teams came to the same conclusion: The dark matter in the cluster isn't spherical but ovoid, like an American football, oriented with one end pointing to the observers. The researchers disagreed, however, on the density of the dark matter across Abell 383. One team calculated that the dark matter increased toward the center of the cluster, while the other measured less dark matter at the center. Even with those discrepancies, the independent efforts proved that dark matter could be detected and successfully mapped.

In January 2012, an international team of researchers published results from an even more ambitious project. Using the 340-megapixel camera on the Canada-France-Hawaii Telescope (CFHT) on Mauna Kea Mountain in Hawaii, scientists studied the gravitational lensing effects of 10 million galaxies in four different regions of the sky over a period of five years. When they stitched everything together, they had a picture of dark matter looking across 1 billion light-years of space – the largest map of the invisible stuff produced to date. Their finished product resembled the earlier computer simulations and revealed a vast web of dark matter stretching across space and mixing with the normal matter we've known about for centuries.

## Dark Matter Particles

Based on the evidence, most astronomers agree that dark matter exists. Beyond that, they have more questions than answers. The biggest question, dare we say one of the biggest in all of cosmology, centers on the exact nature of dark matter. Is it an exotic, undiscovered type of matter, or is it ordinary matter that we have difficulty observing?

The latter possibility seems unlikely, but astronomers have considered a few candidates, which they refer to as MACHOs, or massive compact halo objects. MACHOs are large objects that reside in the halos of galaxies but elude detection because they have such low luminosities. Such objects include brown dwarfs, exceedingly dim white dwarfs, neutron stars and even black holes. MACHOs probably contribute somewhat to the dark matter mystery, but there are simply not enough of them to account for all of the dark matter in a single galaxy or cluster of galaxies.

Astronomers think it's more likely that dark matter consists of an entirely new type of matter built from a new kind of elementary particle. At first, they considered neutrinos, fundamental particles first postulated in the 1930s and then discovered in the 1950s, but because they have such little mass, scientists are doubtful they make up much dark matter. Other candidates are figments of scientific imagination. They are known as WIMPs (for weakly interacting massive particles), and if they exist, these particles have masses tens or hundreds of times greater than that of a proton but interact so weakly with ordinary matter that they're difficult to detect. WIMPs could include any number of strange particles, such as:

Neutralinos (massive neutrinos) – Hypothetical particles that are similar to neutrinos, but heavier and slower. Although they haven't been discovered, they're a front-runner in the WIMPs category.

Axions – Small, neutral particles with a mass less than a millionth of an electron. Axions may have been produced abundantly during the big bang.

Photinos – Similar to photons, each with a mass 10 to 100 times greater than a proton. Photinos are uncharged and, true to the WIMP moniker, interact weakly with matter.

Scientists around the world continue to hunt aggressively for these particles. One of their most important laboratories, the Large Hadron Collider (LHC), lies deep underground in a 16.5-mile long circular tunnel that crosses the French-Swiss border. Inside the tunnel, electric fields accelerate two proton-packed beams to absurd speeds and then allow them to collide, which liberates a complex spray of particles. The goal of LHC experiments isn't to produce WIMPs directly, but to produce other particles that might decay into dark matter. This decay process, although nearly instantaneous, would allow scientists to track momentum and energy changes that would provide indirect evidence of a brand-new particle.

Other experiments involve underground detectors hoping to register dark matter particles zipping by and through Earth.

## Prior to the Big Bang?

It is difficult enough to imagine a time, roughly 13.7 billion years ago, when the entire universe existed as a singularity. According to the Big Bang theory, one of the main contenders vying to explain how the universe came to be, all the matter in the cosmos -- all of space itself -- existed in a form smaller than a subatomic particle.

*Fig.2.8: What came before the beginning? Curtsy - XUANYU HAN/GETTY IMAGES*

Once you think about that, an even more difficult question arises: What existed just before the big bang occurred?

The question itself predates modern cosmology by at least 1,600 years. Fourth-century theologian St. Augustine wrestled with question of what existed before God created the universe. His conclusion was that the Biblical phrase "In the beginning" implied that God had made nothing previously. Moreover, Augustine argued that the world was not made by God at a certain time, but that time and the universe had been created.

In the early 20th century, Albert Einstein came to very similar conclusions with his theory of general relativity. Just consider the effect of mass on time. A planet's hefty mass warps time -- making time run a tiny bit slower for a human on Earth's surface than a satellite in orbit. The difference is too small to notice, but time even runs more slowly for someone standing next to a large boulder than it does for a person standing alone in a field.

Based upon Einstein's work, Belgian cosmologist Rev. Georges Lemaître published a paper in 1927 that proposed the universe started out as a singularity and that the Big Bang led to its expansion.

According to Einstein's theory of relativity, time only came into being as that primordial singularity expanded toward its current size and shape.

Case closed? Far from it. This is one cosmological quandary that won't stay dead. In the decades following Einstein's death, the advent of quantum physics and a host of new theories resurrected questions about the pre-big bang universe.

## Galaxy Types and Parts

They can have as few as 10 million stars or as many as 10 trillion (the Milky Way has about 200 billion stars). In 1936, Edwin Hubble classified galaxy shapes in the Hubble Sequence.

**Elliptical:** These have a faint, rounded shape, but they're devoid of gas and dust, with no visible bright stars or spiral patterns, Figure 2.9. They also don't have galactic disks, which we'll learn about below. Their classification varies from E0 (circular) to E7 (most elliptical). Elliptical galaxies probably comprise about 60 percent of the galaxies in the universe. They show wide variation in size -- most are small (about 1 percent the diameter of the Milky Way), but some are about five times larger than the diameter of the Milky Way.

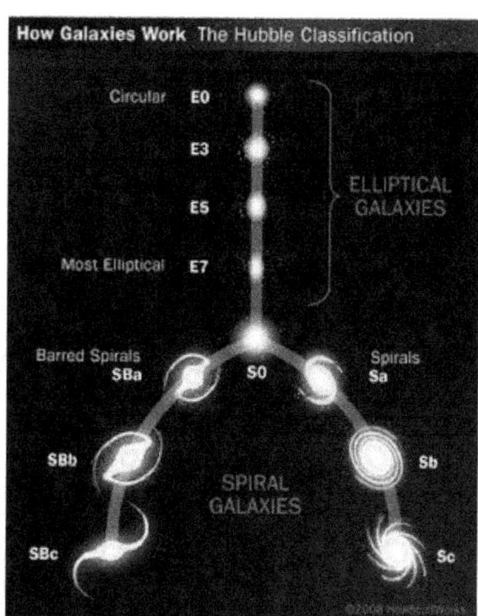


*Fig.2.9: Galaxies come in a variety of sizes and shapes.*

**Spiral**: The Milky Way is one of the larger spiral galaxies. They're bright and distinctly disk-shaped, with hot gas, dust and bright stars in the spiral arms, Figure 2.9. Because spiral galaxies are bright, they make up most of the visible galaxies, but they're thought to make up only about 20 percent of the galaxies in the universe. Spiral galaxies are subdivided into these categories: S0: Little gas and dust, with no bright spiral arms and few bright stars Normal Spiral: Obvious disk shape with bright centers and well-defined spiral arms. Sa galaxies have large nuclear bulges and tightly wound spiral arms, while Sc galaxies have small bulges and loosely wound arms.

**Barred Spiral**: Obvious disk shape with elongated, bright centers and well-defined spiral arms. SBa galaxies have large nuclear bulges and tightly wound spiral arms, while SBc galaxies have small bulges and loosely wound arms (recent evidence suggests that the Milky Way is a SBc galaxy), Figure 2.9.

**Irregular**: These are small, faint galaxies with large clouds of gas and dust, but no spiral arms or bright centers. Irregular galaxies contain a mixture of old and new stars and tend to be small, about 1 percent to 25 percent of the Milky Way's diameter, Figure 2.9.

## The Parts of a Galaxy

Spiral galaxies have the most complex structures. Here's a view of the Milky Way as it would appear from the outside.

**Galactic disk**: Most of the Milky Way's more than 200 billion stars are located here. The disk itself is broken up into these parts:

**Nucleus**: The center of the disk Bulge: The area around the nucleus, including the immediate areas above and below the plane of the disk Spiral arms: These extend outward from the center, Figure 2.10. Our solar system is located in one of the spiral arms of the Milky Way.

**Globular clusters**: A few hundred of these are scattered above and below the disk. The stars here are much older than those in the galactic disk, Figure 2.10.

**Halo**: A large, dim, region that surrounds the entire galaxy. It's made of hot gas and possibly dark matter, Figure 2.10.

All of these components orbit the nucleus and are held together by gravity. Because gravity depends upon mass, you might think that most of a galaxy's mass would lie in the galactic disk or near the center of the disk. However, by studying the rotation curves of the Milky Way and other galaxies, astronomers have concluded that most of the mass lies in the outer portions of the galaxy (like the halo), where there is little light given off from stars or gases.

On the next page, we'll take a walk through the history of galaxies.

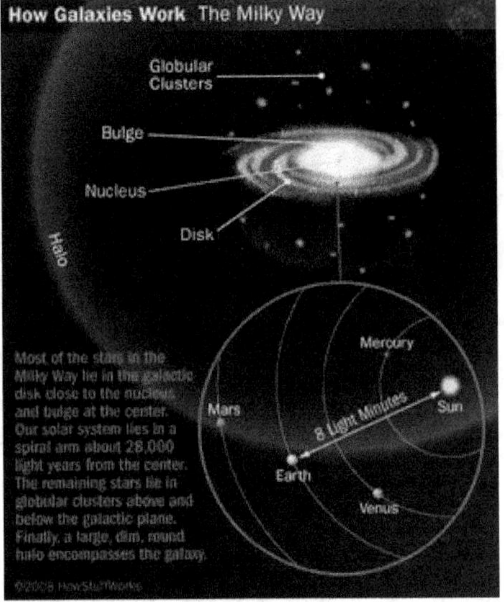


**Fig.2.10**: *The Parts of a Galaxy*

## Chronology of Galaxies

Let's look at the history of galaxies in astronomy.

The Greeks coined the term "galaxies kuklos" for "milky circle" when describing the Milky Way. The Milky Way was a faint band of light, but they had no idea what it was composed of.

When Galileo looked at the Milky Way with the first telescope, he determined that it was made up of numerous stars.

We've known for centuries that our solar system was located within the Milky Way because the Milky Way surrounds us. We can see it throughout the year in all parts of the sky, but it's brighter during the summer, when we're looking at the center of the galaxy. However, to astronomers in the 18th century and earlier, it wasn't clear that the Milky Way was a galaxy and not just a distribution of stars.

In the late 18th century, astronomers William and Caroline Herschel mapped the distances to stars in many directions. They determined that the Milky Way was a disk-like cloud of stars with the sun near the center.

In 1781, Charles Messier cataloged various nebulae (faint patches of light) throughout the sky and classified several of them as spiral nebulae.

In the early 20th century, astronomer Harlow Shapely measured the distributions and locations of globular star clusters. He determined that the center of the Milky Way was 28,000 light years from Earth, near the constellations of Sagittarius and Scorpio, and that the center was a bulge, rather than a flat area.

Shapely later argued that the spiral nebulae discovered by Messier were "island universes" or galaxies (retaining the Greek wording). However, another astronomer named Heber Curtis argued that spiral nebulae were merely part of the Milky Way. The debate raged on for years because astronomers needed larger, more powerful, telescopes to resolve the details.

In 1924, Edwin Hubble settled the debate. He used a large telescope (100-inch diameter, larger than ones that were available to Shapely and Curtis) at Mount Wilson in California and resolved that the spiral nebulae had structure and stars called Cepheid variables, like those in the Milky Way. (These stars change their brightness regularly, and the luminosity is directly related to the period of their brightness cycle.) Hubble used the light curves of the Cepheid variables to measure their distances from Earth and found that they were much farther away than the known limits of the Milky Way. Therefore, these spiral nebulae were indeed other galaxies outside our own.

There are still many mysteries surrounding galaxy formation, but on the next page we'll explain some of the best theories about it.

## Shapes of Galaxies

**Galactic disk:** Majority of the Milky Way›s has more than 200 billion stars are located in the galactic disk, Figure 2.11. The disk itself is fragmented up into the following parts:

**Nucleus:** The center of the disk

**Bulge**: The area everywhere round the nucleus, counting the immediate areas above and below the plane of the disk

**Spiral arms**: These spread external from the center. Our solar system is situated in one of the spiral arms of the Milky Way.

**Globular clusters**: A few hundred of these are dispersed above and below the disk. The stars here are much older than those in the galactic disk.

**Halo**: A huge, blurry, region that environs the whole galaxy. It is made of hot gas and perhaps dark matter.

All of these parts circle the nucleus are detained together by gravity. Since gravity is contingent upon mass, one might consider

*Fig.2.11: From Hypothetical Big Bang to Present Galaxy - Assuming13.8 billion Light Years – (13.8 x 5,55) trillion miles*

that most of a galaxy's mass would lie in the galactic disk or near the center of the disk. Though, through reviewing the rotation curves of the Milky Way and other galaxies, astronomers have decided that the greatest of the mass lies in the outer portions of the galaxy, as the halo, wherever there is little light provided from stars or gases.

## UNDERSTANDING THE CURRENT BIG BANG THEORY

Rendering to the hypothetical present big bang theory, all the matter and energy in the universe laterally, along with space and time, were once confined in a dimensionless point or singularity of infinite density and temperature. Some 13.8 billion years ago, this singularity for a split second substantial sudden expanded appeared as an enormous explosion due to a 'quantum fluctuation', and then endured exponential growth at numerous times the speed of light.

Some of the energy in this sudden expansion shaped subatomic particles, which after ~380,000 years combined to formulate atoms, generally hydrogen with some helium and hints of lithium and deuterium.

After ~700 million years, these atoms merged established by gravity to shape the first stars, called Population III stars, and initial galaxies emerged. More stars shaped, as well as, after ~9 billion years, the sun from a cloud of gas and dust that gravity shrank and fused to provide off light. Then clouds of gas and dust, named the solar nebula, that were revolving

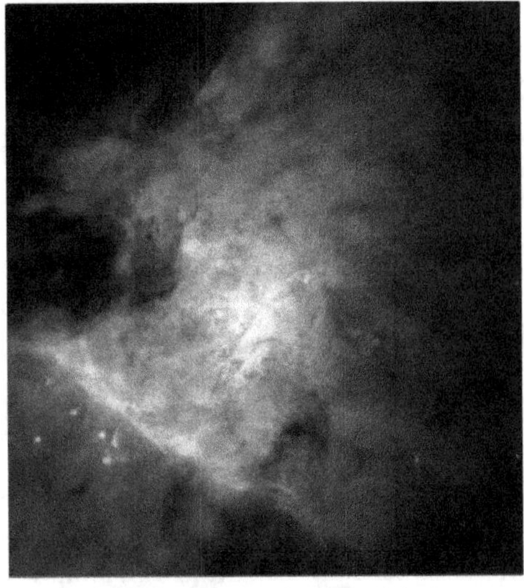

Fig.2.12: *The Great Orion Nebula (M42, NGC 1976). Curtsy NASA*

around the sun in due course advanced into the planets of our solar system, counting our Earth. Over additional millions of years, our Earth progressively cooled to its current state and by some means attained water.

It is located in the 'Sword' part of the constellation of Orion, just below the eastern-most of the three stars that comprise Orion's belt. Approximately 1,500 light years distant, M42 is a very active and turbulent cloud of gas and dust, Figure 2.12.

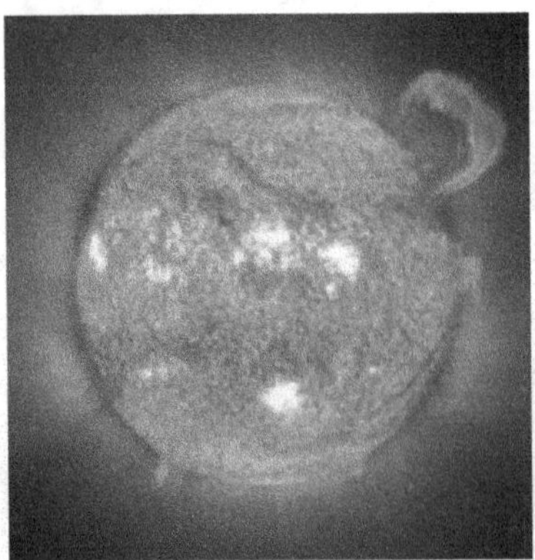

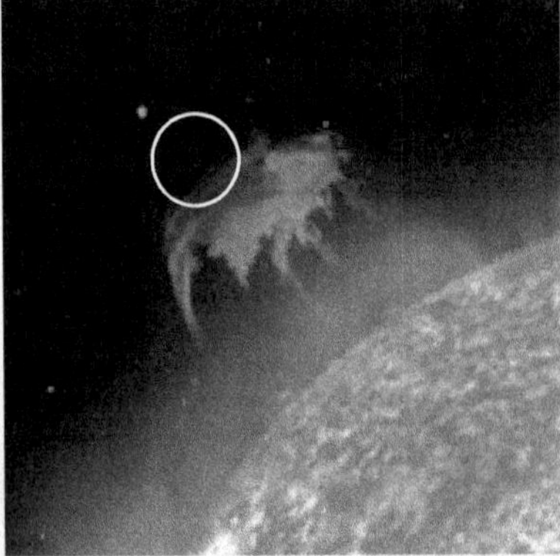

Fig.2.13: *The earth is seen (on the top left) dwarfed here in approximate relative size to the sun - Curtsy: NASA*

The massive fiery plumes (known as coronal ejections) seen here would encompass the earth many times over, Figure 2.13.

A light-year is how far light would travel at its current speed of 300,000 km/s (186,282 mps) in a year—9.5 trillion km or 5.9 trillion miles, Figure 2.14.

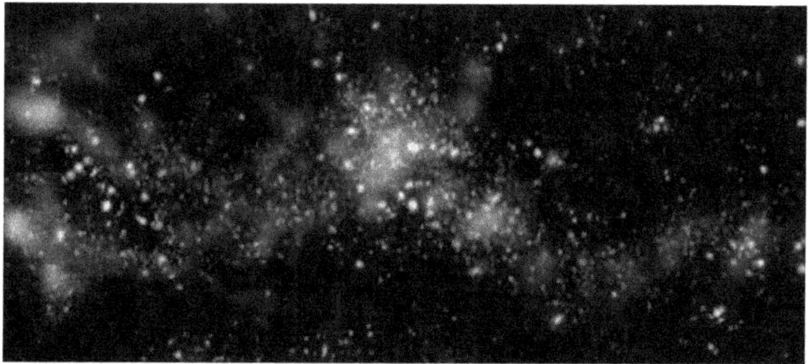

**Fig.2.14:** *The astronomers calculated that the supercluster was 300 million light-years across, and right at the most distant edge of the universe, 10.8 billion light-years away.*

## The Mind of the Infinity and Hawking's Hypothetical "Big Bang"

Andromeda is the closes Galaxy to our Milky Way Galaxy, Figure 2.15

**Fig.2.15:** *Andromeda Galaxy – Curtsy NASA*

'The mind of God' is a term that Christians use to mean 'the reason(s) why God does accomplish an event'. It is also the title of a book by Dr Paul Davies, Professor of Mathematical Physics at the University of Adelaide in South Australia. It is also the concluding phrase in physicist Stephen Hawking's best-selling book *A Brief History of Time*, in which he says, 'If we find the answer to that [i.e., why it is that we and the universe exist], it would be the ultimate triumph of human reason—for then we would know the mind of God.'

## THE 'EVOLUTION' OF THE 'BIG BANG' HYPOTHESIS

In the twentieth century, the first version of the 'big bang' as the explosion of a 'primeval atom' was put forward by Abbé Georges-Henri Lemaître in 1931. Lemaître, who was a catholic monk, already knew of Edwin Hubble's work on the redshift of light from distant stars, which Hubble interpreted to mean that the universe is expanding and, by extrapolating backwards in time, he postulated that the universe originated as a single particle of vast energy but near-zero radius. He argued, erroneously that cosmic rays must have come from such an explosion.

In 1946, one of the Manhattan Project, atomic bomb, scientists, George Gamow, postulated that a universal explosion lasting a few seconds could have produced all the elements we see today. This lost favor after about a decade, when calculations suggested that certain elements could form in stars.

In 1965, a third version of the 'big bang' was put forward by Robert Dicke, P.J.E. Peebles and others, which appeared to receive some confirmation by the accidental discovery by Arno Penzias and Robert Wilson that the universe seemed to be uniformly filled with very even heat at a temperature of about 3 K. "K" is the symbol for kelvin, the base unit of thermodynamic temperature. This was interpreted as being the after-glow in the form of microwave radiation left over from a huge initial explosion.

When Sir Fred Hoyle calculated that a 'big bang' would produce only light elements, notably helium, deuterium, and lithium, it seemed established as the origin of the universe. In the next 20 years, thousands of papers supporting the hypothesis of 'big bang' were produced and *virtually no papers challenging it were accepted*. It became inconceivable that the 'big bang' is considered falsely a theory and – 'could be wrong.' The entire careers in cosmology have been built on the presumption that the unproven hypothesis 'big bang' was fact.

One of the predictions of the 'big bang' is that it would produce large amounts of helium, and, in fact, the galaxies contain about 24 per cent (24%) of helium. However, calculations have shown that the detected matter in the universe is only about 1 per cent (1%) of the amount required to produce the gravitational attraction needed to form all the galaxies and clumps of galaxies, even within the vast time span of a hypothetical 15 billion years. This problem was solved with a stroke of the pen. In the early 1980s, cosmological theoreticians decided that the universe was now made up of nearly 99 per cent of *'cold dark matter'* (*CDM*)—necessarily '*dark*' because no one has ever seen it or detected it, and up to 99 times the amount of the visible matter in the universe. *This CDM could not be composed of detectable elements* like hydrogen and helium, *so hypothetical particles were said to exist*, with names like '*WIMPS*' (*weakly interacting massive particles*) and 'axions.'

## HAS 'DARK MATTER' REALLY BEEN PROVEN?

Dr John Hartnett's research shows that a new physical model explains the observations without recourse to the *fudge factor of dark matter*.

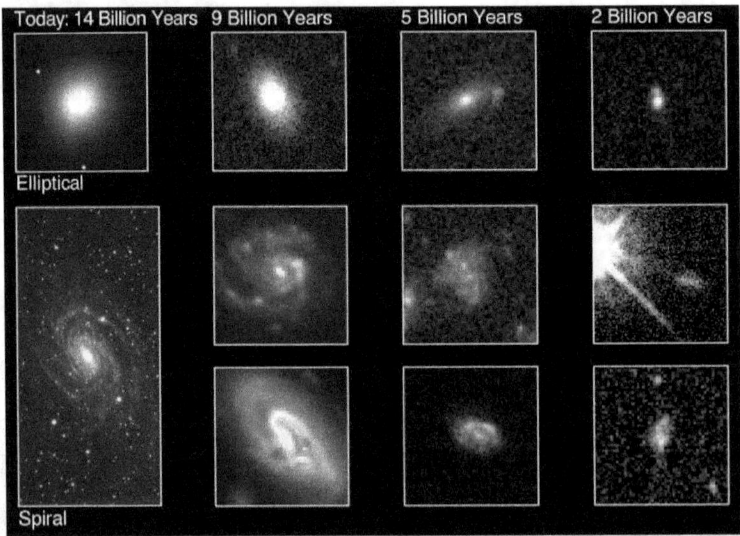


*Fig.2.16: Galaxies Snap Shoy in Time – Curtsy - NASA Hubble Space Telescope*

Another problem was the very smoothness of the so-called background radiation. Large-scale surveys of space have shown that matter is not evenly distributed at all, but exists in the form of huge clusters of galaxies,

Figure 2.16, and even larger-scale clumping including some huge structures which have been given names like the Great Wall, while there are vast empty reaches, one called the Great Void. 'Big bang' theorists decided that if they could find some variation or ripples in the pervasive 3 K radiation, this would be an adequate explanation of the origin of the large-scale galaxy structures. In 1989, NASA launched a space satellite named Cosmic Background Explorer (COBE) to try to detect the needed tiny variations or 'bumps' in the radiation from above earth's atmosphere, Figure 2.17. By 1991, no variation had been detected and the 'big bang' theorists were beginning to panic. Then, in April 1992, a computer program was used to analyze the data, and at last something was detected—hot and cold spots differing in temperature by up to about ***three one-hundred-thousandths of a degree Celsius***.

This background radiation cannot be invoked as conclusive proof of the 'big bang', as there are a number of other explanations for it.

"Russell Ruthen," writing in Scientific American, October 1992, says, 'But controversy has arisen as to whether the COBE measurements have any relation at all to the structure of the universe billions of years ago. Lawrence M. Krauss and Martin White of Yale University argue that the variations in the cosmic microwave background … could be distortions caused by gravitational waves.' And more than two decades ago two Soviet scientists, R.A. Sunyaev and Yakov B. Zeldovich, pointed out that as the background radiation passes through large clouds of intergalactic gas, the resultant change in intensity could cause these 'lumps.'

***Fig.2.17:*** Universe Cosmic Background Radiation – CPBE - -COBE anisotropy maps. The top map contains the dipole and the Galaxy emission on top of the CMB anistropies; the center map had the dipole removed, containing CMB and Galaxy; the bottom map contains CMB fluctuations only.

## BLACK HOLES

*Goldstone 64-metre (210-feet) antenna. Curtsy NASA*

In 1965, mathematician and physicist Roger Penrose, from a consideration of Einstein's general theory of relativity, conjectured that a large star collapsing under its own gravity would continue to do so until all the matter was compressed into a single point of zero volume and infinite density. Such a point of infinite compression is known to mathematical physicists as a '***singularity***' and also as a 'black hole'.

When Stephen Hawking read about this, he worked out a set of mathematical equations reversing the direction in time, so that the collapse into a black hole became instead an expansion from a black hole. In 1970 he published a joint paper with Penrose, supposedly 'proving' that the universe had begun from a 'big bang singularity provided only that general relativity is correct and the universe contains as much matter as we observe.' Since then, Hawking has been trying to deduce a mathematical formula to explain the electromagnetic, nuclear, and gravitational forces in the universe in one 'grand unified theory' or GUT, 'concise enough to be inscribed on a T-shirt'!

Recently quantum physicists have tackled some of the many queries that the 'big bang' theory evokes. For example:

## WHAT HAPPENED BEFORE THE 'BIG BANG'?

### No Place for the Creator

Paul Davies' reply is: 'According to modern physics, the big bang represented the origin of space and time, as well as of matter and energy. This means that *time itself came into existence with the big bang*. Questions like: What happened before the big bang? or What caused the big bang? are therefore ***meaningless***. There was no before.' [Emphasis in the original.] And Stephen Hawking claims that under certain conditions the universe 'would have neither beginning nor end: it would simply be. What place, then, for a creator?'

### How did Nothing Become Something and Then Explode?

Paul Davies' answer is that it happened through *quantum physics* applied to cosmology. He says, 'This "quantum cosmology" provides a loophole for the universe to, so to speak, spring into existence from nothing, without violating any laws of physics.'

This is very significant, as it shows the fallibility of theistic evolution. Theistic evolutionists often urge what is in effect retreat to a '***God of the gaps***' idea. God is invoked as necessary to create the initial particle and to 'light the fuse' as it were—thereafter the rest can evolve more or less by itself. However, ***Hawking says his new theory has no moment of creation and requires no Creator.*** Where does this leave theistic evolutionary compromise?

### The True Definition of Big Bang Hypothesis

So this, more or less, is the current 'big bang' theory—:
- that some 15 billion years ago, at a specific moment
- before which there was no before,
- the entire cosmos created itself
- by suddenly evolving out of nothing
- by means of a quantum fluctuation,
- first as a particle of space/time of zero dimensions and infinite heat, which
- proceeded in a few trillion-trillionths of a second to pass through an

- inflationary stage, and then
- through an incredibly hot 'big bang' stage,
- followed by universal expansion and cooling into its present form.
- The main rationale for this particular 'big bang' scenario is a set of mathematical equations deduced by human reason alone.

## What Should We Believe?

What should Bible-believing Christians think about all this and believe? There are certain contra facts which are indisputable, and certain principles that Christians should always use in evaluating naturalistic theories about origins.

## Regarding the 'Big Bang'

1. Not all scientists agree with the concept of the 'big bang'; in fact, many have never supported it. 'Big bang theory busted by 33 top scientists' (2005), with many more signing this statement] There have been other non-biblical theories about the origin of the universe put forward in modern times—the main ones being the 'steady state' theory and the 'plasma' theory.

   Suggested by Fred Hoyle, Thomas Gold, and Hermann Bondi in 1948, the 'steady state' theory involved the continuous creation of hydrogen atoms to fill the gaps left by the expansion of the universe—to ensure that it remained in a 'steady state'. It assumes that the universe never had a beginning. In 1993, Fred Hoyle, Professor Geoffrey Burbidge of the University of California at San Diego, and one other scientist proposed a new 'steady state' theory in which they explained the cosmic background radiation as being caused by explosions in galaxies that create thin metallic needles which absorb radiation and create the impression that the universe contains heat from a 'big bang'.

   Plasma is high-temperature, ionized hydrogen gas, i.e., comprised of free electrons and protons. The sun and most stars are giant spheres of plasma, and the aurora borealis is due to plasma. The plasma theory of the universe has been promoted by Hannes Alfvén and Tony Peratt for many years, but it took a quantum leap forward in 1991, with the publication of Eric Lerner's 466-page book *The Big Bang Never Happened*. Lerner says that trillions of years ago there was hydrogen plasma, brought into being by unknown evolutionary processes. This plasma 'had motion and energy, thus electrical currents and magnetic fields flowed through it.' And from this the universe supposedly eventually formed.

2. These two theories postulate virtually an infinite age for the universe, both past and future. This rather neatly does away with God both as Creator at one end and as Judge at the other, and thus has some rather obvious advantages for atheists.

3. The 'big bang' scenario involves tremendous (even infinite) energy at the beginning, but supplies no explanation for the source of this energy. Nor is it clear how the gravity of the initial universal black hole can be overcome by a 'quantum fluctuation.' Jargon like this seems to be part of an expanding vocabulary of 'big-bang-speak', masking some formidable, if not insuperable, difficulties. There is also no convincing explanation as to why an outward spray of gas radiating from the 'big bang' should form galaxies, stars, and planets.

4. The evidence propounded for the 'big bang' consists of just three concepts—the alleged expansion of the universe, the microwave background radiation, and the cosmic abundance of helium. However, all of these phenomena are capable of being otherwise interpreted. For example:

   a. No one has ever *seen* the universe expanding; expansion is an *interpretation* of the redshift.

b. There are other explanations for the abundance of helium in the universe. Eric Lerner (op. cit.) happily weaves it into the plasma theory. Bible-believing Christians understand that God created it that way. And since 1921, the lithium content of 20 per cent of 'old' stars studied has been found to contradict 'big bang' predictions.

5. To believe that the known universe was once condensed into a point of zero dimensions takes an unimaginable leap of faith; in fact, ***much more faith than it takes to believe that God created everything in the way He says*** He did in Genesis. The 'big bang' is a major part of the 'creation myth' of the Western nations' major religion—secular humanism.

6. With the passage of time, the lack of proven evidence for the 'big bang' has led some scientists to make such remarks as, 'Never has such a mighty edifice been built on such insubstantial foundations.' And, 'You have to understand that first there is speculation, then there is wild speculation, and then there is cosmology.'

7. In the history of evolutionary theory, from Charles Darwin to the present day, when there has been a conflict between theory and the facts, it is the facts that have been discarded rather than the theory. Thus, in 1993, George Efstathiou, head of astrophysics at The University of Oxford, said, 'In my view, cosmologists should not be too disturbed about small discrepancies between theory and observation. Any attempt to explain cosmic structure involves extrapolating from the Planck era, $10^{-43}$ seconds after the big bang, when strange effects like quantum gravity were important, to the present Universe which is about 10 billion [sic] years old. *So it is not surprising that one or two things don't seem to fit.*' [Emphasis added.]

8. Did God create by means of the 'big bang' or provide the initial energy to kick off the 'big bang'? No! It contradicts the biblical cosmogony. The order of events is different—for example, this theory claims that the sun existed long before the earth, whereas Genesis says the earth was created before the sun. Furthermore, the time-frame is quite different (Exodus 20:11). Also, we have seen that Hawking, Davies *et al*. would repudiate any such threat to the self-creating power of their evolutionary universe. It is sad indeed that theistic evolutionists, including some who would deny such a title, are urging Christians to say Yes.

## Regarding Scientists and 'God'

1. When non-Christian scientists speak or write of 'God', they do not usually mean the God of the Bible, unless they are ridiculing what the Bible says. Otherwise, they may mean something like 'the foundation for existence.' 'a deeper level of explanation,' or something equally vague or esoteric. For Stephen Hawking, 'God' appears to be 'a complete theory of the universe' that 'breathes fire into the equations.'

2. Any theory that dispenses with the true God is itself not true and will not stand the test of time.

3. Although Hawking and Davies say that 'What happened before the "big bang"?' is a non-question, Christians understand that God always existed before He created the universe, and He always was and still is now transcendent to this universe, because He 'inhabiteth eternity' (*Isaiah 57:15*). Christians are able to experience immediate spiritual contact now through prayer with this transcendent holy God.

## Regarding theories of Origins and the Bible

1. Any theory of origins that is contrary to the early chapters of Genesis, Exodus 20:11; 31:17, and the many other references in the Bible that ascribe creation to the work of God, is not true and will not stand the test of time, no matter how well it is promoted by humanistic educational institutions.

2.  In taking the above attitude, are Christians in danger of repeating the error of the seventeenth-century Church when it opposed Galileo for suggesting a heliocentric (sun-cantered) mechanism for our planetary system? Answer: No. Although the Church leaders in Galileo's day mistakenly thought that the Bible supported the Greek idea of a geocentric (earth-cantered) system, there was nothing intrinsically anti-Creator about Galileo's idea that the earth moved.

By contrast, the 'big bang' and all other theories of origins that are based on human philosophies attempt to say how the universe made itself by its own processes and properties, and with no supernatural input.

Accordingly, Christians do not need Hawking's elusive 'grand unified theory' of the universe to know the mind of God or to know who they are, why they exist, and where they are going. We already have access to the mind of God in the Bible. This tells us that through repentance and faith in the atoning work of Christ on the cross we become children of God, that we are here to worship and serve the living Eternal Infinite God, and that one day we who love God and have received His Son the Lord Jesus Christ will go to live with Him forever. In fact, ***the 'grand unified theory' and the 'theory of everything' is the Bible***!

As matters stand at present, there is no better astronomic theory for the origin of the universe than the inspired explanation of the Bible.

- 'In the beginning God created the heaven and the earth' (Genesis 1:1).
- 'By the word of the LORD were the heavens made; and all the host of them, Figure 2.18
- by the breath of His mouth … For He spake, and it was done;
- He commanded, and it stood fast' (Psalm 33:6,9).
- 'All things were made by Him, and
- without Him was not anything made that was made' (John 1:3).
- "Astronomers Just Detected the Beginning of the Big Bang",
- "Big Bang's Smoking Gun Found",
- "Astronomers Discover First Direct Proof of the Big Bang Expansion" and
- "Major Discovery: Smoking Gun for Universe's Incredible Big Bang Expansion Found"

## HAS THE 'SMOKING GUN' OF THE 'BIG BANG' BEEN FOUND?

Media headlines make people think that some astounding scientific 'proof' has been discovered. The reality is far less spectacular:

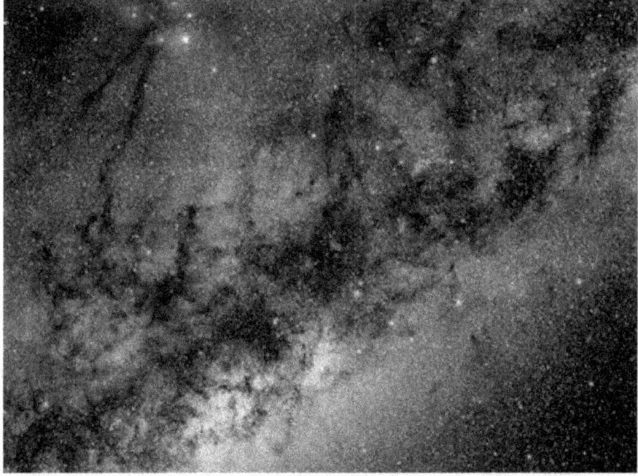

***Fig.2.18:*** *The Universe Came to Existence by the Word of Eternal Infinite God - Curtsy NASA*

These were some of the headlines on Monday 17 March 2014 around the web-based news media.

## One article described it as follows:

Radio astronomers operating telescopes at the South Pole said Monday that they've discovered evidence that the universe ballooned out of the Big Bang due to a massive gravitational force generated by space itself. The discovery is being called the "smoking gun" for the Big Bang theory, and it could have huge implications for our understanding of our universes [*sic*] (and possible others).

Harvard-Smithsonian astrophysicist John M. Kovac and his team detected gravitational waves—tiny ripples in the fabric of space—that could be the first real evidence for the 'inflation' hypothesis of how the universe basically bubbled into being nearly 14 billion years ago. The discovery also suggests that our 14 billion light-years of space aren't all that's out there—our universe could be a tiny corner of something much, much bigger.

In short, the claim is this. The universe began in a big bang nearly 14 billion years ago, but because of various problems encountered with the 'standard' cosmological model for the origin of the universe, the idea of a super-rapid inflation was suggested to have occurred in the first miniscule period of time after the big bang. Wikipedia states:

*"In physical cosmology,* **cosmic inflation**, **cosmological inflation**, *or just* **inflation** *is the expansion of space in the early universe at a rate much faster than the speed of light. The inflationary epoch lasted from $10^{-36}$ seconds after the Big Bang to sometime between $10^{-33}$ and $10^{-32}$ seconds. Following the inflationary period, the universe continued to expand, but at a slower rate."*

The term 'inflation' is used to refer to the hypothesis that inflation occurred, to the theory of inflation, or to the inflationary epoch. The inflationary hypothesis was originally proposed in 1980 by American physicist Alan Guth, who named it 'inflation'. On 17 March 2014, astrophysicists of the BICEP2 collaboration announced the detection of inflationary gravitational waves in the B-mode power spectrum, providing strong evidence for Guth's theory of inflation and the Big Bang.

The problems that inflation was proposed to solve are:

1.  **Why is the universe flat**? That means, why is its geometry Euclidean? We experience such a universe but there is no good reason for that to be so.

2.  **Why is matter (the galaxies) uniformly distributed everywhere** (i.e., homogeneous on the largest scales) and the same in every direction (isotropic)?

3.  Why are there no magnetic monopoles detected, when the Grand Unified Theory (GUT) predicts them in the early big bang universe?

4.  Why is the ***cosmic microwave background (CMB)*** radiation so uniform in all directions in the sky? It is at a uniform 2.72548±0.00057 K temperature, but that is a problem because the radiation has not had an opportunity to mix up bringing the temperature of the universe into equilibrium. This has been called the *horizon problem*. For these and other reasons inflation comes to the rescue.

But what really caused inflation, that rapid initial expansion? What caused the sudden start and smooth stopping of the process? What is the physics behind it? The particle physics mechanism responsible for inflation is unknown. A hypothetical particle or field thought to be responsible for inflation is called the "inflaton," an 'unknown' proposed to solve the above problems. It has been given a name and certain properties; the rest is unknown.

Now it is claimed that from inflation theory they have a prediction that has been verified by observations.

## What have They Detected?

The question is, if you could discover what this "inflaton" field or particle is, would you have evidence for the big bang? Well, no, it is not that simple. Consider what is being observed. The CMB radiation is being studied and a map of a certain polarization mode in the noise from that photon field is teased out using numerous statistical filtering methods.

Did they 'directly' observe the big bang? Not unless you redefine the meaning of the word 'direct.' No, they observed millimeter-wave photons (at about 150 GHz) at the surface of the earth, with a South Pole based telescope.

This is the state of modern cosmology. Because you cannot interact with the universe, you can only observe what it produces, and you have to run statistical arguments based on what you observe.

Is the discovery the 'smoking gun' of the big bang? This implies that it is something similar to just after a crime was committed and you found the guy holding the smoking gun. Well, even if that were so, you could still have the wrong guy, because you were not there to 'directly' see the crime happen. It is circumstantial at best. In this case it would have to be shown that the evidence could not come from any other possible source or mechanism. This is the problem with cosmology in general.

Not only is the claim that inflation has been detected, but also that they have discovered the effects of gravitational waves associated with the initial big bang. They claim they see the ensemble effect of primordial gravitational waves on the CMB photons. This is based on their modelling of what power should be expected in certain CMB fluctuations imprinted by gravitational waves from the big bang.

Gravitational waves, a prediction of Einstein's General Relativity theory, so far have *not* been detected with the very large earth-based interferometric detectors like LIGO. This current claim is that they have detected the signal of these inflationary gravitational waves in the B-mode power spectrum around $l \sim 80$ ($l$ is a measure of the multipole expansion term used to quantify the power in the CMB radiation on different angular scales across the sky).

The paper associated with this release states,

*The power spectrum results are perfectly consistent with lensed-$\Lambda$CDM with one striking exception: the detection of a large excess in the BB spectrum in exactly the l range where an inflationary gravitational wave signal is expected to peak. This excess represents a 5.2 σ excursion from the base lensed-$\Lambda$CDM model. We have conducted a wide selection of jack-knife tests which indicate that the B-mode signal is common on the sky in all data subsets. These tests offer very strong empirical evidence against a systematic origin for the signal.*

It is stated that various methods were used to exclude foreground contamination from the galaxy. In the data they have heavily filtered to extract the result they have modelled what a primordial gravitational lensed signature should look like. In addition, they see an excess effect on the polarization of the CMB photons, over a particular angular scale expected from the theory.

## ONE UNIVERSE

We only observe one universe. Due to this fact there arises the problem of 'cosmic variance,' because we do not know what a typical universe should look like. So even if they were able to separate the foreground contamination by modelling the expected signal from dust in the galaxy how could you know if your model was correct? You can't test it on another universe. They claim to have subtracted "…the best available estimate for foreground dust …" after which a null detection is disfavored at 5.9 σ. This relies heavily on what the modelling says a typical universe without a gravitational wave spectrum should look like.

What you do is divide up the data set into various blocks from different regions of the sky and use that to generate a set of 'different' data sets. This is called 'jack-knife' which is used to give you some sort of estimate of your standard error in the measurements. Of course, it is all from the same universe and really this only gives a measure of the variation across the sky. Next you model what a typical universe should look like and generate mock universes, then compare your observed data to that from the mock universes. You can generate many mock universes and use them to create a statistic.

## Cosmology of Big Bang is not Science

But this belies the problem of 'cosmic variance.' It all depends on the biases you have in your model. They use the dark energy ($\Lambda$), cold dark matter (CDM) cosmology to construct their mock universes and compare the observed data to that universe. It *heavily* depends on the biases built into the model. Those biases also determine what data you accept or reject in the analysis. Besides, there still does not exist any laboratory evidence for dark energy nor dark matter. After that it is just modelling.

'But it must be correct because the big bang happened, we all know that!'

This is the problem:

Cosmology may look like a science, but it isn't a science. —James Gunn of Researchers have measured the temperature variations in the CMB so precisely that the biggest uncertainty now stems from the fact that we see the microwave sky for only one Hubble volume [i.e., only one possible observable universe - James Gunn ], an uncertainty called *cosmic variance*. 'We've done the measurement,' [Charles] Bennett says. (Emphasis added)

That barrier to knowledge, some argue, is cosmology's Achilles' heel. *'Cosmology may look like a science, but it isn't a science, '* says James Gunn of Princeton University, co-founder of the Sloan survey [currently the biggest large-scale survey of millions of galaxies— James Gunn]. 'A basic tenet of science is that you can do repeatable experiments, and *you can't do that in cosmology'.* (Emphasis added)

'The goal of physics is to understand the basic dynamics of the universe,' [Michael] Turner says. 'Cosmology is a little different. The goal is to reconstruct the history of the universe.' Cosmology is more akin to evolutionary biology or geology, he says, in which researchers must *simply accept some facts as given.* (Emphasis added)

## MANY UNIVERSES

These results are also suggested to tell us something about all the other universes out there, because if the inflation of this universe is correct then our universe must be only one of many bubble universes that have bubbled into existence and we just happened to have evolved in this one and now we are able to discover this. This is the multiverse, which solves the problem of *why we are here*. It is just random chance that we live in this particular universe—the 'Goldilocks' universe—not too hot, not too cold, but just right—for life to exist.

This now shows the philosophical nature of cosmology. Only by first assuming the big bang cosmogony to be true, followed by cosmic evolution, does a discussion of a multiverse have any relevance.

So much hinges on the 'unknowns.' And the 'unknowns' result from applying the particular cosmological model—now known as the 'standard' model—to the observational data. The fact that these fudge factors are needed does not seem to disturb the atheist world because this is what they expect from the universe. To me it would engender a lot more confidence in the science if the picture of the universe they obtained made sense.

'It made itself.' We must accept that as a given, and then we can discover how that happened.

## EX-NIHILO – OUT OF NOTHING

The universe began as God said in Genesis chapter 1, with no big bang. He created it out of nothing, by His supernatural omnipotence. God said: "Let there be light." He filled the universe with His glorious light, until the fourth day of creation. Then in the fourth day of creation He created the sun, the moon and the galaxies with stars as source of light. That light might be what we observe today in the CMB radiation. Our own cosmology—a derivative of Moshe Carmeli's own big bang cosmology, but with biblical initial conditions—

has a period of super-rapid accelerating expansion in the 4th Day of Creation, not quite the same as inflation, yet sufficient to adiabatically cool the initial light from a temperature of about 9,000K to nearly 3K today.

Making a prediction from some esoteric quantum theory in regard to the putative inflationary epoch, and then claiming a fulfilment of this in these observations is not the same as a clean prediction in testable, repeatable physics. In the case of the latter, there are ways to interact with the experiment and repeatably test one's hypothesis. In the case of the former, i.e., when the laboratory is the cosmos, we cannot do that. The best we can do is run simulations on what we think the universe should look like and try to quantify the likelihood of the outcome of an observation. Astrophysicist Richard Lieu wrote,

Hence the promise of using the Universe as a laboratory from which new incorruptible physical laws may be established without the support of laboratory experiments is preposterous, and unscientific.

Far from being a definitive proof of either inflation or the big bang, this so-called 'smoking gun' is very 'model-dependent,' which means it depends on unprovable assumptions—including that there was a big bang to begin with. Whereas even the idea that the CMB is the leftover echo of this alleged event has some serious and unresolved problems; for example, if the radiation really is coming from deep space, why is there no 'shadow' in it from objects supposedly in its foreground?

Consider for a moment something else, something consistent with all the observations, including these latest reports; namely, that the universe did not begin in a big bang, because the universe never started in a singularity. It began in time, but, "In the beginning, God created the heavens and the earth."

## Cosmology is Not Even Astrophysics

Physicists, when critical of a theory that is really bad, that is really 'way off base,' will sometimes say that the theory is 'not even wrong.' They mean that some theories stand up to scrutiny and survive, other theories don't and are rejected as being wrong, but nevertheless not scorned at, but some theories are so bad, so ridiculed, so poorly esteemed, that they are way below those that are considered 'wrong.' Therefore, physicists say they are 'not even wrong.'

The phrase was apparently coined by the early quantum physicist Wolfgang Pauli, who was known for his colorful objections to incorrect and sloppy thinking. According to *Wikipedia*, 'An apparently scientific argument is said to be *not even wrong* if it is based on *assumptions that are known to be incorrect*, or alternately *theories which cannot possibly be falsified* or used to predict anything.' [emphasis added]

That seems to be the state of cosmology today. Is it because the unverifiable starting assumptions are inherently wrong?

Some brave physicists have the temerity to challenge the ruling paradigm—the standard big bang LCDM inflation cosmology. One of those is "Richard Lieu" of the Department of Physics, University of Alabama.

According to Lieu, 'Cosmology is not even astrophysics: **all the principal assumptions in this field are unverified (or unverifiable) in the laboratory** … .' In a recent paper, Lieu says that 'because the Universe offers no control experiment, i.e. with no independent checks, it is bound to be **highly ambiguous and degenerate**.'

- This seems a fair analysis, because 'cosmologists' today are inventing all sorts of stuff that has just the right properties to make their theories work, but

- stuff that has never been observed in the lab. They have become 'comfortable with inventing unknowns to explain the unknown.'

Cosmologists tell us we live in a universe filled with invisible, unobserved stuff—74% dark energy and 22% dark matter. But what is this stuff that we cannot detect yet should be all around us? *For 40 years, one form or another of dark matter has been sought in the lab*, the axion, for example, Figure 2.19.

Long before that, scientists invoked dark matter to explain puzzling dynamics in the solar system, such as the planet Vulcan hiding behind the Sun to account for the discrepancy with the orbit of the planet Mercury. But *Einstein solved the problem with his general theory of relativity*. What was needed was new physics and not some unproven hypothesis. Is this the same today?

However, today we have also dark energy that is supposedly driving the universe apart at an even faster pace than in the past. On 30 May 2004 Physicsworld.com reported

'New evidence has confirmed that the expansion of the universe is accelerating under the influence of a gravitationally repulsive form of energy that makes up two-thirds of the cosmos.

'It is *an irony of nature that the most abundant form of energy in the universe is also the most mysterious.* Since the breakthrough discovery that the cosmic expansion is accelerating, a consistent picture has emerged indicating that two-thirds of the cosmos is made of "dark energy," (The Hand of God)—some sort of gravitationally repulsive material.' [emphasis added]

*Fig.2.19*: *74% dark energy and 22% dark matter – It is a Godly Entity –  Curtsy –  NASA, wikipedia.org*

Apparently dark energy is a confirmed fact. But does the evidence confirm the universal expansion is accelerating? They are right about the irony, that although this energy is so abundant it cannot be observed locally in the laboratory. Is this really the state of the Universe today? Or does the emperor need new clothes?

Lieu says '… *astronomical observations can never by themselves be used to prove "beyond reasonable doubt"* a physical theory. This is because we live in only one Universe—the indispensable "***control experiment***" is not available.'

There is no way to interact and get a response from the Universe to test the theory under question, like an experimentalist might do in a laboratory experiment. At most the cosmologist collects as much data as he can and uses statistical arguments to try to show that his conclusion is likely. Lieu again, '*Hence the promise of using the Universe as a laboratory from which new incorruptible physical laws may be established without the support of laboratory experiments is preposterous …* '

Lieu lists five evidences, listed below, where cosmologists use '*unknowns*' to explain '*unknowns*', and hence he says they are not really astrophysicists. Yet these evidences are claimed to be all explained (and in the case of the CMB even predicted) by the LCDM inflation model. None of them are based on laboratory experiments and they are unlikely to be ever explained this way. They are:

1.    **redshift** of light from galaxies, explained by expansion of space,

2.   **CMB**, Cosmic Back Ground) explained as the afterglow of the Big Bang,

3.   **rotation curves** of spiral galaxies, explained by dark matter,

4.   **distant supernovae** being dimmer than they should be hence an accelerating universe, explained by dark energy,

5.   **flatness** and isotropy, explained by Inflation.

As an experimentalist, I know the standards used in so-called 'cosmology experiments' would never pass muster in my lab. Yet it has been said we are now living in the era of 'precision cosmology.'

Recently, Max Tegmark said, ' ... *30 years ago, cosmology was largely viewed as somewhere out there between philosophy and metaphysics. You could speculate over a bunch of beers about what happened, and then you could go home, because there wasn't a whole lot else to do.'* But now they are closing in on a 'consistent picture of how the universe evolved from the earliest moment to the present.'

How can that be true if none of Lieu's five listed items can be explained by '***knowns?***'

They have been explained by resorting to 'unknowns' with the sleight of hand that allows the writer to say 'we are closing in on the truth.' I recall Nobel Laureate Steven Chu speaking to a large gathering of high school children on the occasion of the Australian Institute of Physics National Congress at ANU in Canberra in 2005. He said that we now understand nearly all there is to know about the Universe, except for a few small details like what is dark energy and dark matter, which [allegedly] make 96% of the stuff in the Universe. As Homer Simpson would say, 'Duh!'

Lieu then lists a few counter evidences to LCDM, ΛCDM – Lambda Cold Dark Matter, assumes that the universe is composed of photons, neutrinos, ordinary matter, inflation cosmology, all of which have been, or are in the process of being, published in topmost astronomy journals. They are:

1.   the number density evolution curve in galaxy clusters (there is a massive dark matter problem in them also) does not agree with the LCDM, Lambda Cold Dark Matter, prediction to $7\sigma$ statistical significance,

2.   only 50% of the baryons predicted by the LCDM model seem to exist at low redshifts—called the *missing baryon problem*,

3.   no explanation for the soft X-ray excess from clusters, Abell 3112, for example,

4.   the disparity between the values Sandage *et al.* ($H_0 \approx 62$ km/s/Mpc) and Freedman *et al.* ($H_0 \approx 70$ km/s/Mpc) determine for the Hubble constant from two independent analyses of HST data,

5.   galaxy groups like our own Local group seem to harbor too much matter,

6.   very feeble SZE detected by WMAP, the Wilkinson Microwave Anisotropy Probe (WMAP) is a NASA Explorer mission that launched June 2001 to make fundamental measurements of cosmology -- the study of the properties of our universe as a whole, —no shadow cast in the foreground as expected from a background source; this has now been tested by separate authors respectively on two sets of 31 and 100 rich clusters,

7.   Axis of Evil in the CMB octupole and quadrupole expansion terms correlate with HI clouds in the Galaxy, where Lieu concludes that a significant fraction of the WMAP anisotropy at the primary acoustic peak is not cosmological,

8.   dwarf galaxy rotation curves. The data indicate constant density cores, whereas LCDM halo profiles have central cusps.

These evidences match other cosmological models better than the LCDM, yet no model has a good match to all the evidence. (See Lieu's figure 3 for a comparison with two other models and his paper for details.)

There also remain discrepancies particularly with analyses of the WMAP data. Only recently did the WMAP, (the Wilkinson Microwave Anisotropy Probe (WMAP) is a NASA Explorer mission that launched

June 2001 to make fundamental measurements of cosmology – the study of the properties of our universe as a whole), people at NASA release individual year maps and difference maps of the anisotropy data. They should be a good test of whether the anisotropies are cosmological. If cosmological then the noise levels should be frozen in time, not affected by local effects from one year to the next. The differences should be nulled across the whole sky, they are not. But at the time of writing this, I have not yet had an opportunity to examine them in any detail. However, I have been told that such a 'difference map' between WMAP2 and WMAP3 data was presented at COSMO 06, the International Workshop on Particle Physics and the Early Universe, 25–29 September 2006, Lake Tahoe, California. Apparently it caused quite a stir.

Most of the difference map had very mild (small) temperature fluctuations but 5% of the difference map had an oblong area of very high temperature fluctuations with the highest temperature differences at the area's center. Either there is an error in the data reduction extracting the effects from the Galaxy (which is meant to be minimal at the high frequency band chosen), or the effect is not cosmological but caused by something closer to the earth. My informant who was there he tells me ' ... the audience went into a defense mode with excuses, band-aids and blaming dark matter, whatever. The speaker claimed it really was not part of his formal presentation. The mood was, 'let's not do that again, and we all know that the data looks better if we put all the years together so we can attain our planned objectives.'[18]

Well, the cosmologists may have their planned objectives, to shore up their faith in a model that is based on false and unverifiable assumptions, but it is leaky bucket that cannot hold back the evidence that ultimately will be published against it.

The fact is that the history of the universe cannot be determined from a model which cannot be independently tested. The Big Bang cosmology is verified in the minds of those who already hold to that belief—that the Universe created itself about 14 billion years ago—*ex nihilo*. To me the biblical big picture is far more believable—only we are left to fill in the details.

## Has the 'Big Bang' been Proven?

All around the world, newspaper headlines have exulted about a recent discovery of 'ripples' of temperature in deep space which they claim 'proves' the idea of a 'big bang'. In fact, it could be said that the secular media, in a number of instances, went after this information with what could only be called 'religious fervor'.

The 'big bang' is the evolutionary belief that the universe was once compressed into a tiny dot which exploded, and from this explosion came, unaided, the entire world of stars, galaxies, planets, palm trees and people.

Almost all the newspaper articles mentioned God. Either God was now not necessary, or else the *'big bang' itself was now God*. One English news-paper said about this supposed evidence, *'it has blown away the last shreds of necessity for a supreme being to explain how the universe came into existence ... God is made redundant. Science really does now have "no need of that theory"?.'*

Many theologians did their usual retreat into a *'god of the gaps'* idea by saying that, since nobody knows *'who lit the fuse'*, it must have been God. One scientist said the discovery was *'like looking at God'*. Another called it the *'Holy Grail'*. Some Christians have said that we must now believe that the 'big bang' was God's way of creating the universe.

However, it should be obvious that this is very different from what God has revealed about the origin of all things in His Word to mankind. Not only is the order of events wrong, but the 'big bang' is purely a mechanistic view to explain how the universe made itself *without* any miraculous creation. But what was really discovered?

## Alternative Explanations Possible

First of all, we need to understand a little about the supposed 'evidence' for the 'big bang'. Many may not realize that to date, the 'big bang' has consisted of much speculation based on only three pieces of observation, **all** of which have alternative possible explanations.

One of these 'evidences' is the microwave radiation coming in from all directions. This radiation (the same as would be given off by heat) is interpreted as the 'echo' or 'left-over heat' of the big fireball that started everything off. This radiation has in the past been found to be extremely uniform—it's the same everywhere!

However, because the universe itself has been found to be extremely clumpy (with great walls of galaxies, and great voids in between) then, if this happened as a result of the 'big bang', this background radiation should also be clumpy. In other words, the temperature of this radiation should be uneven—there should be hot spots and cold spots. Because the radiation had been found to be so uniform, even secular scientists were saying that the 'big bang' idea was in trouble. The search for an unevenness in the radiation became intense.

Recently, however, the press was told that the 'big bang' had been rescued because new measurements showed the unevenness the 'big bang' adherents were looking for. The press announced to the world that the 'big bang' had been proved, and God was now redundant.

But, what did they really find? The temperature differences that have been discovered, and which have caused such a worshipful reaction, are around 30 millionths of a degree! However, these differences may not even be real.

At a creation lecture in the USA recently, a man rose and said he was part of the team which designed the instruments used to make these measurements, and he could categorically say that they were **not even that sensitive!** This is confirmed in the journal *Science* of May 1992 ,1 (p. 612): the variations claimed are 'well below the level of instrumental noise'—they have been obtained by statistical methods which still need careful checking.

Also, from the same article, the opinion of George Smoot, the man in charge of the project, is that 'he's pretty sure the effect he's seeing is real but adds that there's always a chance it's wrong'. Even if the measurements are real, Smoot admits 'it can be caused by other effects, such as the motion of our galaxy through the background radiation.'

Thirty millionths of a degree, even if real, is not much to get excited about anyway. Imagine scanning a floor tile with a temperature probe. Even if the whole surface at first appears to be the same temperature, if you make instruments more and more sensitive you will eventually find some patches which are ever so slightly warmer or colder than others, as nothing is ever perfectly uniform.

## BRAINWASHED

Even if real, *Nature* (March 30, 1992, p. 731) concludes that all one can say is that they are 'consistent with the doctrine of the Big Bang', and that it is a 'cause of some alarm' that the media has called it 'proof that "we now know"? how the Universe began.'

Sadly, many will never read the scientific journals for the truth about this information, and will go on believing a lie they have been indoctrinated with by the anti-God secular media. Many people don't realize how brainwashed they have become.

'Big bang' theory also seems to need the associated concepts of 'inflation' (an assumed early rapid expansion) and 'dark matter' (the belief that more than 90 per cent of the universe's mass is made up of a mysterious, unobservable and unknown substance).

The same *Nature* paper also says that neither of these has any 'true independent support, outside the cosmological arena for which they were invented.' It goes on to indicate that those with alternative theories to the 'big bang' will probably be able to 'claim the new data as support for their theories also'. Reading this calm, objective assessment in a leading journal makes it clear that no one has come remotely close to proving that there was a 'big bang'.

Scientific theories are always changing. What seems to fit a few facts in one generation may be replaced by a totally different view which fits those facts and even more. There are already observations which do not presently fit the 'big bang'. One prominent group of astronomers believes that the facts better fit a new version of the steady-state theory—not only no explosive beginning, but no beginning at all—an eternal universe. If the next generation of astronomers adopts that view, what will be the position of those theologians who have been pressured into believing (contrary to God's revelation) that the 'big bang' was 'God's method of creating'?

## The Big Bang and Before

A BBC documentary with this title was aired on SBS-TV in April, 2012. In it, several cosmologists discuss 'the unthinkable'—perhaps the big bang was not the beginning of everything after all. It seems that scientists have discovered a new law. Well, not actually new—just one that has been treated as if it didn't exist for the last half century or so by 'big-bangers' such as Stephen Hawking, Roger Penrose, Paul Davies, Edwin Hubble, *et al*, namely the law of Cause and Effect.

The program explains that the concept of the big bang[1] postulates that "everything we see in the universe today—us, trees, galaxies, zebras—emerged, in an instant, from nothing. And that's a problem. It's all effect and no cause." We are then given five different explanations from five different scientists concerning what this cause may (or may not) have been.

## MICHIO KAKU AND THE MEANING OF 'NOTHING'

Dr Michio Kaku is Professor of Theoretical Physics at City University, New York. He asks: "How can it be that everything comes from nothing?" His solution: "If you think about it a while, you begin to realize it all depends on how you define 'nothing'!"

We are then shown a huge NASA vacuum chamber, the largest in the world—the nearest we can get to a state of nothing, but which still has dimensions ('nothing in 3D'), and through which light can pass. **Kaku tells us: *"I think there are two kinds of nothing. First there is something I call absolutely nothing: no equations, no space, no time, no anything that the human mind can conceive of, just nothing. Then there is the vacuum which is nothing but the absence of matter."***

The host then comments: "Kaku's version of nothing is the perfect vacuum where on the face of it there is only energy. But in a perfect vacuum, energy sometimes transforms itself temporarily and briefly into matter. It is one of these tiny explosions that might have been going on and ended up in the big bang."

Kaku: *"So for me the universe did not come from absolute nothing—that is a state of no equations, no empty space, no time; it came from a pre-existing state—also a state of nothing. Our universe did in fact come from an infinitesimally tiny little explosion that took place giving us the big bang, and giving us the galaxies and stars, we have today."*

The host: *"For Kaku, the laws of physics did not arrive with the big bang. The appearance of matter did not start with the clock of time. His interpretation of nothing tells us there was, in short, a 'before.' If he is right, there is an opportunity for a cause to have an effect, after all."*

## ANDREI LINDE'S 'RADICAL EXPLANATION'—INFLATION

The host continues: *"The idea of the big bang was a very bold idea but it had problems. … Why is the universe as big as it is now? Who made it expand? What caused the explosion? The big bang was clearly a very special explosion. Ordinary explosions are messy. This one produced a universe that wasn't messy at all. Our universe is, more or less, the same in every direction. It was an observation that required a radical explanation."*

According to Dr Andrei Linde, who is Professor of Physics at Stanford University: *"Just after matter first appeared, rather than a messy explosion, there was instead a massive and unprecedented growth in the size of the universe. This is called Inflation. If one assumes there was a period of exponential expansion of the universe in some energetic vacuum-like state, then you can explain why the universe is so large, why the universe is so small at a very large scale, why properties of the universe in different parts are so similar to each other. All these questions can be addressed if one uses inflation."*

The host: *"Inflation was a pre-existing condition that has been there, well, for ever. For Prof. Linde, the big bang wasn't really a starting point at all; he thinks that it was simply the end of something else. The universe appeared out of what he calls eternal inflation. Our universe is not the only one. There are others, all co-existing. He has counted them. There are ten to the power 10 to the power 10 to the power 7. His ideas of a multi-verse, as odd as they seem, are now within the scientific mainstream. For many cosmologists' eternal inflation is in itself a reasonable explanation of what existed before our universe. **For others it's utter nonsense**."* (Emphasis added.)

## Param Singh, the Big Bounce

Dr Singh is a Distinguished Research Fellow at the Perimeter Institute for Theoretical Physics, Waterloo, Ontario, Canada. In the program he tells us: *"The principal mathematical objection [to the universe expanding from nothing] is that as the clock is wound back and Hubble's zero hour is approached, all the stuff in the universe is crammed into a smaller and smaller space. Eventually that space will become infinitely small. And in mathematics, **invoking infinity is the same as giving up, or cheating**."* (Emphasis added.)

**His Solution**: *"Instead of emerging from nothing, our universe owes its existence to a previous one that had the misfortune to collapse in on itself. Then, thanks to some clever math's, rebounded to what we see today. So, the big bang was not a bang at all. It was rather a big bounce. … Of course, it might all be nothing more than a fantasy world of maths and little else, and there's always the nagging question of what started the infinite bouncing in the first place. **It was certainly not the big bang. That is impossible**."* (Emphasis added.)

## Lee Smolin, Natural Selection

Smolin is a researcher at the Perimeter Institute in Canada. We are told that his solution owes more to Charles Darwin than to Albert Einstein. It has been called 'cosmological natural selection'. He believes that the universe had an ancestor which had another ancestor. According to this hypothesis, the universe was born inside a black hole.

**Smolin:** *"There is a bounce inside every black hole. Material contracts and contracts and contracts again, and then begins to expand again, and that is the big bang which initiates the new region of the universe."* The commentator adds: *"Smolin's natural selection idea proposes that for a universe to prosper it must reproduce and for that to happen it must contain black holes that, according to Smolin, spawn offspring universes."*

**Smolin:** *"Before the big bang there was another universe much like our own. In that universe was a big cloud of gases. It collapsed to form a massive star. That star exploded. It left behind a black hole and in that black hole there was a region, if you were misfortunate enough to fall in, you would find it becoming denser and denser and* *denser. You wouldn't survive this but imagine you did—then all of a sudden you would explode again and that would be our big bang."*

## Neil Turok, Membranes Collided

Dr Neil Turok is the Executive Director of the Perimeter Institute in Canada. He says: "There are essentially two possibilities at the beginning. Either time did not exist before the beginning; somehow time sprang into existence. That's a notion we have no grasp of and which may be a logical contradiction. The other possibility is that this event which initiated our universe was a violent event in a pre-existing universe.

His solution requires ten special dimensions plus time. Dr Turok: *"We live on an extended object called a brane (short for membrane). … You can't have only one; there must be at least two, separated by a gap. These two branes collide. When they collide, they remain extended; it's not all of space shrinking to a point. … They fill with a density of plasma and matter, but it's finite. Everything is a definite number which you can calculate, and which you can then describe using definite mathematical laws. That's the essential picture of the big bang in our model."*

The host comments: *"For many cosmologists **this is mathematical sleight of hand**."* (Emphasis added.)

The program then conveniently summarizes these ideas and asks which is correct:

**Michio Kaku**: Stop thinking of nothing as nothing, but rather just the absence of stuff.

**Andrei Linde**: He redefined the big bang as inflationary energy of a mega burst dying out ten to the power 10 to the power 10 to the power 7.

**Param Singh**: No big bang at all; just the big bounce, again and again and again.

**Lee Smolin**: Our big bang was simply the other side of a black hole in a galaxy far, far away.

**Neil Turok**: Colliding branes in another dimension.

**The host**: *"They would be easier to dismiss as the half-baked musings of the lunatic fringe were it not for the fact that some of the very people who constructed the everything-from-nothing big bang model are themselves starting to dismantle it."*

## SIR ROGER PENROSE

Sir Roger Penrose is Professor of Mathematics at the University of Oxford. For many years he spent much of his time dismissing the idea of 'before the big bang.' He now says: *"The current picture of the universe is that it starts with a big bang and it ends with an exponentially expanding universe, where it eventually cools off with not much left except protons. … This very expanded universe is the equivalent to a big bang of another one. … This universe is one eon of a succession of eons. Each expanding universe accounts for the big bang of the next."*

The host adds: "Because of this a nearly infinitely large universe could just as well be the infinitely small starting point for the next one. A simplistic system with a 'before' and an 'after'. Quite a bold thrust for a man who was until five years ago a pre-big-bang denier."

We are then told that "in science ideas are just ideas until they are confirmed or denied by observations" and the program discusses how researchers are investigating gravity waves in an effort to observe the big bang itself.

The program concludes with Prof. Kaku telling us: "My parents were Buddhists. In Buddhism there is no beginning, no end, there is just nirvana. As a child I also went to Sunday school, where we learned that there was an instant where God said, '**Let there be light**.' I kept these two mutually contrasting paradigms in my head, but now we can know these two paradigms together in a pleasing whole. Yes, there was a genesis. Yes, there was a big bang. And it happens all the time."

Discerning viewers of this BBC program will have noticed several things:

1. Almost all the concepts of the big bang are now under attack by secular scientists, e.g., that everything came from nothing, that everything was once contained in a singularity, the beginning of time, the origin of the laws of physics.

2. All the 'solutions' to the problem of what happened before the big bang involve pre-existing universes or conditions, at least one of which is said to have existed for ever, despite the second law of thermodynamics (which says this is an impossibility).

3. The proponents of these 'solutions' provide no physical evidence whatsoever in support of their ideas.

4. Not one says how his postulated first universe came into existence.

5. When atheists reject the eternal truth of the Word of God in the Bible, we should not be surprised to find that their attempts to find alternative explanations fail to stand the test of time and evaluation by true scientific laws.

6. Christians who have tried to write the big bang into Genesis have problems.

    i. First, of course, it contradicts the Bible (e.g., the big bang has Sun appearing long before Earth; Genesis has Earth created before Sun).

    ii. Second, all this rethinking and, frankly, confusion, makes one thing quite obvious. Namely, that building a theology of origins on an allegedly 'assured' and 'scientific' foundation such as the 'big bang' idea is in reality building it on shifting sand.

7. How much better to take Genesis at its face value (as indeed Jesus Christ Himself always did) and build one's scientific models on the rock of Scripture.

## THE UN-SCIENTIFIC BIG BANG

According to most text books promoting evolutionary beliefs, the universe came into existence around 14 billion years ago. Supposedly, it began as a singularity, an unimaginably hot 'point' into which was compressed all the matter and energy now making up the billions of galaxies found throughout space. Following the big bang, this singularity started to expand and cool, enabling quarks and electrons to form, these being the building blocks of atoms. As the universe continued to cool, quarks combined to form protons and neutrons.

These then combined to produce 'light elements', namely hydrogen (including a little 'heavy hydrogen' or deuterium), helium, and lithium. And all this supposedly happened in the first three minutes of the universe!

However, these elements were still nuclei, not atoms. That supposedly needed about 379,000 years of cooling so the nuclei could combine with electrons and form atoms. This hydrogen and helium gas is said to have provided the material from which the first stars formed. (Hydrogen, helium, and lithium are known as 'light elements' because they have very little mass. Evolutionary cosmologists believe that the heavier elements (which they call 'metals') were formed inside stars.)

Biblical creationists, of course, should reject big bang theory because the order of events contradicts the Bible. Whereas in big bang theory, the stars precede the earth, in Genesis, the earth is made before the stars.

There are three main pieces of evidence presented in support of the big bang: background heat, the abundance of 'light elements and the expansion of the universe.

## BACKGROUND HEAT

Space appears to be bathed in a low-level background heat which is said to be the 'after-glow' or 'smoking gun' of the original big bang 'explosion'. This is known as the 'cosmic microwave background radiation' (CMBR).[1] It has been measured very accurately, and in great detail, and is said to have a temperature consistent with what would be expected from the big bang.

Creation scientists, however, point out that this background heat is actually a major *problem* for big bang theory. **This is because its temperature is virtually the same across the universe and this would *not* be expected from a big bang**. A conventional 'explosion' would leave behind an uneven pattern of heat, not the extremely even pattern actually observed.

Evolutionists, of course, are well aware of this difficulty, called the '***horizon problem***', and some claim to have a solution. Supposedly, very soon after the initial 'bang', and for only a brief period, the universe expanded at a much higher rate—in fact faster than the speed light—and this enabled the background heat to be smoothed out. This very rapid expansion is known as 'inflation'.

However, according to Paul Steinhardt, Albert Einstein Professor in Science at Princeton University, "inflation is very flexible … [and] can be adjusted to give any result … any outcome is possible." Hence, he says, "it is not possible to find evidence to support or refute inflation." Creation scientists would agree, and argue that it is really no more than a story called upon to explain away facts that point to the big bang theory being wrong.

## ABUNDANCE OF LIGHT ELEMENTS

Big bang theory is said to have accurately predicted the amounts of light elements that we actually find in the universe. Hydrogen would be expected to be the most common, followed by helium, deuterium and lithium—which is what is observed. Particularly, it's said, the amount of helium (25%) is consistent with calculations based on big bang theory. Historically, however, this claim has been very controversial. As pointed out by Professors Burbidge and Hoyle,

"We have now reached the stage where it is argued that the existence of helium … is taken, together with the microwave background radiation, as primary evidence in favor of the … big bang … . However, this argument is only powerful if there is no other way to explain the helium abundance and the microwave background radiation."

They then argued that this helium was produced by the burning (fusion) of hydrogen in stars (rather than the big bang) and that the light emitted by the stars and absorbed by dust clouds produced the cosmic microwave background radiation.

According to Professor Burbidge, neither the observed abundance of helium, nor the level of the cosmic microwave background, were really predictions of big bang theory. Scientists had already measured the abundance of helium and the theory was adjusted so that it would give the 'right result'. Referring to the parameter (the number in the big bang theory) which would result in it 'predicting' the amount of helium in the universe, he wrote:

"*It is chosen to make things come out right … . This is why the big bang theory cannot be argued to explain microwave background or to explain a cosmic helium value close to 0.25 [i.e. 25%]. … if you really believe in a big bang, you can choose parameters which will make observation and theory agree, but the argument is not based on basic theory*" (emphasis original).

**Evolutionary cosmologists** sometimes claim that big bang theory can predict the amount of the light elements using information (data) obtained from satellite measurements. This, however, cannot be so because their theory relies on the existence of 'dark matter', a form of matter that cannot be observed. (This 'dark matter', which many creationists and some secular cosmologists do not believe exists, is also needed to explain how galaxies and stars form by natural processes.)

## THE EXPANDING UNIVERSE

It is thought that expansion of space would cause light waves to be stretched, resulting in light being reddened. This is known as 'redshift' and is observed when we view galaxies. Moreover, in general, the more distant a galaxy, the greater is the redshift. This is understood to indicate that the universe is expanding and that, the more distant a galaxy, the faster it is moving away. Hence, it is argued, winding the clock back, we would see the universe getting smaller and smaller until it reached its original state as a singularity.

While many creation scientists accept the evidence that the universe is expanding, this does not mean that it must have started off as a singularity—it could have started expanding from a fairly large state.

**Creationist scientists** argue that the atheists' claim that our universe arose from a random 'explosion' is absurd. For example, the rate of expansion would have needed to be just right, as even a tiny deviation from the required rate would have been catastrophic. If just a little faster, particles would have simply flown away from each other, never coming together to form stars and planets. If just a little slower, gravity would have pulled everything back together resulting in a violent 'great crunch', with no planets and no life.

According to Nobel prize-winner, Professor Steven Weinberg, the number determining the required expansion rate (known as the 'cosmological constant') would have had to be just to right to within 120 decimal places.

But what does this mean? Well, let's use as an illustration the mixing of concrete. Here the amount of water added has to be controlled quite carefully, otherwise the concrete will not have the required strength. Typically, for every 100 kg of cement, around 40 kg of water should be added. For a particular building project, there might be generous leeway, where an error of 1 kg in the amount of water added could be tolerated and the concrete still have the required strength. Working with a margin of error of 0.1 kg might still be practical; however, if the allowable error was only 0.000001 kg this would clearly be impractical. 0.000001 has the decimal point six positions to the left of the 1:

6 positions

0.000001

But a number with 120 decimal places has the decimal point 120 positions left of the 1:

120 positions

0.00000000000  .........  0000000001

How realistic is it to believe that an 'explosion' just happened to produce an expansion rate this critical?

The expansion rate, however, is just one of many factors that would have had to be 'fined-tuned' for the big bang to have produced a universe like ours in which life could exist. For example, unless the masses of the particles that make up atoms, the forces that hold atoms together, and the force of gravity had all had the right values, the big bang would have produced a lifeless universe.[11,12] Creation scientists argue that a process that is this critical could not have occurred by chance.

Christians need not be intimidated into accepting secular accounts of origins. Big bang theory only appears to be scientific because people are exposed only to the evidence that appears to support it. At the same time, nothing is said about its major scientific problems. Big bang theory contradicts the account of creation in Genesis, and Bible believing creationists should reject it on the authority of God's word.

*Fig.2.20: B*

## The Big Bang Flaws

In the big bang theory, foremost points of inconsistency to the Scripture comprise:

1. **The sun** is supposed to come to existence before the earth, conflicting to Genesis 1, which declares God created the earth on Day One, but then the sun and the stars on Day Four of Creation Week. Correspondingly, when God first created the earth, it was imperturbable and dark, whereas, the big bang hypothesizes that the commencement was immensely hot and bright.

   The big bang theory encompasses a gradual progressive construction of astronomical objects stretch over billions of years. This runs in conflict to one of the details for its plea to Christians who have not understood it through carefully and so mistakenly consider of it as an unexpected one-time creation event. Such a gradual progression over vast ages is not an brand that learnt minds of the past, former to the popularity of the idea of huge ages, in geology first, ever once collected from the Bible, Psalm 33:8–9 - *Let all the earth fear the Lord; let all the inhabitants of the world stand in awe of him! For he spoke, and it came to be; he commanded, and it stood firm*

2. God may well have created the **universe** in an **expanding** condition, but not over eons—The scripture in fact, precisely describes *against* this long-age idea.

3. This opinion declares that the **days of creation** were extensive ages of time instead of **6 Days of creation**. This idea was unidentified until enthusiastic Anglican theologizer "Stanley Faber (1773–1854)," wished-for that the days of Genesis 1 were actually eons of time. This was not broadly acknowledged until it was promoted by the Scottish geologist and confessing evangelical, "Hugh Miller (1802–1856)," who abandoned the "gap theory" and began encouraging the day-age interpretation in his book "*Testimony of the Rocks*." This was published in the year after his untimely death, by suicide. He risked that the days were actually long ages. Miller thought that Noah's Flood was a limited to a local flood and that the rock deposits were arranged down over extensive periods of time. The best recognized contemporary advocate of this opinion is "Hugh Ross (1946– )," and his group "Reasons to Believe," whose defenses for the opinion have been methodically humiliated.

4. As we have repeatedly illustrated, the days of Genesis 1 essentially is normal-length days, as they have together a number and **an evening plus morning**. Correspondingly, the Fourth Directive plainly quotes

Creation Week as the outline for our working week (Exodus 20:8–11) - *"Remember the Sabbath day, to keep it holy. Six days you shall labor, and do all your work, but the seventh day is Sabbath to the Lord your God. On it you shall not do any work, you, or your son, or your daughter, your male servant, or your female servant, or your livestock, or the sojourner who is within your gates. For in six days the Lord made heaven and earth, the sea, and all that is in them, and rested on the seventh day. Therefore, the Lord blessed the Sabbath day and made it holy. —we don't work for six eons and rest for one eon!*

5.  Even the Order is Wrong!

6.  Some are attracted to the incorrect Day-Age interpretation as they think that if the time scale of Genesis were overextended out, then it would tie the long-age/evolutionary order. Nonetheless, this interpretation is naive, as will be illustrated. There are foremost illogicalities between a straightforward reading of Scripture and the order claimed by uniformitarian/evolutionary "science." Day-agers agree to take the evolutionary order, nonetheless, just repudiate biological transmutation of kinds, Figure 2.21.

Contradictions between Genesis and Uniformitarian 'science' in the order of events

| Day | Biblical creation | | Big bang, evolution | Evolutionary sequence |
|---|---|---|---|---|
| 1a | Heavens and Earth | | Light | 1 |
| 1a | Darkness | | Expanse ('space') | 2 |
| 1a | Water, oceans | | Stars | 3 |
| 1b | Light | | Water | 4 |
| 2 | Expanse | | Sun | 5 |
| 3a | Dry Land | | Earth | 6 |
| 3b | First life: land plants, trees | | Dry land | 7 |
| 4 | Sun, moon, stars, other planets | | Oceans | 8 |
| 5a | Fish, whales (including dolphins) | | First life: single-celled organisms | 9 |
| | Aquatic reptiles (including ichthyosaurs, mosasaurs, plesiosaurs, pliosaurs) | | Death | 10 |
| | | | Fish | 11 |
| | | | Trees | 12 |
| 5b | Birds, bats, pterosaurs (flying reptiles) | | Land reptiles | 13 |
| 6a | Land mammals | | Aquatic reptiles and flying reptiles, all from land reptiles. | 14 |
| 6a | Land reptiles ('creeping things') | | Land mammals (from land reptiles) | 15 |
| 6b | One man from dust | | Birds from land reptiles | 16 |
| 6c | One woman from the man's rib | | Bats and whales (from land mammals) | 17 |
| A few days after Creation Week | Death (from Adam's sin) | | A population of ape-like creatures evolved into a population of humans. | 18 |

***Fig.2.21:*** *This is explained in the table above some of the major highlights are:*

7.  Big bang theory seats the Big Bang at 13.8 billion years ago, at that point for billions of years later, stars were appeared and died, and from "stardust" our particular sun was born—all essential's elements above helium were allegedly freed by exploding stars. Then the earth and the rest of the globes shaped about 4.5 billion years ago by some means condensing out of a churning cloud of gas and dust (nebula). Therefore, the sun came to existence prior to Earth, and numerous stars came billions of years prior to the sun. *Nonetheless, Genesis explains that God made the earth on day 1, and the sun and stars on Day 4.* And this earth was created sufficiently calm and cool enough for water to be liquid, and was no light before God *said: let there be light.* Nonetheless, the big bang theory declares that the beginning of creation was enormously hot and bright.

Some scientists assert that the reality of what occurred on this fourth "day" was the sun and other celestial bodies "gave the impression" that a compressed cloud layer degenerated after millions of years. This is not only imaginary science but corrupt exegesis. The Hebrew word *"asah"* means "make" through Genesis 1, and is occasionally utilized interchangeably with "create" (*bara'*)—e.g., in Genesis 1:26–27- *26 Then God said, "Let us make man in our image, after our likeness. And let them have dominion over the fish of the sea and over the birds of the heavens and over the livestock and over all the earth and over every creeping thing that creeps on the earth. "So, God created man in his own image, in the image of God, he created him; male and female he created them.*

8.  It is pure desperation to apply a *different meaning* to the *same word* in the *same grammatical construction* in the *same passage*, just to fit in with atheistic evolutionary ideas like the big bang. If God had *meant* 'appeared', then He presumably would have *used* the Hebrew word for appear (*ra'ah*), as He did when He said that the dry land 'appeared' as the waters gathered in one place on Day 3 (Genesis 1:9).

9.  Evolution assumes that the first existing organism was a single cell, which ascended from an oceanic primordial soup by chemical evolution. After that, living beings evolved in the seas long before land plants and animals, and longer still before trees. Nonetheless, *Genesis reveals that God created land plants, including trees, first.*

10. Evolution imparts that ichthyosaur (marine creatures or fish-lizard) and the other marine reptilians evolved from land reptilians, and that whale evolved from land-dwelling mammals, which had evolved from other land reptilians. Day-agers obediently assert that the former was created after the latter. Similarly, evolutionists have faith in birds and pterosaurs evolved from land-dwelling reptilians, while bats evolved from land-dwelling mammals. Nevertheless, *Genesis explicitly teaches that God made the sea and flying creatures on Day 5, a day before He made land creatures, Day 6.* This has led to very imaginative Scripture-twisting: that the eon-"days" were overlapping rather than consecutive.

    Any kind of evolution must decently repudiate a precise first solitary man and woman who are the sole ancestors of all other humans who are ever being in existence. Somewhat, they state that a populace of ape-like creatures evolved into a *populace* of humans. Genesis in contrast explains that the first man was formed not from other existing beings but from lifeless matter, "dust of the ground," which did not turn out to be living up until God breathed upon it (Genesis 2:7) - *then the Lord God formed the man of dust from the ground and breathed into his nostrils the breath of life, and the man became a living creature...* And the first woman similarly had no mother, but was made from the man's rib (Genesis 2:21–24) - *So the Lord God caused a deep sleep to fall upon the man, and while he slept took one of his ribs and closed up its place with flesh. And the rib that the Lord God had taken from the man he made into a woman and brought her to the man. Then the man said, "This at last is bone of my bones and flesh of my flesh; she shall be called Woman, because she was taken out of Man.*

11. Day-age creationists understanding a little better than evolutionists, as they sightlessly receive the evolutionary dating. This "dates" fossils of undoubted *Homo sapiens* at almost 200,000 years ago, distant ancient than Adam could be, even with the most illogical stretching of the scripture timeline given in Genesis 5 and 11 - *This is the book of the generations of Adam. When God created man, he made him in the likeness of God; Thus, all the days of Enosh were 905 years, and he died.*

12. All billions-of-years interpretations place death before sin. It's hard to overstate the significance of this. These interpretations teach that virtually as soon as living things ascended, they also died. Though, *the first recorded death of a biblically living creature, Hebrew* nephesh chayyāh, *occurred after Adam and Eve sinned,* when God sacrificed an animal to make skin clothes for them (Genesis 3:21).

# PROBLEMS IN THE PRESENT

If what God the Infinity in Perfection, All Knowledge (Omniscient) and all Wisdom states in the scripture about creation is deemed to be wrong or to need modernizing, this unlocks the door for non-believers to regard as wrong or needing modernizing. Either the whole Bible truly is the Word of God or it is not; therefore, it is not a combination of bits that are not accurate together with other bits of it that are accurate.

## Big-Bang-Believing Christians – Identified

1. "Progressive Creationists – Reject biological evolution and accept a real Adam and Eve, nonetheless, have no difficulty with cosmological or geological evolution, or as
2. "Theistic Evolutionists" Believe God not only utilized cosmological and geological evolution, nonetheless, also believe biological evolution to create living things.

In each instance, enormous problems ascend if one enhances billions of years to the timeframe of the Scripture.

## Problems Regarding the Past

With regard to man's appearance on Earth, millions of years puts death, suffering, and diseases like cancer, arthritis and abscesses (visible in the fossil bones) all *before* Adam sinned, instead of as the *result* of Adam's sin, which led to that perfect original world becoming cursed (Genesis 3:17–19).

But the Apostle Paul wrote that death was the *result* of sin, and that sin came into the world through one man, namely Adam (Romans 5:12–14).

## Problems Regarding the Future

Another aspect for big-bang-believing Christians is what the Bible says about the future, namely that God will create "new heavens and a new earth" (2 Peter 3:13; cf. Isaiah 65:17; Revelation 21:1). So how long do long-agers allow for God to create the new heavens and the new Earth, if they maintain that it took Him billions of years to produce the present heavens and the present Earth?

## What the Bible Clearly Teaches

According to Genesis Chapter 1, God created everything by His word of command, over a period of six days that were chronological in order and 24 hours in duration, about 6,000 years ago,

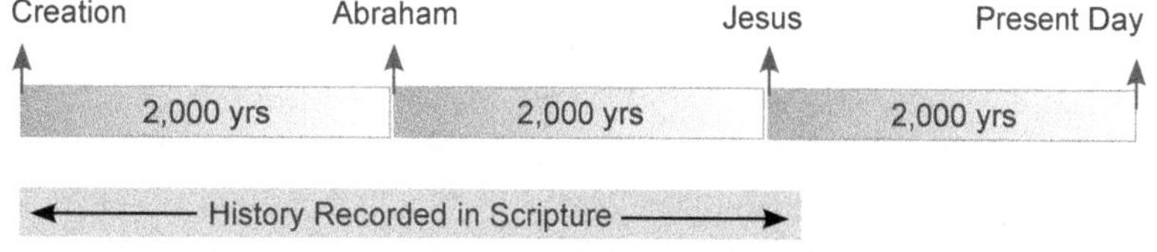


Various strategies have been used to evade this obvious meaning. However, because Genesis means what it so obviously says, which is how the Lord Jesus and the NT writers obviously understood it, it is doubtful if any theological institution would raise these alternatives today if it were not for their desire to be 'scientifically respectable'. For detailed refutations see the classic commentary *The Genesis Account* and/or use 'search' in creation.com.

# The Big Bang Fails as Objective Science

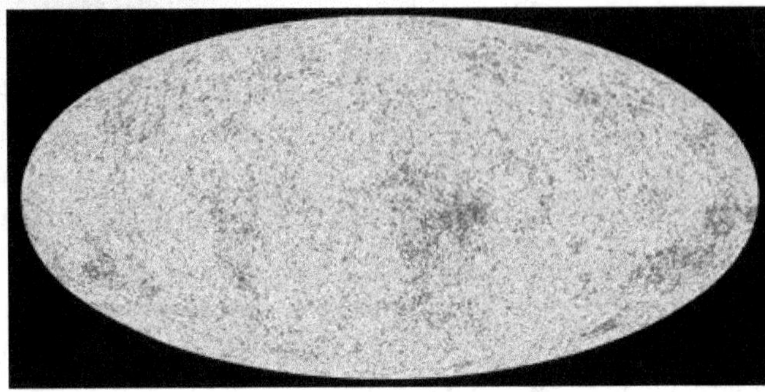

**Fig.2.23:** *Cosmic Back Microwave Ground  CMB: Planck Collaboration –
Curtsy European Satellite Agency*

The chief 'proof', for the big bang theory is the unevenness's (of 1 part per 100,000) in the cosmic background radiation of the universe, as per this so-called 2013 Planck 'map' thereof. But this involves circular reasoning. The evidence is interpreted assuming the big bang theory to be factual, and then it is used to support the theory.

The big bang is one of the most non-scientific narratives ever propounded. It has evolved considerably over the last 30 years or so, not so much because of new evidence in support of it, but because more and more problems have needed to be answered.

Here are some of the scientific problems of the big bang:

- Energy can be converted into matter according to Einstein's equation $E=mc^2$, but when this happens exactly equal amounts of matter and antimatter are produced. So where are the 200 billion galaxies of antimatter that had to form to balance the 200 billion galaxies of stars, for the big bang theory to 'work'?

- The big bang is supposed to have begun by means of a quantum fluctuation. But what was it that quantum fluctuated? And how could this have happened before there was any time or space for anything to quantum-fluctuate in?

- The big bang depends on early inflation of the universe that lasted from $10^{-36}$ to between $10^{-33}$ and $10^{32}$ seconds, at many times the speed of light, with no known mechanism either to cause this, or to un-cause it once it began.

- The big bang has a light-travel time problem, arising from the fact that the cosmic microwave background (CMB), Figure 2.23, has the same temperature over the entire sky, namely $2.726 \pm 0.001$ K (i.e. above absolute zero). However, there hasn't been enough time for radiation to travel between widely separated regions of space at the speed of light, to produce the consistent CMB temperature over the whole sky. This is technically known as the big bang 'horizon problem'.

- Big bang theory only produces an expanding cloud of gas. Expanding clouds of gas do not spontaneously reverse their expansion and condense into the objects we see in the real universe around us.

- Big bang theorists require the universe to be composed of ~27% 'dark matter' in order to supply the gravitational force needed to cause the outer edges of galaxies to rotate at the same rate as their centers. The word 'dark' means 'unseen', i.e. such dark matter is not in the form of stars or planets or indeed any particles that have ever been seen or that emit radiation.

- Another conundrum for big bang theorists is that they also need the universe to be composed of ~68% 'dark energy'. This is an alleged form of anti-gravity that is supposedly supplying the force needed to cause the expansion of the universe to accelerate. This too has never been identified in any laboratory experiment. These dark entities are best described as 'fudge factors' conceived only because without them the big bang theory would not 'work'. **All this leaves only ~5% of the universe as identified matter!**

- It may not be widely known that 33 scientists signed a letter published in *New Scientist* No. 2448, p. 20, 22 May 2004, itemizing these hypotheticals, never observed 'dark' fudge factors on which the big bang depends, and saying: "Without them, there would be a fatal contradiction between observations made by astronomers and the predictions of the big bang theory. **In no other field of physics would this continual recourse to new hypothetical objects be accepted as a way of bridging the gap between theory and observation.**"[7](Emphasis added.) Of course, the big bang in its non-theistic form also contradicts the First Law of Thermodynamics, i.e. energy can neither be created nor destroyed, only changed from one form into another. So energy could not have been created by the big bang, as secularists maintain. To get around this, it has been proposed that the laws of science themselves have not always been the same—i.e. science itself may be jettisoned if necessary.

## SATAN'S STRATEGY

On the matter of the importance of heeding what the Word of God actually says, the very first temptation of anybody by Satan was his suggestion to Eve to doubt what God had said: "Did God actually say, 'You shall not eat of any tree in the garden?'" (Genesis 3: 1). And Satan followed this sowing of the seed of doubt about what God had said by his outright denial of what God had said: "You will not surely die" (v. 4). This in turn led to Eve's (and Adam's) disregard for the imminent judgment of God.

Today Satan continues this strategy of initiating doubt and denial regarding the Word of God. People today not only do not believe in the judgment of God—they do not even believe in the existence of God, whose chief communication to us is through His Word, the Bible. Satan's strategy, which worked with Eve, has proved to be no less effective with modern man.

Sadly, this reverence for the evolutionary paradigm (with its intrinsic denial of the historicity of Genesis) is now advocated by many theological colleges. This is despite the fact that in the Bible God repeatedly warns believers not to add to or change what He has said.[8] The result is that today many theological college graduates who are now church pastors do not believe that what God says in Genesis is historical fact.

Surely God does not want ministers or missionaries to tell converts that God does not mean what He says in the early chapters of Genesis. Jesus warned that those who reject Moses' writings eventually reject His words too (John 5:46–47). Christian organizations should have a question in their application papers asking what an applicant or prospective member believes about creation and Genesis.

Accordingly, the answer to the question in the title of this article is No! Acceptance of the big bang (with or without accepting biological evolution) undermines biblical authority in crucial Gospel-related ways. It is thus a major reason for the increasing impotence of the church in Western culture. The remedy is self-evident.

## SCIENTIFIC EVIDENCE THAT THE UNIVERSE IS YOUNG

Here are some of the strong arguments for a 'young' age of the earth and the cosmos, composed by Dr Jonathan Sarfati's *"The Genesis Account."*

- Soft, still stretchy tissues, cells, and proteins are now found in dinosaur fossils that should, by physical laws, have decayed long ago if millions of years old.

- Coal and diamond often contain carbon-14, but this decays so quickly that it would be undetectable in much less than a million years.

- The earth's magnetic field is decaying exponentially, as a giant resistance/inductive circuit. This decay has been accompanied by rapid field reversals, recorded in thin, quickly-hardening lava flows.
- Salt in the ocean is accumulating too rapidly.
- Comets lose so much mass every time they pass the sun (that's why we see their tails) that they would have evaporated after a few hundred passes. This means they could not have been orbiting for millions of years. Proposed hypothetical replenishing sources, such as the Kuiper Belt and Oort cloud, have numerous problems.
- Polystrate fossils (which span multiple sedimentary layers), show that the layers they span must have been deposited quickly, before the organism had time to decay.
- Flat gaps: smooth boundary lines between geological layers shows that there has been no time for significant erosion between such layers allegedly separated by millions of years.
- Mutations in humans and other 'higher' creatures are accumulating so rapidly that they (and we) would have become extinct in well under 100,000 years.

Additionally, Dr Don Batten's article Age of the earth: 101 evidences for a young age of the earth and the universe; a shorter version is titled Evidences for a young age of the universe.

## Universe Singularities—Evolution Singularities

Evolutionists claim their findings of origins is founded upon science only. This is not factual. The evolution story of the origin of life and the cosmic origins, entails of a long succession of sole one-off action that are best labelled as singularities.

Science may only deal with repeatable measures, so singularities are ordinarily outside its reach. Though, when evolutionary events are proclaimed to be naturalistic, repeatable, and depending on chance, then we can utilize probability philosophy to calculate the prospect that they only occur once.

The inference is that a long succession of singularities is blurry from a long series of miracles. As it turns out, all claims of origin comprise of a series of singularities, so all of them are outside the scope of scientific analysis. However, the causes and consequences of origin claims are exposed to scientific investigation.

The big events in the evolution claim all lack trustworthy causes. The scriptural claim does have a trustworthy cause—***Almighty Infinite God***—and its significances have plentiful peer-reviewed backup evidences in creation scientist literatures. Notwithstanding of evidence, nevertheless, the singularity problem is a cue that origin claims are not constructed mainly upon science, but on history, particularly the imaginary eons of time in evolution, philosophy, and a subsequent worldview.

## Origin of Species - Charles Darwin

Four years after publishing the first edition of the Origin of Species on March 29, 1863, Charles Darwin corresponded with his friend and counsellor Dr Hooker stating, in part,

*"I have long regretted that I truckled to public opinion, and used the Pentateuchal term of creation, by which I really meant 'appeared' by some wholly unknown process. It is mere rubbish, thinking at present of the origin of life; one might as well think of the origin of matter."*

Astonishingly, what Darwin supposed to be the tougher of the two problems—the origin of matter—was the more readily explained.

## Albert Einstein "Special Theory of Relativity

In 1905 Albert Einstein printed his "Special Theory of Relativity," counting on his celebrated equation:

$$E = mc^2.$$

Matter can be consequent from energy, in agreement with the laws of *quantum physics*, and vice versa. It was the origin of life that turned out to be the tougher problematic. Today, the origin of life, combined with the finely tuned universe that provisions it, has to be equally traced back to the *quantum world*. Many claim that universes can arise from *quantum fluctuations* within "nothing," a quantum vacuum.

## Singularities, Utilize Laws of Probability

If universes and life can ascend from any naturalistic cause whatever, then they will continue to ascend each time those identical causes continue operating. If they don't continue to ascend it must mean they only occurred once. Things that happen just once are called "*singularities*," and we can utilize the laws of probability to hunt for them. The evolutionary worldview depends on from start to finish on singularities, and when we apply probability theory to look for them, we discover that they don't exist.

## Stephen Hawking's Probabilities Error

Physicist Stephen Hawking said in his "masterpiece." A Brief History of Time that it is conceivable, though unlikely, for the molecules of gas in a sealed box to all move down one end and occupy only one half of the box. "The probability … is many millions of millions to one, but it can happen." This is not factual.

## Dr Hugh Ross Greatest Errors

Dr Hugh Ross, founder of "Reasons to Believe," fell into the same trap so let us utilize the setup to show some elementary values of probability theory. Hawking's first error was to put probability in apposition to an event in a manner that suggests the small but finite probability caused or gave purpose for the event to occur. But chance is not a force that can do things, and probability is nothing more than a set of *hypothetical tools* that humans have established to help them make decisions about ambiguous events. A respectable decision is one that evades "false positive" outcomes: a Type I error, and "false negative" outcomes: a Type II error.

For instance, if a medical exam declares you have cancer when you don't, a false positive, it can cause needless worries and expensive, wasteful, medical treatment, with possible associated negative side effects. But if the test declares you don't have cancer when you do have it, a false negative, it may place your life in peril. Doctors utilize the history of such exams to perform probability calculations to assist them make the best decision. If a self-assured decision cannot be made, they will recommend further testing.

A basic rule of probability is that:

$$p = 1 - q,$$

where p is the probability that an event will happen, and

q is the probability that it will not happen.

The Null Hypothesis, meekest postulation in statistical testing is that there is no change between some test measure and zero. If this supposition is verified false, at some calculated level of confidence, then the Alternate Hypothesis is recognized that it is different from zero. As the value of $p$ becomes smaller, the value of $q$ becomes larger, so a confident decision must strike a balance between Type I and Type II errors. The tables in the back of statistical textbooks carry a set of **p values** that optimize these risks, usually

$$p = 0.05, 0.01, \text{ and } 0.001.$$

Hawking's own probability statement gives us a value of

$$p < 0.000000000001,$$

and we will see below that it is not pointedly different from zero. We should place no confidence in his reasoning, even by his own standards. But shockingly, Hawking massively miscalculated the scope of his problem. To comprehend the reasons, we require to make simpler his setting.

Assume his box has just one molecule of gas in it and with the purpose of observing the insides we addition a divider at the half-way point so the molecule is either in the **left (L)** or **right (R)** hand end. There is a 100% probability that the gas molecule is either in L or in R. Add a second gas molecule, and to compute combinations we must now label them, say **A and B**.

The possible combinations are: Underlined groups signify all in one end

$$[A \text{ \& } B \text{ in } L],$$
$$[A \text{ \& } B \text{ in } R],$$
$$[A \text{ in } L \text{ \& } B \text{ in } R], \text{ and}$$
$$[B \text{ in } L \text{ \& } A \text{ in } R].$$

Consequently, the probability of both molecules being in one end only is two cases out of four, or **50%**. Add a third molecule, C, and we get these combinations:

$$[A, B \text{ \& } C \text{ in } L],$$
$$[A, B \text{ \& } C \text{ in } R],$$
$$[A \text{ \& } B \text{ in } L \text{ \& } C \text{ in } R],$$
$$[A \text{ in } L \text{ \& } B, C \text{ in } R],$$
$$[A, C \text{ in } L \text{ \& } B \text{ in } R],$$
$$[B \text{ in } L \text{ \& } A, C \text{ in } R],$$
$$[B, C \text{ in } L \text{ \& } A \text{ in } R],$$
$$[C \text{ in } L \text{ \& } A, B \text{ in } R].$$

The probability is now two cases in eight, or **25%**. Add a fourth molecule and the probability drops to 2 cases in 16, or **12.5%,** and so on.

The outcomes are plotted in Figure 2.24 below. The pattern that arises is that there are constantly only two likelihoods of Hawking being correct **(all in L or all in R),** while the number of possible combinations rises as

**2n** where

**n = the number of gas molecules**.

The Binomial distribution describes this situation where one of two outcomes is probable. It illustrates that:

with 10 molecules **p = 0.001**, and

with 20 molecules **p = 0.000001**.

However, a shoe box of ordinary air would contain something in the order of **1023**, a hundred thousand million trillion, molecules of gas. Consequently, the likelihood of Hawking being correct is not "many millions of millions to one," but

just two chances in **2100,000,000,000,000,000,000,000,000**, or about

$$1 \text{ chance in } 1030,000,000,000,000,000,000,000.$$

Subsequently, there are only about 1080 atoms in the universe, Hawking's decision to say "it can happen" is an excruciating Type I error! Nonetheless, there is a significantly more essential error in Hawking's scenario. The activities of gas particles are not chance events; they are resolute by the laws of motion.

Probability theory can be practiced to such events only as an approximation to physical reality. Chance cannot achieve what the laws of physics forbid. Being a physicist Hawking should have questioned himself whether the laws of physics would allow such a scenario.

## The First Law of Motion

The first law of motion states a body that is moving or at rest will remain in that state except a force act upon it. At ordinary room temperature and pressure the gas molecules—mostly nitrogen ($N_2$) and oxygen ($O_2$)—zip around at in excess of a thousand miles/hour, bouncing about as billiard balls on a billiard table but in 3-dimensions.

What if that gas molecules did start to concentrate down one end of Hawking's box. Molecules in the transition region between the dense region and the empty region will face abundant problems if they move in the direction of the dense region. They will bounce around among lots of other gas molecules accumulating to the pressure and temperature in that end.

On the other hand, molecules that move towards the empty region will face no such resistance, so trillions of them will continuously be zipping back into the empty region at over a thousand miles per hour. Hawking's scenario could never eventuate.

People often implement probability theory improperly and an amusing impression is offered in David Hand's book "The Improbability Principle:" Why coincidences, miracles, and rare events happen every day. Nevertheless, in attempting to clarify everything with chance, David Hand falls into the identical trap as Hawking and Ross by applying probability theory to make-believe events that are unconnected from physical authenticity.

## Employing Chance as a Substitute for Ignorance

Astrobiologists make their living doing this very same thing. They "guesstimate" an assembly of probabilities and supplement them into the terms of the famous "Drake equation" to calculate the number of extraterrestrial intelligent civilizations in our galaxy and universe. Such submissions are virtuously imaginary—employing chance as a substitute for ignorance—deprived of appropriate respect for whether the wished-for events are physically possible.

## The Singularity Problem

As specified, a singularity is an exclusive event that only occurs once. Singularities must have distinct causes, not common causes. Things that have common causes harvest common events, not singularities.

For instance, rain is a common event, in most places on Earth, caused by specific features of the hydrological cycle. Rain usually falls downwards, not upwards, since gravity generally pulls things towards the center of the earth. If rain is ever seen to move upwards, then it can typically be clarified by an updraft in air currents—alternative aspect of planetary climate.

In physics, a gravitational singularity happens in the heart of a **"black hole."** When a big star burns-up to consume all its nuclear fuel, it breakdowns in on itself, as it is a prediction of Einstein's General Relativity.

The internal self-gravitational attraction of the burnt-up star could no longer withstand the gravitational pull of the Back Hole. Thereby, the Black Hole sucks everything nearby, as well as light, with a singularity at its center.

## Hawking's Probability Error

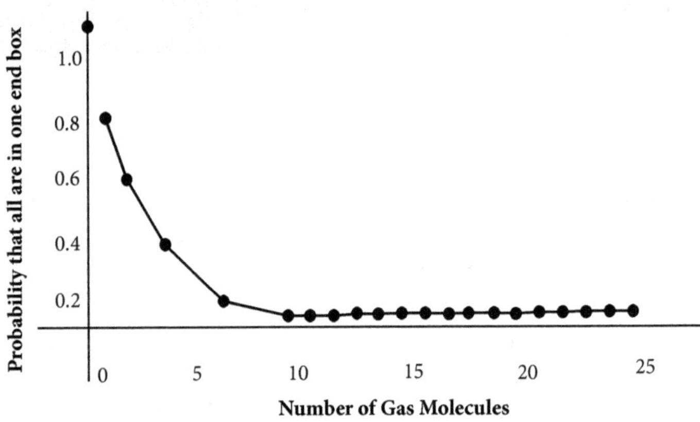


*Fig.2.24: Error in Hawking's Probabilities –*

Probability that all gas molecules will accumulate in one end of Hawking's box, plotted against the number of molecules in the box

This is a single point with density and temperature approaching immensity in an infinitesimally small volume, conferring to the theory. When this happens there is no possibility of moving back it. This type of event can occur numerous times as there are many huge stars, but it can only occur once to a specific star as it is irreversible, on the timescale commonly presumed for the universe.

In mathematics, a singularity is a point in a given algebraic function where the derivative, which is the rate of change, is undefined, but every near-neighborhood point does have a derivative. This type of singularity is to some extent as a sheet of metal along, which the algebraic function defines straight lines, but at a particular point there is a hole, which is a singularity across.

## In The Beginning

***Genesis 1:1 In the beginning God created the heavens and the earth.*** If we can't trust this verse, then nothing else in the Bible makes sense. Since this verse is so foundational, it is not surprising that atheists have feverishly attacked this concept. Some of the attacks are childish, while others have the veneer of philosophy or advanced science.

## World Religions and Christian Faith

Below is Table 2.1 providing distinction between world religion philosophies and the Christian faith as means of life among believers.

**Table 2.1 :** Definition of Various Doctrine of Religious Sects Vs. Christan Doctrine

| World Religion | Christian Doctrine |
|---|---|
| Atheism: there is no God. | • God exists, and was present 'at the beginning'.<br>• God created the universe; the universe neither spontaneously appeared nor has it existed forever. |
| Agnosticism: it is impossible to know whether God exists. | • God has revealed Himself in Scripture as Creator. |
| Dualism: Good and Evil are eternally co-existent (as Zoroastrians believe). | • God was alone when He created.<br>• God is perfectly good.<br>• Beings who became evil are part of the created order. |
| Finite-god views (e.g., Open Theism and Process Theology). | • God created the space-time universe.<br>• Thus, He is not limited by anything in the universe, including the future, since God created time itself. |
| Evolutionism: man evolved from other species of animals | • God created all things. |
| Humanism: man is the measure of all things. | • God is the ultimate reality.<br>• Man is part of the created order.<br>• God created us so *He* is the measure of all things. |
| Materialism: Matter (or mass-energy) is the only reality. This is a synonym of: Naturalism: natural laws describe all things. | • God created matter (and mass-energy); or, God created nature.<br>• God is thus sovereign over the natural world.<br>• Thus matter (mass-energy) are not eternal or self-existent. |
| Pantheism: all is god; god and creation are the same thing. | • God created the universe.<br>• Thus, God is distinct from His creation. |
| Panentheism: "all is in god". | • God *transcends* what He created. |
| Polytheism: there is more than one god | • Only *one* God created all things. |
| Unitarianism (that God is an absolute unity, e.g., Islam, modern Judaism, Jehovah's Witness doctrine, classical Unitarianism). | • *Elohim* is a plural noun with a singular verb, teaching a plurality in the Godhead.<br>• The NT reveals this further as the Trinity. |

# APPENDIX

- Elements heavier than iron are believed to have formed via a complex network of nuclear reactions known as r and s processes. These involve a competition between neutron-capture and beta-decay.
- Feser, E., Not Understanding Nothing, A review of *A Universe from Nothing, First Things*, June/July 2012; firstthings.com.
- That is, a follower of Thomas Aquinas and his Aristotelian approach to metaphysics and apologetics.

- Ostriker, J.P. and Gnedin, N.Y., Reheating of the universe and Population III, *Astrophysical Journal Letters* 472:L63, 1996. Also see Bromm, V., Exploring the physics of primordial star formation, *American Astronomical Society Meeting* 195, #125.05, 12/1999 as cited at the NASA ADS Astronomy abstract service <adswww. harvard.edu>, 13 July 2000.

- Barnes, L., Out of nothing, *letterstonature.wordpress.com*, 1 April 2011.

- Hear also this radio episode, Militant anti-Theism, hijacking science and Dawkins & Krauss on Redemption Radio, *Redemption Radio*, 22 February 2012, where I was interviewed for about two hours to critique a misotheistic love-fest between Krauss and Dawkins at Arizona University, redemptionradio.podbean.com/2012/02/22/militant-anti-theism-hijacking-science-and-dawkins-krauss-on-redemption-radio.

- Hume, D., letter to John Stuart, 1754.

- Other apologists have used different arguments. For example, the leading medieval theologian and apologist Thomas Aquinas (c. 1274–1225), while he believed that the universe began in time because the Bible said so, didn't think this could be proved philosophically. Instead, in his famous 'five ways' (*Summa Theologiae*, Question 2, The existence of God) he argued that even if the universe had no beginning, then it would *still* not exist or undergo change *here and now* unless there were a necessary being, 'prime mover', and 'first cause', who upholds the universe in its *moment-by-moment existence*. "This all men speak of as God." That is, while the *Kalām* argument concerns God's *creative* work, Aquinas's arguments are for God's *sustaining* work since He finished His creation on Day 7 (Genesis 2:1–3). Then Aquinas spends hundreds of pages arguing that the 'God' with these properties would be perfectly good, all-powerful, and all-knowing.

- See also *Christianity for Skeptics*, ch. 1. This is called the *Kalām* Cosmological Argument. It goes back to the church theologian Bonaventure (1274–1221), and was also advocated by medieval Arabic philosophers. The word *kalām* is the Arabic word for 'speech', but its broader semantic range includes 'philosophical theism' or 'natural theology'. The *kalām* argument's most prominent modern defender is the philosopher and apologist Dr William Lane Craig (1949–): *The Kalām Cosmological Argument*, Barnes & Noble, New York, 1979. Unfortunately, Dr Craig compromises on the plain meaning of Genesis—see William Lane Craig's intellectually dishonest attack on biblical creationists.

- Actually, the word "cause" has several different meanings in philosophy. But in this section, we are referring to the *efficient cause*, the chief agent causing something to happen.

## REFERENCES


1. Similar lists are in Morris, *The Genesis Record, A scientific and devotional commentary on the Book of Beginnings*, p. 38, Baker Book House, Grand Rapid, MI, 1976, p. 38; and Fruchtenbaum,A.G., *The Book of Genesis*, p. 35, Ariel's Bible Commentary, Ariel Ministries, San Antonio, TX, 2009.

2. See this comprehensive analysis and refutation: Ammi, K., Atheism, creation.com/atheism, 11 June 2009; and *Christianity for Skeptics*, ch. 3.

3. This is the 'hard' form of 'agnosticism', as coined by 'Darwin's Bulldog', T.H. Huxley (see also Grigg, R., Darwin's bulldog—Thomas H. Huxley, *Creation*, 31(3):39–41, 2009; creation.com/huxley). A 'soft agnostic' merely claims himself not to know that there is a God.

4. Theologians first used the word 'omnipotent' (all-powerful) to mean that *God has no limitations from **outside** Himself*. See Sarfati, J., If God can do anything, then can He make a being more powerful than Himself? What does God's omnipotence really mean? creation.com/omnipotence, 12 January 2008.

5. Anscombe, G.E.M., "Whatever has a beginning of existence must have a cause": Hume's argument exposed, *Collected Philosophical Papers*, Volume 1, Basil Blackwell, 1981.

6. Anscombe, G.E.M., Times, beginnings and causes, in her *Collected Philosophical Papers*, Volume **2**, Basil Blackwell, 1981.

7. Feser, E., *The Last Superstition: A Refutation of the New Atheism*, Kindle Locations 5253-5254, St. Augustine's Press. Kindle Edition, 2012.

8. Lemley, B., Guth's grand guess, *Discover* 23(4): 32–39, April 2002; discovermagazine.com.

9. Krauss, M., *A Universe from Nothing: Why There is Something Rather than Nothing*, Free Press, 2012. See also review by Reynolds, D.W., *J. Creation*, 27(1):30–35, 2013.

10. Davies, P., *God and the New Physics*, p. 215, Simon & Schuster, 1983.

11. Albert, D., On the Origin of Everything: review of *A Universe From Nothing*, by Lawrence M. Krauss, *New York Times*, 23 March 2012; nytimes.com.

12. *National Audubon Society Field Guide to the Night Sky*, Chanticleer Press, p. 22, 1998.

13. Faulkner, D.R., The role of stellar population types in the discussion of stellar evolution, *CRSQ* 30(1):8–12, 1993.

14. Rose, W.K, *Advanced Stellar Astrophysics*, Cambridge University Press, pp. 16–17, 1998. Also see Low, C. and Lynden-Bell, D., The minimum Jeans mass or when fragmentation must stop, *Royal Astronomical Society, Monthly Notices* 176:367–390, August 1976 as cited at the NASA ADS Astronomy abstract service <adswww. harvard.edu>, 29 June 2000. This abstract shows that for normal molecular gas clouds, it is about 0.007 $M_\odot$.

15. Silk, J., On the fragmentation of cosmic gas clouds. I—The formation of galaxies and the first generation of stars, *Astrophysical J.* 211(1):638–648, 1977 and also Silk, J., On the mass range of the first stars, *ESO Workshop on Primordial Helium*, Garching, West Germany, 2–3 February 1983, Proceedings (A83-50030 24-90) as cited at the NASA ADS Astronomy abstract service <adswww.harvard.edu>, 6 July 2000.

16. Nakamura, F. and Umemura, M., On the mass of Population III Stars, *The Astrophysical J.* 515(1):239–248, 1999 as cited at the NASA ADS Astronomy abstract service <adswww. harvard.edu>, 6 March 2000.

17. Talking Back, Water, water (almost) everywhere, *Astronomy* 27(6):16, 1999.

18. Adams F.C. and Laughlin, G., The future of the universe, *Sky & Telescope* 96(2):34, 1998.

19. NewsNotes, Free-floating planets in the Orion Nebula? *Sky & Telescope* 100(1):18–19, 2000. Also see Lucas, P.W. and Roche, P.F., A population of very young brown dwarfs and free-floating planets in Orion, *Monthly Notices of the Royal Astronomical Society* 314(4):858–864, 2000 as cited at the NASA ADS Astronomy Abstract Service, <adswww. harvard.edu>, 11 July 2000.

## Rod Bernitt

by *Dr Jonathan D. Sarfati*

by: *Robert Lamb & Patrick J. Kiger* | Updated: Apr 15, 2021

by *Russell Grigg*

by *John G. Hartnett*

# 3

# THE ORIGIN OF THE UNIVERSE

## SOMETHING CAME OUT OF NOTHING

Ancient times, even before time existed, there was nothing at all. There was no universe, no sky, no sun, no moon, no stars, and no Earth. There was no water. There were no plants, no trees, no animals, no Adam, and no Eve. Nothing existed. Nothing at all. Except, of course, God – the Glorious Infinity. He has always existed from Infinity Everlasting to Infinity Everlasting

### Miraculous Week

God decided that He wanted all these things to come into being. He would take "six whole days" to make them all, and He would create something different every day for a week. We call this week "Creation Week.

'In the beginning God created the heavens and the earth. And the earth was formless and void, and darkness was over the surface of the deep; and the Spirit of God was moving over the surface of the waters. Then God said, "Let there be light;" and there was light. And God saw that the light was good; and God separated the light from the darkness. And God called the light day, and the darkness He called night. And there was evening and there was morning, one day.' *Genesis 1:1–5.*

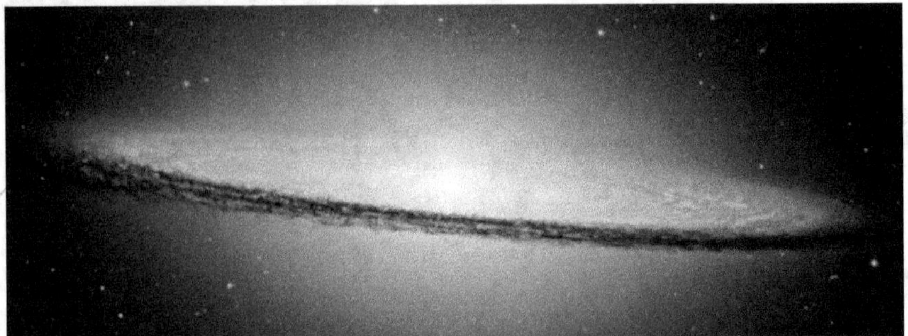

**Fig.3.1:** In the Beginning God Created - *Curtsy NASA/Hubble Heritage Team*

On "Day 1" of Creation Week, God made four important creations!

1. **Time.** You may be astounded to learn that God had to create 'time' itself. 'Before' then, even time did not exist. Only God existed. He is eternal. Accordingly, 'before' the beginning (Genesis 1:1), Figure 3.1, there was no time or …

2. **Space.** God had to create space. Space is like unique peaceful compartment.  This was so there would be somewhere for God to put everything after He had created it.

3. **Matter.** God also had to create the earth. To begin with He created the stuff which makes up the earth, and tons and tons of massive cold refreshing water. God created all this material, and everything else too, from nothing. How could that be? Well, the Infinite power God simply commanded it all to come into being, and it did. His Word is so powerful that this was all that was needed.

   At first, the earth was all covered by water, so it was not a molten blob. God's Spirit was there. Over the next few days infinite His power and infinite energy would make the earth into a beautiful place for animals and people to live. However, first God brightened the place up with …

4. **Light.** In the beginning, it was very, very dark—pitch-black, inky black—everywhere. So, early on "Day 1," having made time, space and matter, God then said **let His own glorious light glow**. How did He do this? He just said, 'Let there be light!' and there was light. No problem for the Eternal Infinite God. Again, His Word was all that was needed.

This glorious light was neither pixels of photos, nor the sun, as God had decided that He would not make the sun until the fourth day. So where did this light come from? The Bible doesn't tell us what this light source was, but we know it was shining onto the earth to begin with, as the earth rotated. In the new heavens and new earth, God Himself will be the source of light (Revelation 22:5).

If the light was on one side of the earth, then, as the earth turned around, part of it would have been in the light for a while, and part of it would have been in the dark. God gave names to these two periods of time. He called the light 'day', and He called the darkness "night." He had made the first 24-hour day, and a way for us to measure time by.

God said that the light was "good." This means it was a perfect part of God's plan for creation, and God was pleased with it. He knew that we would need light to see our way around. He knew too that we would also need darkness to help us sleep at night.

# CREATION NEEDS A CREATOR

God the Infinity did not have a beginning. He created time, space, matter, and sources of light to serve normal day and night. We can trust this, because He says His thoughts are higher than our thoughts in *Isaiah 55:9*

Some people who believe in evolution try to tell us in books and TV programs that time, space, matter, energy, and everything else all made themselves, and by accident. But how could anything make itself before it existed?

Others say that the universe has always existed. This cannot be so. If the universe had existed for ever, our sun and all the stars would long ago have used up all their energy and stopped shining. This condition would make everything so cold and lifeless that it has been called 'heat death.' But this hasn't happened, has it? The sun and stars are all still here, so the universe could not have existed forever.

Some scientists, cosmologists, and mathematicians, but not nearly all, suggest that there was once a "big bang" that made everything. Nevertheless, when did you ever see 'an explosion' produce anything other than a huge mess? And *we know from God's Word,* --the Bible Scriptures, that God did not use evolution or the big bang to make everything. He did not use evolution or the big bang to make *anything*! He did not need to. He is almighty. This means that He is absolutely powerful. Powerful enough to do anything He wants to do. Like creating time, and space, and matter, and energy, and everything else. He created everything—just by His Word.

God didn't need to experiment either. Everything He did, He did right the very first time.

## Who Created God?

Some people ask: 'If everything needs someone to make it, then who made God?' The answer is that everything *that has a beginning* needs someone to make it. But God did not have a beginning. He has existed forever. Therefore, He did not need anyone to make Him!

Evolutionists like to say that the universe is billions of years old, but the Bible denies this. Adding up people's ages and other time periods in the Bible, --for example Genesis 5 and 11, 47:9; 1 Kings 6:1), we find that "Creation Week" was about 6,000 years ago.

The creation account that God has given us in Genesis is neither a fairy story, nor poetry, nor a parable. It is a reliable record of what actually happened.

It is the history of the earth and the universe from their very beginning, and of how everything came into being. It is scientifically accurate, and it shows us the power, wisdom and goodness of our almighty Creator God.

Who created God? Atheists frequently position this question to validate their disbelief. "Bertrand Russell (1872–1970)," Figure 3.1, a famous British philosopher, in his important little essay, *Why I am not a Christian,* put this forward as his first objection. Noonday's atheists repeat the opposition, including Richard Dawkins (*The God Delusion*)

**Fig.3.2:** Portrait of Bertrand Russell (1872-1970), Oil on Canvas - *Curtsy Duce, Alberto (b.1915)*

and Australia's Philip Adams, who said, "the great argument for God was that there had to be a Creation, a beginning. … But my objection was simple. If God was the beginning who began God?"

The universe had a beginning; almost no one in disagreement, since the laws of thermodynamics mandate it: the universe is running down and it cannot have been running down forever, *or it would have already run down.* No stars would be still whipping out energy and we would not be here.

Some have projected one universe giving birth to another, but again, there cannot be an infinite succession of such births and deaths, as each cycle must have less energy available than the last and if this had been happening for eternity, the death of everything would have already transpired.

## God Created the Universe – Who Created God?

A number of skeptics ask this question. But God *by definition* is the *uncreated* creator of the universe, so the question 'Who created God?' is illogical, just like 'To whom is the bachelor married?'

Therefore, a more sophisticated questioner might ask: 'If the universe needs a cause, then why doesn't God need a cause? And if God doesn't need a cause, why should the universe need a cause?' In reply, Christians should use the following reasoning:

1. Everything **which has a beginning** has a cause.
2. The universe has a beginning.
3. Therefore, the universe has a cause

The universe requires a cause because it had a **beginning,** as will be shown below. God, unlike the universe, had **no beginning**, so doesn't need a cause. In addition, Einstein's general relativity, which has much experimental support, shows that *time is linked to matter and space. Therefore, time itself would have begun along with matter and space.* Since God, by definition, is the creator of the whole universe, he is the creator of time. This. He is not limited by the time dimension He created, so has *no beginning* in time—God is 'the high and lofty One that inhabits eternity' (Is. 57:15). Therefore, He does not have a cause.

In contrast, there is good evidence that the universe had a beginning. This can be shown from the *Laws of Thermodynamics,* the most fundamental laws of the physical sciences.

- 1st Law: The **total** amount of mass-energy in the universe is **constant**.
- 2nd Law: The amount of energy **available for work** is running out, or *entropy* (decaying) is increasing to a maximum.

If the total amount of mass-energy is limited, and the amount of usable energy is decreasing, then the universe cannot have existed forever, otherwise it would *already* have exhausted all usable energy—the 'heat death' of the universe. For example, all radioactive atoms would have decayed, every part of the universe would be the same temperature, and no further work would be possible. Therefore, the obvious corollary is that the universe began a finite time ago with much of usable energy, and is now running down.

Now, what if the questioner accepts that the universe had a beginning, but not that it needs a cause? But it is self-evident that things that begin have a cause—no-one really denies it in his heart. All science and history would collapse if this law of cause and effect were denied.

So would all law enforcement, if the police did not think they needed to find a cause for a stabbed body or a burgled house. Also, the universe cannot be self-caused—nothing can create itself, because that would mean that it existed before it came into existence, which is a logical absurdity.

Accordingly:

- The universe (including time itself) can be shown to have had a beginning.
- It is unreasonable to believe something could begin to exist without a cause.
- The universe therefore requires a cause, just as Genesis 1:1 and Romans 1:20 teach.

- God, as creator of time, is outside of time. Since therefore He has no beginning in time, He has always existed, so does not need a cause.

There are only two ways to refute an argument:

1. Show that it is logically invalid
2. Show that at least one of the premises is false.

# GOD IS THE CAUSE!

The big bang is an atheistic worldview that is invalidated by the major laws of science. For atheists to uphold their worldview they must bid naturalistic explanations for everything which the Bible attributes to the power and authority of Almighty Infinite and Eternal God. Hence the modern prerogative that the universe originated as the result of a "big bang" some 13.8 billion years ago, rather than being created by God in the way that Genesis records, and in the time frame and with the order of events that the Bible indicates.

## God Being the Cause

Dr Michael Shermer, Executive Director, Skeptics Society, asks: "Why would a deity make a universe that's 13.8 billion light years in radius, in which almost none of it is useable? It's just a waste of stuff?"

There are, at the minimum, two reasons why God made the universe so huge:

1. According to Psalm 19:1, "The heavens declare the glory of God, and the sky above proclaims his handiwork." The fact that God created billions of stars and located them in the billions of galaxies that contain them validates, for those willing to see it, the omnipotence and omniscience of the Almighty infinite God.
2. It also validates that the worldview that says that all the mass and energy of the billions of stars in each of the 200 billion galaxies were once contained in a point of zero dimensions is scientific nonsense.

Atheist, Sam Harris chips in with: "It is the most agonizingly wasteful system we could devise. Nowhere in that mayhem is there the suggestion that man is somehow central."

## Wrong on two counts!

1. A Fellow of the Royal Astronomical Society, clarifies that, at least within our solar system, Jupiter is significant for life on Earth because Jupiter's gravity sweeps up comets, meteorites and all kinds of debris which could perhaps destroy Earth.
2. Earth is indeed the prime focus of God. According to the record in Genesis that God Himself has given us, God first prepared the earth to be a habitation for mankind on Days 1, 2, and 3, of Creation week, and this was *before* He made the stars and galaxies on Day 4. Moreover, it was on Earth that the Son of God, the Lord Jesus Christ, was incarnated and died to pay the penalty for mankind's sin; not to pay for the sin of demons or imaginary aliens.

Additionally, it's tiresome to see atheists point to the vastness of the universe as if it were news. Though, this truth has been well recognized for almost all the history of the Church. E.g., the Roman Christian philosopher Boëthius (AD c. 480–524/525), in prison awaiting trial and execution for an undeserved charge of treason, wrote *The Consolation of Philosophy*, an imaginary dialogue between himself and "Lady Philosophy." She points out that just as the earth is just a point in space, how much more trivial is any glory of any of its inhabitants:

As you have perceived from the demonstrations of the astronomers, in comparison to the vastness of the heavens, it is decided that the whole extent of the earth has the value of a mere point; that is to say, were the earth to be compared to the vastness of the heavenly sphere, it would be judged to have no volume at all.

This was one of the most widely read and influential books in the West during most of the Middle Ages. So, churchmen were well aware of how tiny the earth is, without considering it the slightest threat to faith. Furthermore, it was exactly this consideration that led some medieval scientists to propose that it was the tiny earth that rotated rather than the immense cosmos revolving around it.

Theist Prof. William Lane Craig makes the point: "Since something cannot come out of nothing, since being does not come from non-being, there must exist some sort of transcendent Cause which brought the universe into existence, and this is the traditional concept of what theists have meant by God."

Atheist Michael Shermer asks: "If you posit a god that started it, I can just say, who created God? … If you say that God is that which does not need to be created, why can't the universe be that which does not need to be created?"

The answer is that everything that has a beginning needs a creator. God did not have a beginning, so He *did not* need to be created. On the other hand, the universe did have a beginning, so it *did* need to be created. See:

## GENESIS—FACT NOT FANCY

William Lane Craig then takes it upon himself to attack the 'Young Earth' Creationist (Genesis-means-what-it-says) biblical position, when he says:

Faith and science ought not to conflict, if both are means of discovering truth about reality. I think that the impression that there is a conflict has largely arisen because of literalistic interpretations of the opening chapter of the book of Genesis in the Hebrew Bible. … People have taken this to describe six consecutive 24-hour days, and that viewpoint has been exploded by modern science.

Not so! Conflict is not between faith and science, but between faith and scientists (both atheists and theists) who do not believe God's Word. Genesis is the true account of the world's history, which began with God creating the earth and everything in it in six consecutive days. The Bible clearly teaches:

- The days were 24 hours long.
- The whole universe was created during creation week, which was the same length as our working week.
- Mankind was created on Day 6, "from the beginning of creation" (Mark 10:6).
- The genealogies imply that Adam was created about 6,000 years ago.
- Death arose from Adam's sin, not before.
- There was a worldwide Flood that left so much evidence that scoffers are "willingly ignorant" (2 Peter 3:3–7),

This viewpoint is rejected by atheists, but it has certainly not been "exploded by modern science" as claimed by Craig. We believe all this because the text of the Bible says so, and Jesus and the New Testament writers took it to mean this. The proper way to describe this hermeneutic is 'historical grammatical,' or perhaps 'classical literal,' rather than 'literalistic.'

## QUANTUM MECHANICS – MULTIPLIED UNIVERSES OUT OF NOTHING

To comprehend how a universe might be created out of nothing, we must turn to the theory of "quantum mechanics." It describes how in the tiny world of the atom small particles seem to appear and disappear for no obvious reason. Seemingly absurd! Imagine if in our everyday world people appeared and disappeared *without obvious cause*. And yet the theory of "quantum mechanics" describes how this can materialize at the atomic level, and that might suggest an alternative to a creator.

Nonetheless, why should a universe as fruitful as ours just pop out of nothing and somehow work?

Envisage not just one tiny universe popping into existence but millions. Envisage that, among the millions, one of them by chance had the properties to support life. Maybe that universe was ours. This is the multiverse concept, which suggests that out of millions of options we by chance are in the one universe that works.

The question was 'how could our universe pop out of nothing?' And the answer is: 'envisage millions of universes popping out of nothing!' That's not an answer; that's millions more 'how' questions!

And here are a few other explicit ones: For "quantum mechanics" to work, the laws of quantum mechanics must already be in existence, so how did they form out of nothing? Another problem is: how does what seems to happen at the sub-atomic level in a laboratory experiment (called a quantum fluctuation) apply to the formation of all the stars in the 200 billion galaxies in our visible universe – 2-3 trillion Galaxies altogether in the universe visible and invisible? If they all appeared *without obvious cause* because of quantum mechanics (Fluctuation), why aren't they also disappearing *without obvious cause* by reason of quantum mechanics (as per the equivalence given above)?

In laboratory experiments, virtual particles do appear within the vacuum of space. However, in the primordial singularity there was no space and so no vacuum.

## Four Reasons to Reject the Big Bang Theory

1.  **The theory lacks a credible mechanism**.
    a.   The big bang universe begins with all matter, energy, space and time compressed into a point allegedly of infinite density.
    b.   There is no known mechanism to start the universe expanding out of this singularity—equations in the theory only work *after* the expansion has begun.
2.  **It depends on 'fudge factors.'**
    a.   It has to violate physical laws and appeal to unknown forces (Dark Energy) and unknown substances (Dark Matter) to explain what we observe.
    b.   Dark Matter/Dark Energy exists to hold the universe together and maintain it stability
3.  **The Multiverse concept is excluded.**
    a.   Appeals to chance, e.g., via an infinite number of bubble universes, do not work because chance cannot accomplish what the laws of nature do not allow.
4.  **Science cannot produce any final answers on the subject of origins**.
    a.   Present-day observations, when applied to the past, are always based on the belief system of the person(s) making the claims, and hence express their worldview,
    b.   not facts about how the past came to be that way.

## THERE WAS A BEGINNING

One of the most recognized principles of logic / science / reality is the principle of "causality:" something that has a beginning has a sufficient cause. The principle is not, 'Everything has a cause'; Bertrand Russell, Figure 31), misstated it. No, the principle is,

'Everything *that has a beginning* has a sufficient cause'.

Just a moment's thought confirms this—something which had no beginning has no need of a cause. Moreover, a cause has to be sufficient, or adequate.

'You were found in a cabbage patch'

is not a sufficient explanation for your existence.

This principle of causation is so vital that if I said that

the chair you are sitting on, which must have had a beginning, just popped into existence without any cause,

you might justifiably think I need a psychiatric assessment!

Today's atheists, who like to use words like:

'rational', 'reasonable' and 'scientific' in describing their beliefs,

believe that the greatest beginning of all—that of the universe—had no cause at all! Some acknowledge it is a problem, but they claim that saying:

'God did it' explains nothing because you then have to explain where God came from.

But is this a valid opposition?  What must the cause of the universe have been like?

The cause of the universe must have been non-material because if the cause was material / natural, it would be subject to the same laws of decay as the universe. That illustrates it would have to have had a beginning itself and you have the same problem as cycles of births and deaths of universes. Consequently, the cause of the universe's beginning must have been:

*super*-natural, i.e., non-material or spirit—a cause *outside* of space-matter-time.

Such a cause would not be subject to the law of decay and so would not have a beginning. That is, the cause had to be eternal spirit.

Moreover, the cause of the universe had to be extremely powerful; the sheer *size* and *energy* observed in the universe together voice of that power; there had to be a *sufficient* cause.

That sounds like the God of the Bible to me. The Bible makes known the Creator of the universe as:

## Eternal

Before the mountains were brought forth, or ever you had formed the earth and the world, from everlasting to everlasting you are God. (*Psalm 90:2*)

## All-Powerful

Yours, O LORD, is the greatness and the power and the glory and the victory and the majesty, for all that is in the heavens and in the earth is yours. Yours is the kingdom, O LORD, and you are exalted as head above all. Both riches and honor come from you, and you rule over all. In your hand are power and might, and in your Hand, it is to make great and to give strength to all. (*1 Chronicles 29:11–12*)

## Spirit - Non-Material

God is spirit, and those who worship Him must worship in spirit and truth. (*John 4:24*)

Notice that the Bible says, "In the beginning God created the heavens and the earth" (*Genesis 1:1*). Here God created time itself. Only One who is outside of time, that is, timeless, or eternal, could do this.

Now to ask where someone who is eternal, someone who had no beginning, came from (**Who Created God?**') is like asking,

'To whom is the bachelor married?' As we stated previously, it is an irrational question. The Bible matches reality, which is not surprising when we consider that it claims to be from the Creator Himself.

*Two* 'great beginnings'—without any cause!

Those who reject the Creator not only have to believe that **matter** came into being deprived of any cause; they also have to believe that **life** itself burst into existence deprived of a satisfactory cause.

## SINGLE CELL LIFE

Even the modest single-celled life is impressively intricate. A humble bacterium is jam-packed with extremely sophisticated nano-machines that it desires to live. A cell requires at least over 400 dissimilar proteins to make the machines that are unquestionably indispensable for life, Figure 3.3. How could these protein-based machines create themselves, even if all the precise components, 20 dissimilar amino acids, but many of each, could make themselves?

The amino acids, frequently thousands of them, have to be combined together in the precise order for each protein to work well.

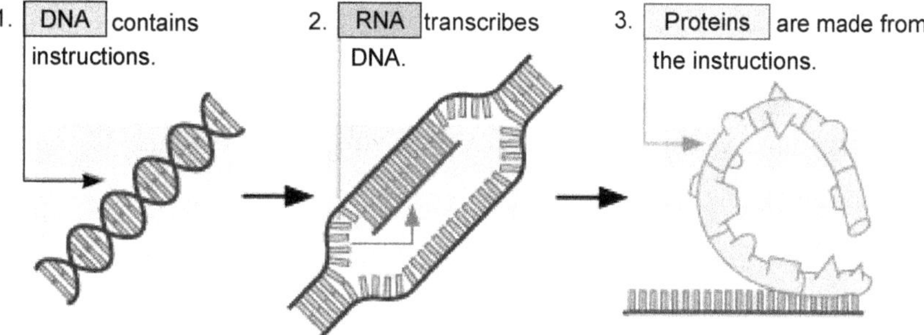


**Fig.3.3:** *Living Cells Deigned the Desire to Live*

Just ponder about one indispensable machine that copies the DNA directives for making each protein. Then let's obtain just one protein element of that machine, less than 10% of the total. This protein span is 329 amino acids in length. What would be the coincident of obtaining this one protein by chance, assuming that the precise, and *only* the precise, amino acid ingredients were present? Calculate it this way:

$$1/20 \times 1/20 \times 1/20 \dots 329 \text{ times!}$$

This is a probability of 1 in $10^{428}$ ... a number with 428 zeros after the 1! Even if every atom in the universe ($10^{80}$—a number with 80 zeros) signified an experiment for every molecular vibration possible ($10^{12}$ per second) for the supposed evolutionary age of the universe (14 billion years=$10^{18}$ seconds), this would allow 'only' $10^{110}$ experiments—a long, long way short of the number needed to have a ghost of a chance of getting just this one protein to form, let alone the over 400 others needed.

## RICHARD DAWKINS ACKNOWLEDGES

It's no wonder that Richard Dawkins admits that scientists might never work-out in what way life could ascend by natural processes. Yet, he throw-outs the creation account for the erroneous reason above.

Now what sort of cause is adequate to explain the origin of life?

The cause must be unbelievably intelligent—far outside our intelligence. We marvel at the scientists who are discovering the nano-technology in living things—and it is an astounding innovativeness. But then again, what of the One who invented these things?

How much more intelligent is He?

This reminds us of another characteristic of the God whom the Bible reveals:

He is **omniscient** (all knowing). *Psalm 139:2–6; Isaiah 40:13–14.*

We know enough about the Creator from His creation to be "without excuse". *Romans 1:18–22* says,

*"For the wrath of God is revealed from heaven against all ungodliness and unrighteousness of men, who by their unrighteousness suppress the truth. For what can be known about God is plain to them, because God has shown it to them. For his invisible attributes, namely, his eternal power and divine nature, have been clearly perceived, ever since the creation of the world, in the things that have been made. So, they are without excuse. For although they knew God, they did not honor him as God or give thanks to him, but they became futile in their thinking, and their foolish hearts were darkened. Claiming to be wise, they became fools … "*

And here the Scripture enlightens the reasons, otherwise intelligent people select to believe impossible things—that firstly the universe, then life, just popped into existence deprived of any sufficient cause. They select to irrationally agree to take that their two 'great beginnings' had no sufficient cause, rather than recognize and honor their Creator.

## BIG BANG – SOMETHING CAME OUT OF NOTHING

Almost all the concepts of the big bang are now under attack by secular scientists, e.g., that everything came from nothing, that everything was once contained in a singularity, the beginning of time, the origin of the laws of physics.

All the 'solutions' to the problem of what happened before the "big bang" encompass pre-existing universes or conditions, at least one of which is said to have existed for ever, despite the second law of thermodynamics, --which says this is an impossibility.

The proponents of these 'solutions' provide no physical evidence whatsoever in support of their ideas. Not a single scientist says how his postulated first universe came into existence.

When atheist's cast-off the eternal truth of the Word of God in the Bible, we should not be astonished to find that their attempts to find alternative explanations fail to stand the test of time and evaluation by true scientific laws.

Some Christians who have tried to write the big bang into Genesis have substantial problems. First, of course, it contradicts the Bible, e.g., the "big bang" has Sun appearing long before Earth; Genesis has Earth created before Sun.  Second, all this rethinking and, bluntly, confusion, makes one thing quite obvious. Explicitly, that building a theology of origins on an allegedly "assured" and "scientific" foundation such as the "big bang" idea is in reality building it on shifting sand.

How much better to take Genesis at its face value, as indeed Jesus Christ Himself always did, and build one's scientific models on the rock of Scripture.

Usually, Christians comprehend that the original process that God utilized, involved His Word, the creation of matter from nothing on "Day 1," and the forming of that matter into the objects that fill our physical universe.

The Bible clearly reveals that God created this universe:

### I. By His Word

"And God said, 'Let there be light'; and there was light" (*Genesis 1:3*). "Then God said, 'Let the waters below the sky be gathered into one place, and let the dry land appear!'; and it was so" (*Genesis 1:9*). "In the beginning was the Word, and the Word was with God, and the Word was God. He was in the beginning with God, Figure 3.3. All things came into being through Him [i.e., the Word], and apart from Him nothing came into being that has come into being" (*John 1:1–3*). " ... by the word of God the heavens existed long ago and the earth was formed out of water and by the water" (2 Peter_3:5). "By faith we understand that the worlds were prepared by the word of God ... "(Hebrews 11:3).

**Fig.3.4:** *And God Said Let There Be Light – The Glory of God – Curtsy - Zetu Quality AI Art*

## II. Out of Nothing – Ex-Nihilo

"By faith we understand that the worlds were prepared by the word of God, *so that what is seen was not made out of things which are visible*" (Hebrews 11:3). ' ... God, Who gives life to the dead and calls into being that which does not exist' (Romans 4:17).\

## III. Creative Acts

Although the universe, heavens and earth, were created out of nothing, much of the contents of this universe were created out of the material which God had previously created out of nothing. Thus, the Bible says, 'God separated the light from the darkness' (v.4) or 'God made the expanse' (v.7), or 'God made the two great lights' (v. 16), or 'God formed man of dust from the ground, and breathed into his nostrils the breath of life, and man became a living soul' (Genesis_2:7). Nonetheless, all these creative acts, still, not by fiat and not *ex nihilo*, are nevertheless done in less than a day. They are not accomplished over a long period of time, such as thousands or millions of years, but according to the numerals used, in less than one solar day.

To extend this thought, Henry Morris in *The Genesis Record* provides the following commentary:

"The earth itself originally had no form to it (Genesis 1:2); so [Genesis 1:1] must speak essentially of the creation of the basic elements of matter, which thereafter were to be organized into the structured earth and later into other material bodies. The word is the Hebrew "***erets***" and is often also translated either 'ground' or 'land'. Somewhat similarly to the use of 'heaven', it can mean either a particular portion of earth (e.g., the 'land of Canaan'—Genesis 12:5) or the earth material in general (e.g., 'Let the earth bring forth grass'—Genesis 1:11)" (p. 41).

Morris goes on to describe that "Initially there were no stars or planets, only the basic matter component of the space-matter-time continuum. The elements which were formed into the planet Earth were at first only elements, not yet *formed* but nevertheless comprising the basic matter—the 'dust' of the earth [emphasis in original]" (p. 50).

There are numerous places in Scripture which talks about God being the "potter" and "forming" his creation (e.g., Isa. 64:8, Rom. 9:21, Psa. 95:5, Psa. 103:14, Psa. 104:26). Like a potter who works with clay, God used the material he created on "Day 1," to form/create the plants on "Day 3,"

The birds and fish were not described as coming from the "dust." However, Gen. 2:19 states "Now the LORD God (YHWH) had formed out of the ground all the wild animals and all the birds in the sky". Thus, the Bible does say that birds were created from the same 'dust' material.

## Time-Machine

Dr Carson is professor of New Testament at Trinity Evangelical Divinity School in Canada; he utilized a time-machine as an illustration to the instant functionality of God's creation.

Let us envisage that you and I have at our disposal a neat, comfortable, and very precise time-machine which looks like a large telephone booth with two armchairs in it. We turn the dials all the way to the left, marked "**creation**—sixth day, morning." Within a few moments we are present just as God has finished creating -- the Bible says "fashioned" -- Adam, the first man.

There Adam stands in all his beauty and splendor. Accordingly, we get out of our time-machine, walk over to Adam and ask: **"How old are you, Adam?"** Adam smiles and answers: "God created me just two minutes ago." Then, we look Adam up and down—yet what we see is certainly not a human being just two minutes after conception, nor two minutes after birth. Here stands a full-grown man of perhaps 30 years of age.

What a inconsistency! Adam claims he is two minutes old, yet his appearance shows he may be 30 years old!

Who is correct? Is Adam, right? Or are we, the scientists, right?

Adam informs us of his age on the basis of a fact: the very simple, straightforward fact that he came into existence only about two minutes ago. How can anyone challenge this fact? But you and I, with the advantage of our great scientific information (after all we have just arrived in a time-machine), know exactly how to determine the chronological age of a human being: we measure the skull, the size of nose, ears, mouth, teeth; we measure the length of his arms and feet and the size of his torso. And we compare Adam's body with our charts on anatomy—and, on the basis of the facts before our eyes, we come to the deduction that Adam is some 30 years old, give or take a few years. How can anyone argue with these facts? Really, does it matter what Adam tells us?

## Created Mature - Means Two Ages

Facts against facts? Yes, in one way. But in another way, no. All of us must agree that any person or thing created mature has to have **two ages**: an actual:

(Real, or revealed) age (two minutes) and: an apparent (or scientific) age (30 years—apparent because it is determined by its appearance, and scientific – because scientific methods and tools are used to determine the age), Figure 3.4. This is, first of all, not an inconsistency—but a necessity! In other words, one cannot even speak about "creation" without immediately necessitating the postulation of two ages.

***Fig.3.5:*** *Adam created a Mature Man – Curtsy VoVatia*

What I am claiming is simply this: **Creation, by definition, always results in two ages.**

On a certain way, byway in a suburb of Bergen County, New Jersey, I discovered a curious sign outside a shop. It read: "**New Antiques**." Now I ask you, what does that mean? If the things inside are antiques, they cannot, by definition, be new. And if they are new, then by definition, they cannot be antiques.

There is a way out of this dilemma: the manufacturer may make "antiques," that is, pieces of furniture which appear to be very old but which are actually brand new.

But is this not deception? Are all these pieces not just fakes?

Well, that depends on the manufacturer, doesn't it? If he sells you that Louis XIV chair for $10,000, claiming that the King of France himself used it in his palace in Versailles, then he would be swindling you. But if the manufacturer sells you the chair for $250, claiming he copied it out of a "collectors" guide, and that he has attempted to make it look as genuine as possible—then there is no deception.

## IV. A Week-Old Universe – Fully Functioned

Back to our question of what we mean when we say that God created this universe. If He created this world, as we believe, out of nothing—in other words, He spoke the Word and it came into being—then this universe and our earth was an entirely grown, mature, fully functional creation. On the seventh day, it was only a week old.

But if the evolutionary scientists of today had been present on the seventh day, they could have measured the distance of the stars from the earth and would have claimed that the universe had to be at least hundreds of millions of years old because that is the time the light of the stars needed to reach earth, Figure 3.5.

They could have investigated the mountains and allotted ages to them in the millions of years. They could have cut a few trees and measured the rings, declaring that they were perhaps 75 to 100 years old. And so forth. **Not one scientist could have measured scientifically any of the evidence and come to the conclusion that the earth and all in it was one to six days old.**

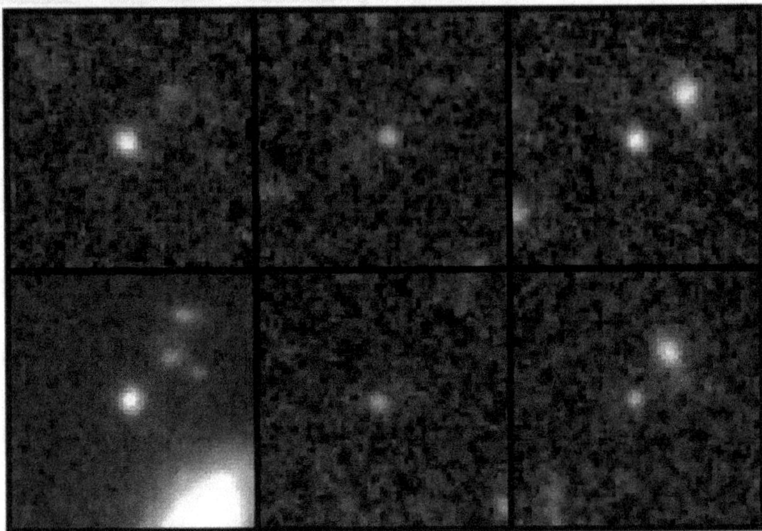

***Fig.3.6:*** *Images of the six objects thought to be massive galaxies Proves Young from Universe – Curtsy NASA, ESA, – (Swinburne University of Technology).*

## DID GOD CREATE WITH THE 'BIG BANG'

What do you really mean when you say God created this universe? Did He create the universe by causing a 'big bang' and then permit everything to take its course, perhaps infrequently lending a helping hand with the evolution of things? Or did God create the universe **out of nothing**—*ex nihilo*? Figure 3.6.

We could philosophize about this and discuss what is most likely and what is somewhat probable and what would be the most elegant model considering who God is. We expect more.

*We want the manufacturer to tell us the truth.*

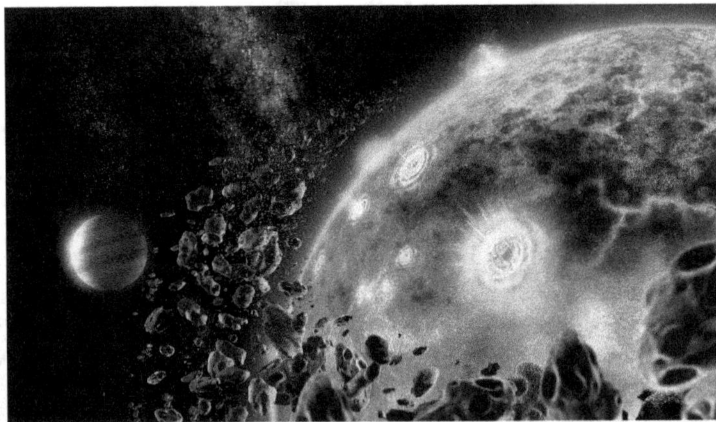

***Fig.3.7:*** *Science Does Not Disprove God – Curtsy Dorling Kindersley - Getty Images/Vetta*

That's really all that counts. Actually, there is more involved, as we showed already with the illustration of the antique-manufacturer. If the manufacturer does not disclose, *reveal*, to us that this genuine-looking antique is actually brand-new, just off the conveyor belt so to speak, he is deceiving us. If God the Creator, in other words the Manufacturer, had not informed us about who created this universe and how it was created, all of us would merely be guessing. There would be many speculations, many theories, many varied and colorful opinions, but absolutely no certainty, Figure 3.7.

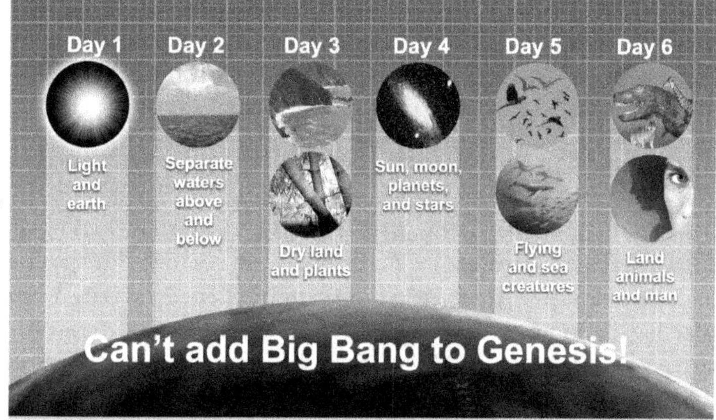


*Fig.3.8: Did God use the "Big Bang" to create the universe. No, the Big Bang model has the stars before the sun & then the earth forms as a hot molten blob. God's Word records that God made the earth first covered with water, and the sun, moon & stars weren't made until the 4th Day*

## THE CREATION MISCONCEPTION

However, God has disclosed all the pertinent information. He has told us how He has created this universe. It is a fallacy, propagated in some Christian circles, to claim that God has not told us the how, only the who: that He only told us that He did it but not how.

This fallacious claim relegates the first two chapters of Genesis to the literary category of saga, myth, or legend (man's philosophical speculation about origins) but does not believe that God's Holy Spirit oversaw the writing of these chapters so that only what actually happened was recorded, constituting—to remain within our illustration—the manufacturer's manual which informs.

In other words, man, employing his common sense, looks at the stars or solar system or galaxies in the sky; he looks at the sun and moon; he looks at the mountains and the seas; he looks at the trees and plants and birds and animals; he looks at the human body—the eye, the food-processing plant, the brain—and he comes to the conclusion that all this cannot have come into being by *chance*, but that there is a Creator, an intelligent Maker behind it all.

## Is Evolution an Option Today?

Evolution, and we are pondering macro-evolution, one kind of creature's allegedly evolving into an entirely diverse kind, not minor changes such as diverse skin color, ear shapes, and so on, has never really been an option for the Christian, nevertheless, it always has been a strong option for the scientist. Evolution is the effort of man to evaluate the evidence without turn to the manual. Thus, evolution is a system which contradicts the necessity for and the existence of God. Evolution has carefully chosen to ignore some of the evidence; it has elected to ignore the Manufacturer's Manual. Evolution has sifted through the evidence and chosen what it could use and ignored that which spoiled the model.

Even as a scientific hypothesis evolution has advanced rather badly, so much so that evolutionism is no more a well-thought-of as hypothesis in some scientific circles (cf. among others Michael Pitman, *Adam and Evolution*, Rider & Co., London, 1984; Michael Denton, *Evolution: A Theory in Crisis*, Burnett Books, London, 1985).

We conclude that evolution is not a option for a Christian and not a resounding choice for a scientist. We believe the Manufacturer's Manual, the Word of God:

'In the beginning God created the universe [*ex nihilo*— out of nothing-- creative acts—as a mature creation which, of inevitability, had to have an actual/revealed/young and an apparent/scientific/ old age.

## ETERNAL UNIVERSE

Much has been written about the universe, with its alleged big bang origin 13.8 billion years ago, with its expansion forcing all galaxies away from each other. And about two decades ago it was 'discovered' that the expansion is accelerating, driven by some very strange form of energy—Dark Energy—that acts like an antigravity force. Yet the big question remains. What is the ultimate fate of the universe? Secular cosmology does not have a precise answer. I felt I should provide several of their scenarios below. However, I believe that the Scripture has the answer to this question. That answer may seem too many to be contrary to known science, but the same could be said of the creation of the universe from nothing and without a cause, whether it be by the action of the Creator God or by secular physics invoking some quantum fluctuation of a metastable false vacuum.

## Day-Night Cycle - No Sun till Day 4

We know today that all it takes to have a day-night cycle is a rotating Earth and light coming from one direction. The Bible tells us clearly that God permitted *light* of His Glory to shine on the first day, as well as on the Earth. Thus, we can deduce that the Earth was already rotating in space relative to this permitted light.

God can, of course, create anything without a secondary source. We are told that in the new heavens and Earth there will be no need for sun or moon (Rev 21:23). In Genesis, God even defines a day and a night in terms of light or its absence.

"Progressive creationists," sometimes use the argument that the days are really long periods, although God could have used words for that if He had really meant that. The creation of the sun after the Earth undermines progressive creationists' attempts to harmonize the Bible with billions of years. Therefore, they must explain this teaching away.

Some "Progressive creationists," declare that what really occurred on this "4th day" was that the sun and other heavenly bodies 'appeared' when a dense cloud layer dissipated after millions of years. This is not only fanciful science, but bad exegesis of Hebrew. The word *'asah* means 'make' throughout Genesis 1, and is every now and then used interchangeably with 'create' (*bara'*), e.g., in Genesis 1:26–27. It is pure desperation to apply a *different meaning* to the *same word* in the *same grammatical construction* in the *same passage*, just to fit in with atheistic evolutionary ideas like the "big bang." We know that the Holy Spirit oversees every written word in the scripture. If God had *meant* 'appeared', then He presumably would have *used* the Hebrew word for appear (*ra'ah*), as when the dry land 'appeared' as the waters gathered in one place on "Day 3" (Genesis 1:9).

We have checked over 20 major translations, and all clearly teach that the sun, moon and stars were *made* on the "4th day."

The evidence that ordinary days are being referred to is so overwhelming that even liberal Hebrew scholars acknowledge that the author can have had no other intent — predominantly when the words "evening" and "morning" are used from the first day.

On the "4th day" the present system was instituted as the Earth's temporary light-bearers were made, so the glorious light of the Infinity God from the first day was no longer needed.

## Pagans Worshipped the Sun

1. This would have been very substantial to pagan world views which inclined to worship the sun as the source of all life. God appears to be making it deliberately clear that the sun is secondary to His Creator-hood as the source of everything. He doesn't "need" the sun in order to create life.

2. This unusual, counter-intuitive order of creation, light before sun, actually adds a hallmark of authenticity. If the Bible had been the product of later "editors," as many critics allege, they would surely have adapted this to fit with their own understanding. It is only recently that the astronomical fact has been comprehended that a day-night cycle needs only *light plus rotation*. Having "day" without the sun would have been generally inconceivable to the ancients.

## THE EARTH IS SPECIAL

Those who believe the Bible should are ecstatic—after all, the earth was created first, before the sun, moon and stars, and was particularly designed to house millions of diverse kinds of living things.

One must comprehend that evolutionary astronomers have omitted a Creator by decree, and in its place believe that our solar system formed by itself:

'Astronomers agree that the planets and moons of our Solar System shaped in a swirling disc of dust and gas around the Sun. In the inner regions, dusty particles melted and bonded together, forming hot blobs of rock that cooled and compounded to make Mercury, Venus, Earth and Mars.'

That is, according to evolutionists, the solar system was born in a crumpling cloud of dust and gas called a *nebula*, hence the term *nebular hypothesis*. Most of this collapsed into the sun, while the inner planets were shaped from fragments that collided and fused together.

Though, the more scientists have examined this, the more they have realized that there is a problem. There was no reason for the rocky particles to melt—what would have heated them? If anything, back then the sun would have been cooler than today. Therefore, only a small and very close planet like Mercury could conceivably have become hot enough. But further from the sun, they have admitted a problem:

'While asteroid-sized rocks would have aggregated in the inner Solar System, they would not have melted and clumped together to form planets. The solid rocks would just zoom past each other or collide and recoil like billiard balls.'

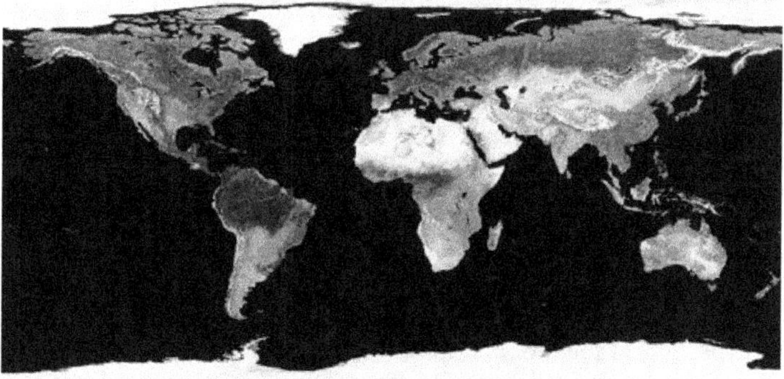

*Fig.3.9: The Earth at Present Time*

Evolutionary astronomers propositioned that a supernova explosion within 50 light years from Earth blasted and provided the nebula with radioactive aluminium-26, which produced heat as it decayed. But this necessitates a highly improbable set of chances, which is why the chances are 'remote'.

Nevertheless, Dr Clarke couldn't bear the thought of Earth being in a preferred place in the universe. It has a great deal to do with the humanistic/atheistic belief that life on Earth, as well as humanity, just 'happened'. Therefore, they would assume our earth to be neither especially equipped nor to possess a special location in the universe. Therefore, he favors the above 'speculative' idea, Figure 3.8!

## Earth facts

- Mass: $5.976 \times 10^{24}$ kg ($1/333,400$ sun, $81 \times$ moon)
- Radius: 6,378 km (equator) 6,356 km (pole) ($1/109$ sun, $3.66 \times$ moon)
- Mean distance from sun: 149.6 million km (1 astronomical unit (AU))
- Land area: 148 million km$^2$ (29% of total area)
- Highest point: 8,848 m (Mt Everest)
- Lowest points: sea-11,034 m (Marianas Trench); land-397 m (Dead Sea)
- Gravitational acceleration at surface: 9.8 m/s$^2$ ($1/27$ sun, 6x moon)

# HIDDEN ASSUMPTIONS OF SECULAR COSMOLOGY

This repudiation of Earth's special place goes even further, and it motivates the "big bang" theory itself. Most people don't comprehend that this theory depends on a philosophical supposition called the Cosmological Principle—that there is no special place or direction in the universe. That is, the universe has no center and no edge to it.

Though, consistent with the Bible, we can start from a different assumption—that humanity *is* distinct in God's sight, and the earth, as our home, *does* display indications of exceptionality, counting its setting. Both a centered and a centerless universe are *consistent with* the observation that almost all galaxies are receding from Earth, so the choice is purely *philosophical* on those grounds. However, only the idea that our galaxy is near the center of the universe fits *all* the evidence, making sense of the quantized redshifts that a centerless universe has great difficulties explaining.

Of course, there is nothing low-spirited about God creating Earth particularly for life, as the Bible says! The sincere scientific evidence, as contrasting to unscientific assumptions, is just another validation that nothing in real science contradicts the Bible—on the contrary it sanctions the Bible's history over and over again.

## What is Dark Matter – Dark Energy

Dark matter is consisted of particles that do not absorb, reflect, or emit light, so they cannot be distinguished by observing electromagnetic radiation. Dark matter is material that cannot be observed directly. We know that dark matter exists because of the outcome it has on objects that we can see directly. Dark matter is stuff in space that has gravity, but it is different to anything scientists have ever seen before. Together, dark matter and dark energy make up 96% of the universe. That only leaves a small 4% for all the matter and energy we observe and comprehend.

Dark Energy is a hypothetical form of energy that exerts a negative, repulsive pressure, acting like the opposite of gravity. It has been hypothesized to account for the observational properties of distant type "Ia supernovae," which illustrate the universe going through an accelerated period of expansion. Dark energy is the name given to the mysterious force that's causing the rate of expansion of our universe to accelerate over time, rather than to slow down. That's opposing to what one might predict from a universe that allegedly started in a "Big Bang." It causes the universe to expand. The author strongly believes that Dark Matter/Dark Energy is the power provided through the Lord Jesus Christ, as He "holds" the Universe together Colossians 1:17.

The discovery of supernova 1997ff, located about 10 billion light-years away (Distance), provided evidence for dark energy. About halfway into the universe's history — several billion years ago — Dark Energy turn out to be dominant and the expansion accelerated.

Distant galaxies seemed to be moving away from us at high speed. Dark Energy is caused by energy intrinsic to the fabric of space itself, and as the Universe expands, it's the energy density — the energy-per-unit-volume — that stays constant. Consequently, a Universe full of dark energy will see its expansion rate stays constant.

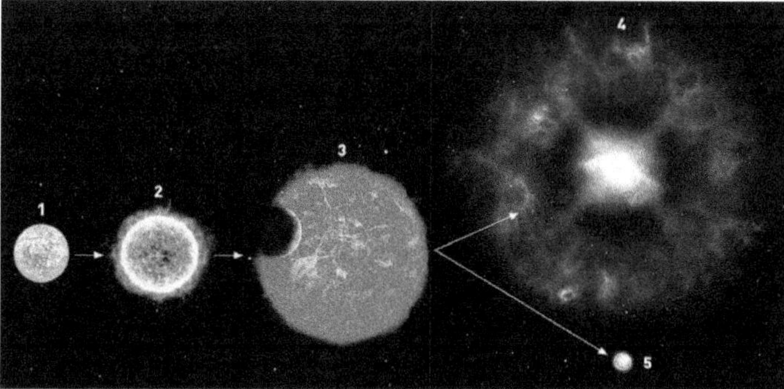

*Fig.3.10*: Steps Making a White-Dwarf as Dead Star – Curtsy NASA/Hubble Heritage Team

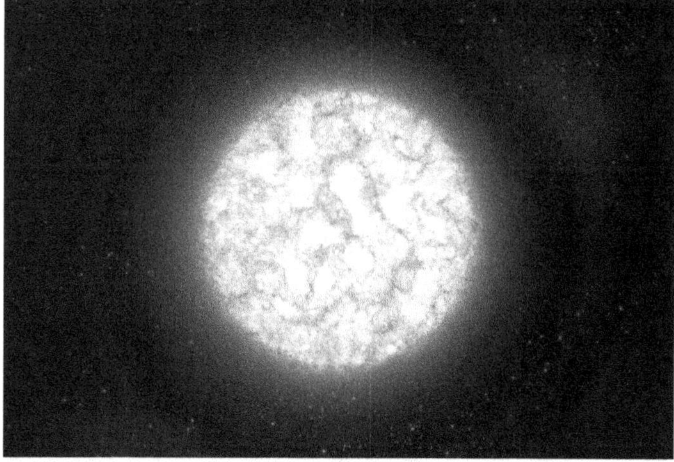

*Fig.3.11*: The Core of White Dwarf – Dark Matter – Remnant of Dear Star – Curtsy NASA/Hubble Heritage Team

## V. Burning Hydrogen

A Sun-like star is powered by nuclear fusion, which turns hydrogen into helium with the release of energy. This can go on for around 10 billion years. During this time, it is said to be a 'main sequence' star.

## VI. Shell burning

A star is said to move off the main sequence when it transitions from burning hydrogen to fusing helium. This often happens in layers surrounding the dead core and so astronomers refer to it as 'shell burning.'

## VII. Red giant

The star gets through its helium one hundred times faster than it did its hydrogen. This creates huge amounts ofoutwards pressure and it surges out into its planetary system, Figure 3.9. Its surface would reach out to about where Earth is now.

## VIII. Planetary Nebula

Helium burning is unstable. It sends shockwaves through the star and convulsions shake it apart. A few of these events, each 100,000 years apart, create nested shells called a **planetary nebula**.

## IX. White dwarf

A core of carbon and oxygen – the products of helium fusion – is left at the center, Figure 3.7. Astronomers refer to this as a white dwarf and it contains about half the mass of the original star packed into an Earth-sized ball, Figure 3.11.

## Dark Matter – White Dwarf

Dark matter could be white dwarfs, Figure 3.11, the remnants of cores of dead small- to medium-size stars. Or dark matter could be neutron stars or black holes, the remnants of large stars after they explode. Most people don't think of dark energy—the all-permeating force driving apart galaxies—as particularly weak. But based on arguments from "quantum mechanics" and Albert Einstein's equations for gravity, scientists estimate that dark energy ought to be at least 120 orders of magnitude stronger than it actually is.

Dark energy may not be destroyed. It may stop to act by itself willingly. It may convert into baryonic matter or even birth a brand-new particle. Because we don't understand how it is made, we also do not know how to eliminate it. It is indestructible.

The author considers the Dark Energy/Dark Matter to be absolute entity, Godly! Scientists, cosmologists and others cannot interact to alter its nature or contents. The universe could be ripped apart by dark energy if its influence steadily increases. Eventually there would be nothing left: no particles, not even space itself, and no universe.

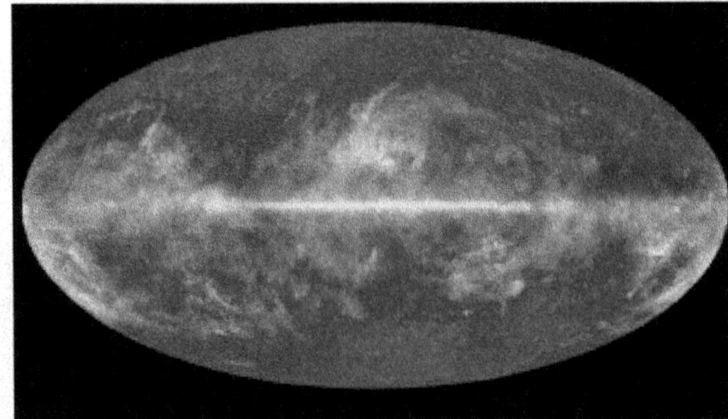

*Fig.3.12: This all-sky image of the cosmic microwave background CMB), created from data collected by the - European Space Agency's Planck satellite's*

For this reason, most experts think that dark matter is abundant in the universe and has had a strong influence on its structure and evolution. The primary evidence for dark matter comes from calculations showing that many galaxies would behave quite differently if they did not contain a large amount of unseen matter.

The presence of dark energy, in whatever form, is needed to reconcile the measured geometry of space with the total amount of matter in the universe. Measurements of cosmic microwave background (CMB), Figure 3.11, anisotropies *(Anisotropy characterize substances that exhibit physical properties with different values when measured in different directions)* indicate that the universe is close to flat. It has been established that dark energy is the cause why time continuously moves forward. The Dark Energy is accelerating the expansion of the universe supports the asymmetrical nature of time.

Dark matter makes up most of the mass of galaxies and galaxy clusters, and is responsible for the way galaxies are prearranged on impressive scales. Dark energy, in the meantime, is the name we give the mysterious effect driving the accelerated expansion of the universe, Figure 3.13.

In a Universe without dark matter, we may still have stars and galaxies, but the only planets would be gas-giant worlds, with no rocky ones, no liquid water, and insufficient ingredients for life as we know it, and no human being.

Dark matter makes up about 85 percent of the total matter in the universe, accounting for more than five times as much as all ordinary matter. Dark matter played an important role in the formation of galaxies. Dark energy makes up approximately 68% of the universe and appears to be associated with the vacuum in space. It is distributed evenly throughout the universe, not only in space but also in time – in other words, its effect is not diluted as the universe expands.

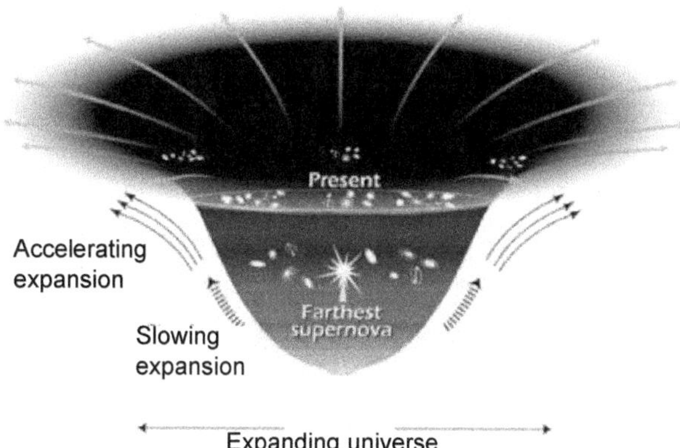


*Fig.3.13: Dark Energy the Cause Time Moves forward. The Dark Energy Accelerating Universe Expansion*

On Earth, there may be more than 10 trillion dark matter particles in each cubic centimeter of the planet's crust.

## Will the Big Rip happen?

If dark energy remains unchanging, space will expand indefinitely while increasingly isolated stars will slowly fade away and go cold, a phenomenon referred to as "Heat Death." And if dark energy keeps accelerating the expansion of the universe, space itself will eventually be torn apart in the Big Rip, Figure 3.13. This will not happen, as the Infinite God the Creator cares about His creation. The universe will go through periods of travail and turmoil, but not to the point of annihilation. He will rearrange the Universe to equip it to last for eternity, absent of any decay and death. Revelation 21 – *"Then I saw a new heaven and a new earth, for the first heaven and the first earth had passed away, and there was no longer any sea. I saw the Holy City, the new Jerusalem, coming down out of heaven from God, prepared as a bride beautifully dressed for her husband.*

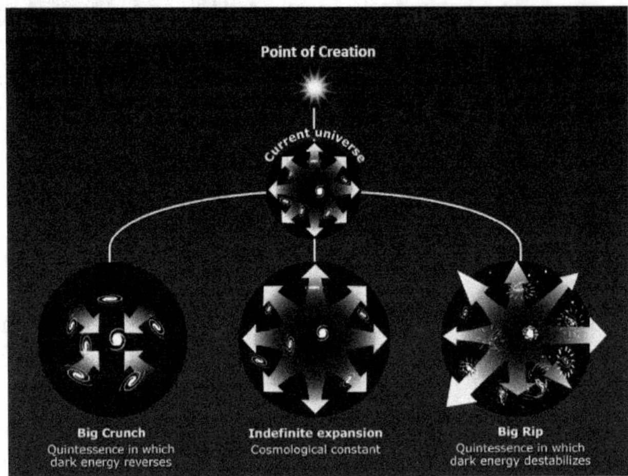


*Fig.3.14: The Big Rip Will Never Happen - NASA/ A. Feild / Space Telescope Science Institute*

## Dark Matter - Dark Energy Complete Mystery

While shadows can move faster than the speed of light, even though absolutely nothing can move faster than the speed of light.

What would occur if dark energy increased? If the hypothetical dark energy continues to increase and dominate the Universe's energy balance, then the present expansion of space will continue to accelerate, exponentially. bodies which are not already gravitationally secured will eventually fly apart, Figure 3.15.

*Fig.3.15: Insecure Gravitational Bodies will Fly Away – NASA Representation*

Is gravity Dark Energy? It is accelerating, and the term "Dark Energy" is a procurator for the mysterious force driving this acceleration. As a result, Dark Energy is, in effect, working against the force of gravity. It is considered as negative gravity — pushing cosmic objects apart as gravity draws them together, Figure 3,14.

Dark Energy in essence is balancing the forces in the universe, without it, galaxies may collide leading to the destruction of the universe. Dark Energy is the hand that stable the entire universe structures, and keeps it intact together. It is considered the Hand of Balancing the universe, it is caring God's entity.

## Dark Matter does not Enter Our Bodies

It is believed that the nuclear forces that grip your nuclei and protons together would become extinct; the electromagnetic forces that caused atoms and molecules to stay together (and light to interact with you) would disappear; your cells and organs and entire body would cease to embrace together.

Dark energy was revealed in 1998 with this method by two international teams that included American astronomers Adam Riess, and Saul Perlmutter and Australian astronomer Brian Schmidt, Figure 3.16.

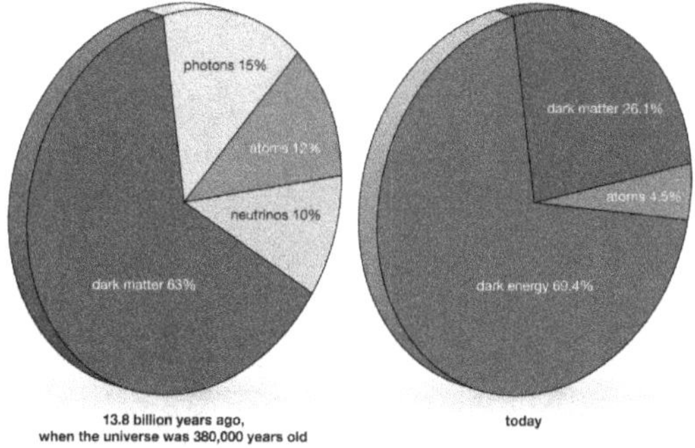


***Fig.3.16**: Hypothetical and Unproven Dark Matter/Dark Energy Then. It is a Well-known Fact Now*

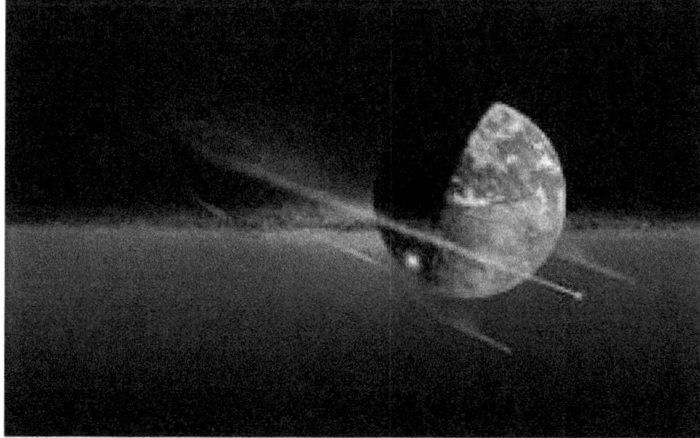

***Fig.3.17**: Dark Matter Effect on Earth*

How does Dark matter affect Earth? The new source of periodic heating by Dark Matter in our planet's interior could lead to periodic occurrences of mantle-plume action and changes in convection patterns in Earth's core and mantle, which could affect global tectonics, volcanism, geomagnetic field reversals, and climate, such as our planet has experienced, Figure 3.17.

What is the problem with Dark Energy? Since it is the energy of space itself, then by means of space expands more Dark Energy comes into the universe, causing the expansion to accelerate ever faster. If Dark Energy is left unimpeded, it is a situation that could eventually cause a "Big Rip" that would tear the fabric of space–time apart.

## The True Identity of Dark Matter/Dark Energy

It is established that scientists, cosmologists, and astronauts collectively have come short to comprehend Dark Matter/Dark Energy, in spite of their diligent efforts and exceptional knowledge of the laws of physics and scientific principles. Certainly, they are well experienced in knowing how to drive mathematical equations to pronounce physical phenomenon within the universe surrounding us. They have deeply studied Eistein understanding of the universal physics and his unique theory of relativity of Time/Space, Matter, and Gravity. They have thoroughly applied Isaac Newton's laws of dynamics, kinematics, and the second law of thermodynamics to balance the universe. All investigations were ended in vain.

A single area they have overlooked to investigate. Sadly, they have not even attempted to cursory read of it. It is the Biblical perception of what the Dark Matter/Dark Energy might be! The invisible power of the Dark Matter/Dark Energy is found in the book of (Colossians 1:15-17), Christ is the visible image of the invisible God. He existed before anything was created and is supreme over all creation, for through him God created everything in the heavenly realms and on earth. He made the things we can see and the things we can't see— such as thrones, kingdoms, rulers, and authorities in the unseen world. Everything was created through him and for him. He existed before anything else, and He holds all creation together.

## The "god-Particle"

Happily, some scientists and astrologists were not far off when they named the unknown entity of the Dark Matter/Dark Energy the "**god-particles**" as a symbol, unknowingly referring to the invisible God's Power to sustain the universe.

Since all creation, visible and invisible, have come to existence through the Infinite God. He has the supremacy of all creation as he sustains it, by holding it together, according to His perfect plan. The plan that was established much before the creation took place. The cause of Dark Matter/Dark Energy, which are invisible substance intended to hold the universe together and keep it advancing in a precise manner till the moment He planned the universe to change its structure to accomplish His perfect will, as planned.

Dark Matter/Dark energy is not a substance to be weighed, seen, nor it is a substance to be physically touched. It is exactly as the scientists unknowingly, described it to be; It is "power." It is the part of the Infinite power of the Lord Jesus Christ, the visible representation of the inviable Infinite God Himself, Figure 3.18.

*Fig.3.18: Dark Matter/Dark Energy Keeps the Universe Held Together – God's Entity*

## Example: Illumination Surrounded by Darkness

Naturally, dark is repelled by the spread of any glimpse of light. This is not a debatable matter, as darkness is the absence of light. However, what if a phantom plane of darkness was erected to prevent light from passing through an illuminated side? I would consider the statement preposterous! Nonetheless, what if all sides of a city were surrounded by phantom walls of darkness to prevent light to pass through darkness. I would say, this is beyond human perception! Rightfully said, it is indeed beyond human perception. The truth sometimes, is more astonishing than fiction.

In the 13th century BC, the Israelite lived in Goshen, Egypt during the reign of king Ramses the II. Goshen was the residence assigned to Jacob and his family, and it was there that the Israelites lived in Egypt (Genesis 45:10; Ex. 9:26). It is currently assumed that the name is derived from the Semitic root, i.e., compact, solid, and fertile land, suitable for grazing and certain types of cultivation. It is believed to have been located in the Egyptian eastern Nile Delta, lower Egypt; perhaps at or near Avaris, the seat of power of the Pharos king.

The pharos king had slaved the Israelite and treated them harshly. As Moses was the leader of the Israelites, he pleaded with the king to let the Israelite depart Egypt. The Pharos was very stubborn to let the Israelite depart. Accordingly, God plagued Egypt with 10 devastating plagues. No plague had fallen upon the land of Goshen, while the rest of Egypt suffered the severe consequences of the plagues.

One of the plagues was "Darkness," no one could see anyone else or move about for three days, all over the land of Egypt. Yet all the Israelites had light in the places where they lived, (Exodus 10:23), Figure 3.19.

*Fig.3.19: Goshen was in Light while the Rest of all Egypt was in Total Darkness for 3 Days and Three Nights*

It is the power and the will of the Infinite God who somehow allowed darkness to surround Goshen preventing light to escape to the neighboring cities of the land of Egypt. We may call this type of Darkness, "The god particles" of Dark Matter!

## Example: Phantom Wall of Dark Matter Separating Light

Another similar incident took place at the edge of the Red Sea in Egypt. When the Israelite had been finally released to depart Egypt, they traveled on foot carrying all their belonging and dragging their animals leaving Goshen, heading to the Red Sea, which is about 75 miles. Josephus stated that it took only three days of journeying for the Israelites to reach the Red Sea. "But as they went away hastily, on the third day, they came to a place called Baalzephon, on the Red Sea," (Antiquities, 2.15. 1).

Also, it is recorded that the Pharoh's King had change of heart and decided to chase the Israelite with his army geared with their chariots and armories to return them back to Egypt.

The infinite God who had been traveling in front of Israel's army, withdrew and went behind them. The pillar of cloud also moved from in front and stood behind them, coming between the armies of Egypt and Israel. Throughout the night the cloud brought darkness to the one side and light to the other side; so, neither went near the other all night long.

Then Moses stretched out his hand over the sea, and all that night the LORD drove the sea back with a strong east wind and turned it into dry land.

The waters were divided, and the Israelites went through the sea on dry ground, with a wall of water on their right and on their left.

The Egyptians pursued them, and all Pharaoh's horses and chariots and horsemen followed them into the sea. During the last watch of the night the LORD looked down from the pillar of fire and cloud at the Egyptian army and threw it into confusion. He jammed the wheels of their chariots so that they had difficulty driving.

And the Egyptians said, "Let's get away from the Israelites! The LORD is fighting for them against Egypt." Then the LORD said to Moses, "Stretch out your hand over the sea so that the waters may flow back over the Egyptians and their chariots and horsemen." Moses stretched out his hand over the sea, and at daybreak the sea went back to its place. The Egyptians were fleeing toward it, and the LORD swept them into the sea.

The water flowed back and covered the chariots and horsemen—the entire army of Pharaoh that had followed the Israelites into the sea. Not one of them survived. But the Israelites went through the sea on dry ground, with a wall of water on their right and on their left. That day the LORD saved Israel from the hands of the Egyptians, and Israel saw the Egyptians lying dead on the shore. And when the Israelites saw the mighty hand of the LORD displayed against the Egyptians, the people feared the LORD and put their trust in him and in Moses his servant. (Exodus 14:19-31).

A phantom wall of darkness separated the Pharoh's army from the helpless Israelites, that neither side was able to see the other. We may call it "The god particles." of Dark Matter!

## Example: Fire does Not Burn

Another incident took place in 1391 BC when Moses was tending the flock of Jethro his father-in-law, the priest of Midian, and he led the flock to the far side of the wilderness and came to Horeb, the mountain of God. There the LORD GOD APPEARED TO HIM IN FLAMES OF FIRE FROM WITHIN A BUSH. MOSES SAW THAT THOUGH THE BUSH WAS ON FIRE IT DID NOT BURN UP. So, Moses thought, "I will go over and see this strange sight—why the bush does not burn up." (Exodus 3:1-6), Figure 3.20.

*Fig.3.20: Moses at the Burning Bush*

The Infinite Eternal Power caused the fire and light to be integrated with a green bush without burning the bush or quenching the fire! This is the same power and the author of the Dark Matter/Dark Energy.

### Example: The Sun Stood Still – Moon Stopped

So, the sun stood still, and the moon stopped, Until the nation avenged themselves of their enemies. Is it not written in the Book of Jashar? And the sun stopped in the middle of the sky and did not hurry to go *down* for about a whole day, (Joshua 10:13).

This is another incident that the Infinite Eternal God has the power to alter the elements within the Miky Way without disturbing the rest of the galaxy, or breaking His physical laws, Figure 3.21.

**Fig.3.21:** *Hoshua – The Sun Stood Still – Moon Stopped*

Since the Infinite God the Creator had created everything in the beginning, we can be confident that He knows how to change or freeze the laws of physics if necessary, and perform a miracle without destroying the universe or creating havoc. God can do whatever He wants to His creation. (Revelation 19:6) reminds us that He is all powerful.

Job 38-39 reveals that God knows the details of His creation. He put the laws of physics in place. Colossians 1:17 states that He continues to **hold** the universe together. He continues to maintain the subatomic particles and the laws of the universe. He is the glue of the universe! We can be confident that God knows how to change the laws of the universe any time He desires without causing a disaster.

"The Greek historian Herodotus . . . wrote that when he visited Egypt, the priests there showed him an ancient manuscript which told the story of a day which lasted about twice as long as a normal day." Ancient Egyptian hieroglyphics also tell a story of Joshua's long day, according to "Fernand Crombette."

According to Gill's Commentary, there is a Chinese account from an ancient manuscript now lost, that recounted a period of time during the reign of Emperor Yau (concurrent with Joshua's account), when the sun stood still for ten days, (since all their sun clocks halted, which is quite possibly an overestimate, subjective guess as to the actual amount of time,).

J. G. Frazer recounts a story from the Fiji Islands, in which the sun stopped from setting.

Since God is all-powerful, we conclude that He can stop the sun and moon, freeze the laws of physics and still maintain the universe. The infinite God does not need our help to explain away a divine miracle with a humanistic and rationalistic explanation.

Joshua 10:12-14 describes a miracle performed by the God of miracles.

## Scientific Hypotheses of Dark Matter/Dark Energy

First there was Dark Matter, then came Dark Energy, then Dark Photons and now there is talk of Dark Stars, Dark Planets and even Dark Intelligent Life, in a whole Dark Galaxy within our Milky Way galaxy.

However, physicists essentially know nothing about Dark Matter and Dark Energy. These terms and the nebulous concepts they represent, were actually invented by astrophysicists because they assumed materialism (matter and energy is all there is). They then inflexibly insisted on rigorously applying this to the origin and structure of the universe.

When such physicists observe the rotation speeds of stars—not only in our own galaxy but also in thousands of other spiral galaxies—they discover that the stars in the spiral disks are moving 'too fast'. They are moving so fast that in their assumed lifetimes, which is the lifetimes of their galaxies, of the order of 10 billion years, the galaxies should have disintegrated because their stars should have flown away from the galaxies, which could not hold onto them.

Because astrophysicists cannot explain these high rotational velocities with standard tried-and-tested Newtonian physics, they have come up with the concept that galaxies really comprise 80% to 90% 'Dark Matter'—stuff that is everywhere but we cannot see or detect it by any method.

## Scientists Abandoned the Word of God

Beginning about 200 years ago, scientists started to abandon the "Word of God" as authoritative in such matters as the creation of the universe, and so today many believe in materialism—that there is no Creator and the universe just created itself from nothing. But when that view encounters data that contradicts it, They could then instead consider the possibility that the universe is not as old as they imagine, that it was created about 6,000 years ago, as the Bible indicates. That being the case, it is no wonder those fast stars have not had time to fly apart, Figure 3.21.

# DARK MATTER EVERYWHERE

## A product of the Dark Side

For this reason, most experts think that dark matter is abundant in the universe and has had a strong influence on its structure and evolution. The primary evidence for dark matter comes from calculations showing that many galaxies would behave quite differently if they did not contain a large amount of unseen matter.

*Fig.3.22: The Fast Stars have not had Time to Fly Apart – Yong Universe*

Why is dark matter supposed to exist in the cosmos? From reading news headlines you would think it has been clearly recognized and that we now know so much about this once elusive stuff. It has been pursued in many different laboratory experiments for more than four decades now, but never found. Why then are astronomers so confident it is out there? As described earlier Dark Matter/Dark Energy is a Godly phenomenon that could not be measured by human hands or human intellect. However, it was identified in, (Colossians 1:17), He holds His creation together. It undoubtedly the entity that hold the universe together and continually guides it to its destiny at a speed equal to the speed of light, Figure 3.22.

## Two Types of Physics

There are really only two types of scientists:

1.  Experimental physicists carrying out experiments in laboratories,

2.  Astrophysicists (or cosmologists) who employ the universe as their 'laboratory'.

Both construct mathematical models to define their observations. Both test their models against those observations.

Nevertheless, the experimentalists (type 1) can interrelate with their experiments in a way the astrophysicists cannot. For example, they can send in a light signal and measure the response in the system, i.e., see what comes out. But the astrophysicists (type 2) cannot interrelate with what they are observing in the universe. The universe is just too large to do that, they claim.

Within our solar system we have been able to send probes to make observations. For example, NASA's Deep Impact probe shot a 370 kg copper bullet into a comet and measured the spectra of the expelled material. And ESA's (European Space Agency) Rosetta spacecraft landed a robotic lander, Philae, on a comet and made, for the first time, direct measurements of the surface constituents. These types of measurements, you could consider, are very parallel to what the experimentalists do in their laboratories. But the latter mission's purposes, extracted from the ESA website, highpoint the type of science involved.

Rosetta's major objective is to aid comprehend the origin and evolution of the Solar System. The comet's structure echoes the composition of the pre-solar nebula out of which the Sun and the planets of the Solar System shaped, more than 4.6 billion years ago. Consequently, an in-depth examination of comet "67P/ Churyumov-Gerasimenko" by Rosetta and its lander will deliver indispensable information to comprehend how the Solar System shaped.

There are straightforward underlying assumptions. The declaration above makes it clear that the scientists who executed the mission trust that the solar system evolved out of a solar nebula originating more than 4.6 billion years ago. That is the unverified main assumption. It is not testable by what they excavate out of the surface of the comet, but rather they trust the measurements of that material will help them comprehend the origin of the solar system *within their original assumption.*

But no matter how much indication they accrue they cannot directly observe the past. Undoubtedly, not without assumptions. They always require to apply interpretations to the evidence, which is the materials they excavate out of those comets.

Even in the case of astrophysics, you might think that the astronomer is detecting the past, because the light entering his telescope allegedly took millions or billions of years to navigate the vast universe to earth. But even this has its restrictions to what we can know.

The astronomer receives light into his telescope on earth and he must make the uniformitarian hypothesis that the light has been travelling at a constant speed (of about 300, 000 km/s – 186,282 mi/s) for the past millions or billions of years to touch earth, and with no relativistic time dilation properties. Only after making that hypothesis can, he *makes the assumption*, not know, that what he observes is coming from some past epoch millions or billions of years past. But how could you test that assumption? You can't! And for that reason, this aspect of astrophysics/cosmology is not directly demonstrable by any experiential test.

In the case of all observations outside the solar system the problem is beyond dispute. You cannot go there. The magnitude, distances to and assumed age of galaxies, and other cosmic sources, is so great that even what we measure is as though we are taking a single motionless photograph; it is just a moment in time.

Astronomers only observe, they cannot interrelate with their experiment as the experimental physicist in the laboratory can do. And what makes matters even more problematic for the astrophysicist or cosmologist is that there are many conceivable descriptions for the same observations. But because they cannot interact with the sources under investigation (which might even be the whole universe) their science is very feeble indeed. For this very cause James Gunn, co-founder of the Sloan survey, said:

"Cosmology may look like a science, but it isn't a science. … A basic tenet of science is that you can do repeatable experiments, and you can't do that in cosmology."

## What do we recognize about gravity?

Now let's examine this statement. *There exists a force of attraction between any two masses in the universe.*

## Can that statement be confirmed?

You might answer, yes. We can locally test gravity and it works, in fact, it works extremely well, and it has been empirically confirmed even down to sub-centimeter distances. Scientists discover strong indication in local laboratory experiments. In fact, new physics is even sought at distance scales less than this because it is trusted that ultimately the gravitational force law must break down, since quantum theory and Einstein's gravitational theory are fundamentally mismatched. But all of that research is done via repeatable experimental physics. Diverse theories can be, and are being, verified by the experimentalists.

Evidence has accrued that supports the law of gravitation. It has been tested repeatedly, and no flaw found, to Einstein's formulation of the law anyway. That is why it is now called a *law*. The law is often called the *universal* law of gravitation.

Can an experimentalist then without harm extrapolate his deductions about gravity, in a laboratory experiment, to the whole universe? No, he can't, not without assumptions. *Therein lies one major part of the problem.*

Next, we must resolve on what we mean by 'evidence.' Typically evidence is data collected. But that data must be interpreted. And models are constructed that make predictions. In a laboratory the experimentalist can test those predictions. In the cosmos this is more problematic. It is likely, though. A model might predict the existence of certain behavior and then the astrophysicist looks for that. But it is more like 'stamp-collecting' than laboratory science. Because he cannot do an experiment he accumulates as many observations as possible and tries to classify the results. He sorts his objects into families, or recognizes a common trend among those in the same family. By collecting a lot of such data, he argues for his model. But because he cannot distinguish by an empirical test his deductions are consistently and inevitably weak.

Now, what if you were to read the headline, "Evidence for dark matter in the inner Milky Way"? What evidence could this be referring to? And how could you distinguish that dark matter really exists?

The 2015 article that bore this headline went on to say (emphases added):

"The existence of dark matter in the outer parts of the Milky Way is well recognized. But factually it has proven very problematic to establish the presence of dark matter in the deepest regions, where the Solar System is located. This is due to the trouble of measuring the rotation of gas and stars with the required accuracy from our own location in the Milky Way.

"In our new study, we attained for the first time a direct observational proof of the presence of dark matter in the innermost part of the Milky Way. We have created the most complete compilation so far of published measurements of the motion of gas and stars in the Milky Way, and equated the measured rotation speed with that expected under the assumption that only luminous matter exists in the Galaxy. The observed rotation cannot be explained unless large amounts of dark matter exist around us, and between us and the Galactic center," says Miguel Pato at the Department of Physics, Stockholm University.

Without going into the particulars of the physics of gravitation, and why bodies like stars orbit the Galaxy center the way they do, we can learn by critically examining these statements.

Dark matter is called dark because we cannot see it. No dark matter particle has ever been observed in a lab experiment in spite of more than 40 years of searching. I even spent a few years doing that myself looking for Para photons, which are classed as "WISPs (disregarded)," a putative dark sector particle. (In *Star Wars* terminology it is a particle from **the Dark Side**.)

If the so-called dark matter particles could be observed by light, or by X-rays, or by some other electromagnetic radiation, it would make their proof of identity easy. But how can the author of the above article claim "direct observational proof"? How can it be claimed that their existence is "well established" in the external parts of our galaxy? They don't observe dark matter, nor do they do an experiment where they send in some radiation into a cloud and get a reply back.

An experimentalist might do something like that to detect a particle he cannot otherwise 'see.' Therefore, the performance of seeing means a response to some radiation. It does not mean a human has to be capable to see it with his own eyes. For example, we know electrons exist. That is not in disagreement, and their existence has been repeatedly long-established by many experiments. (Interestingly, though, we don't know how small they are. It is still an open question.)

In the galaxy, how can the entitlement be made that *the observations cannot be explained unless large amounts of unseen dark matter are assumed*? To make such a entitlement, you would have to know that you have *ruled out all other possibilities*. In such a case—remember this is not a laboratory experiment—you would have to be an all-knowing god.

**Fig.3.23:** *The image shows the rotation curve tracer gases from a photograph of the disc of the Milky Way Galaxy as observed from the Southern Hemisphere. The tracers are color-coded in blue or red according to their relative Doppler motion regarding the Sun. The spherically symmetric blue halo shows the dark matter distribution inferred from the analysis.*

The light approaching from the gases and stars in the Galaxy is detected with a telescope, Figure 23, but more precisely what is observed are the spectral lines in the light from those sources. And they are perceived to be red-shifted or blue-shifted. (Spectral lines shifted toward either the red end or the blue end of the spectrum, as compared to a laboratory sample of the same type of gas.) These properties are interpreted as ascending from the well-established Doppler Effect, where the motion of the gas particles (or the stars) causes this effect in the light. Then that is understood as meaning the gases and stars are moving around the galaxy center at certain speeds, which are typically 100 km/s to 300 km/s.

That interpretation (the Doppler Effect) necessitates some expectations. But all are sensible and within recognized laboratory physics, **except one**. That one is that *the law of gravitation is true out in the Galaxy where these gases or stars are.* That is, the law of gravitation, which has been tested extremely well in the solar system, also applies without alteration out in the cosmos, both in the Milky Way Galaxy and outside it.

But we have by now understood that it cannot be known if that law is *universal*. It is *assumed to be universal* and therefore, a model of the galaxy constructed. If the detected speeds (and they are an interpretation of the meaning of the red-shifted and blue-shifted light, which we will agree to here) shadow the predictable trend then we say all is well and Newton's law (of gravity) works fine. But the tricky is they do not. The stars and gases move too fast around the Galaxy to obey Newton's law. If that condition sustained for hundreds of millions of years the stars in the Galaxy disk region (where our solar system is located) would fly apart and the Galaxy would collapse over longer time scales. But that couldn't be right because (of the fundamental assumption that) the galaxy is steady and has been around for 10 billion years or so.

So, the deduction is, either that the law of gravitation is wrong or that there is more matter in the Galaxy, which we cannot see. Nearly always it is assumed that there is missing, hence 'Dark', Matter prowling out there, which comprises 80–90% of a galaxy's mass. But if it wasn't for Einstein's discovery where he added to and improved on Newton's law of gravitation, we might still be thinking that the Dark Planet Vulcan (also called Dark Matter at the time) was essential to clarify an irregularity of Mercury's orbit in the inner solar system. So it is just as rational to consider that new physics rather than new matter is needed to add to Einstein's improvements the way these did to Newtonian physics. Yet nowadays news headlines tend to speak of discoveries of Dark Matter as though it is being directly imaged.

"Dark Matter observed in the heart of our galaxy," says one news headline, and the article states, "But up to now it has proven very problematic to establish the presence of Dark Matter in the innermost regions." You get the impression it is all now well-established science. This is illustrated in Figure 3.17 above of the Milky Way galaxy with the red-shifted and blue-shifted sources illustrated on either side of the central core of the Galaxy.

Another headline: "Milky Way has half the amount of dark matter as previously thought, new measurements reveal". The story is that by looking at how many Dwarf satellite galaxies our galaxy has around it, and their motions, you can find the mass of our galaxy. This is significant because according to the standard cosmogony the mass of the Galaxy regulates its formation process, and that process is determined from the cosmology that is presumed.

In this story these so-called measurements resolve a mystery. Remember the "big bang" model is presumed. That is called the Lambda (Dark Energy) Cold Dark Matter theory, which predicts that there should be some big satellite galaxies around our Milky Way Galaxy that are noticeable to the naked eye. But that is not what we observe. Though, the new measurements allegedly resolve the problem:

"When you utilize our measurement of the mass of the dark matter the theory foresees that there should only be three satellite galaxies out there, which is exactly what we see; the Large Magellan Cloud, the Small Magellan Cloud and the Sagittarius Dwarf Galaxy."

## Light from the Dark Matter Sector

But some may claim science has distinguished radiation from Dark Matter particles in the cosmos. It was previously reported on the idea that the intergalactic medium has too much light coming from it, where no sources could be recognized, and it was hypothesized as the result of the degeneration of some hypothetical Dark Matter particles. This caused by a mismatch between theory and observations, and henceforth, Dark Matter was proposed as the solution.

Already, you must have understood how opportune it is to have Dark Matter particles, or actually, anything from the dark sector. It can fill in what is missing in the theory deprived of a necessity to reject the underlying theory itself.

Dark Matter is vital in the formation of both stars and galaxies. Without it they would not form naturally. If you don't know how stars and galaxies shaped, you don't know much about how the universe, which we observe, shaped.

Galaxy development is a seriously big problem for "big bang" cosmology. In the computer simulations, modelling the development of the large-scale structure (super-clusters, filaments of galaxies etc.) in the universe, Dark Matter is presumed from the beginning. For the condensation of individual galaxies, it is a comparable story. By starting with a critical density of Dark Matter the models are capable to illustrate galaxy development under gravity where the Dark Matter attracts the normal matter into the central region to form a galaxy.

The Dark Matter must exist in as a spherical halo around the spiral galaxies, which have a thin disk of glowing normal matter. This state is obtained from the studies of the speeds of gases and stars in the disks of thousands of spiral galaxies. But there is even a problematic here too, called the Dark Matter cusp problem.

The problem is since the unobserved made-up stuff does not fairly act as you would expect matter to behave under the influence of the law of gravitation.

Since Dark Matter is meant to be like typical matter under gravity's influence it should pile up in the center of galaxies. Its density should be extreme in the core—hence, there should be a cusp or peak in the density distribution there. But to accurately model the motions of the stars and gases, Dark Matter is not required in the central cores, only in the disk regions. Newtonian gravity alone easily accounts for the visible matter in the central nuclei of these galaxies, Figure 3.24.

*Fig.3.24: Dwarf Galaxy – Dark Matter*

So even the research that infer the existence of the dark matter contradict what matter should do under the influence of gravity. One research on dwarf galaxies highpoints this problem:

"Our measurements contradict a basic prediction about the structure of cold Dark Matter in dwarf galaxies. Except or until theorists can adapt that prediction, cold Dark Matter is inconsistent with our observational data."

"Dwarf galaxies are consisted of up to 99 percent Dark Matter and only one percent normal matter like stars. This disparity makes dwarf galaxies ideal targets for astronomers looking for understanding Dark Matter, Figure 3.23.

Astrophysics and cosmology are by their very nature encumbered with philosophical reinforcements. In principle there is nothing wrong with that. You could not do any kind of science short of a basis to build your model. I would call these philosophies worldviews. And we all have a worldview. We form that based on what we have faith in about the world around us and how it all began. The modification here is that my worldview is founded on the biblical truth that Infinite God, the Creator, created the universe about 6000 years ago. It was not the result of an accident or a "quantum fluctuation" of some dishonest vacuum or a "big bang" of any sort. If it were, God would have said so in the Bible.

The worldview that motivates modern cosmology, and cosmogony (on the origin of the universe) is an atheistic one. It has no place for a Creator, and only depend on what man can discover for himself. As a result, he has had to resort to all types of "fudge factors" to make his model fit the observational data, the evidence from the cosmos. Dark Matter has arisen from this. But even when assumed to fix such problems the supposed dark matter does not perform like traditional matter under the influence of gravity. It is stranger than fiction and I am afraid it is no more real than the 'Emperors' new clothes.

## Stars Need Dark Matter – The God of the Gap

"God of the gaps" is a **theological viewpoint in which gaps in scientific acquaintance are taken to be evidence or** proof of God**'s existence**. Dark Matter' is an indispensable ingredient to form stars naturally specified only standard known physics. 'Dark Matter' is a hypothetical exotic form of matter, unknown to laboratory physics, which does not interact with or emit light in any way, hence it is invisible to all forms of detection within the electromagnetic spectrum, from radio-waves to gamma radiation. 'Dark Matter' itself, therefore, is outside of standard known physics. It a superior entity. It has been given one special property, which is that it gravitates, that is, unlike normal matter, it is a source of gravity power only. It is Godly.

## Detection of 'Dark Matter'

Has dark matter been revealed by any direct measurement? That is separate from deducing its existence due to anomalies like galaxy rotation curves where the motions of stars and gases in the arms of spiral galaxies do not follow the expected Keplerian law in line with standard Newtonian physics?

No, it has not and that is after 40 years of searching in laboratory experiments. Yet it is believed to exist—a 'God of the Gaps'—and is essential, otherwise many astrophysical observations just do not agree with those expected by application of standard laws of physics. It is outside the human realms of experimentations. It is there are a caring balancing power of our created universe.

## Six Days Creation

- Are the days of creation ordinary days – 24 hour-day?
- Could they be long periods of time?
- Why six days? Is Genesis poetry?
- Does the length of the days really affect the Gospel?
- How can there be 'days' without the sun on the first three days?
- Does Genesis 2 contradict Genesis 1?
- What about the framework hypothesis?

## The Importance of Six Days Creation

Does it really of any substance if the days of creation in Genesis 1 are real, 24-hour days? Many may say it doesn't matter. In effect, the view that the days should be understood as 'ordinary' days is sadly, a minority view in churches today, although in the past this was not the case, Figure 3.25.

Some may consider that the days can be understood as eons of time, but that God stepped in to do some of the more unbelievable things at numerous times—like making pine trees and people. This so-called 'progressive creation' understanding has God creating *progressively* over eons of time.

Others claim that Genesis is a mere literary device, an outline upon which suspends significant theological teaching—like clothes hanging on a clothesline. They contend that the clothes are the important thing, not the clothesline, so we should not be concerned about attempting to attach Genesis to the history of the world, this is the 'framework hypothesis.'

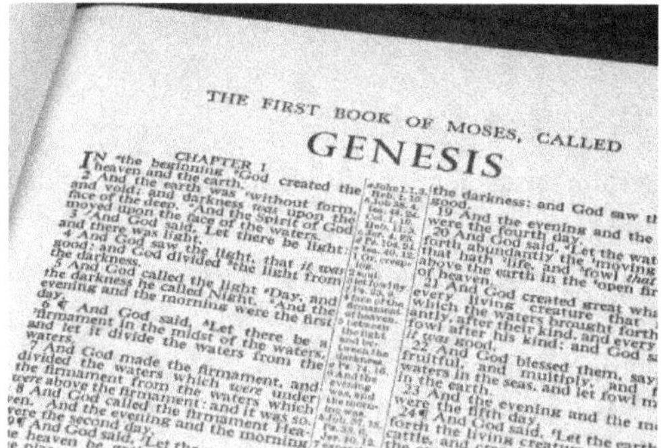

**Fig.3.25:** *In the Beginning God Created – Genesis 1:1*

Yet others may say God employed evolution to brand everything, 'theistic evolution,' and that Genesis has no significance to understanding the history of the universe; it is some sort of 'myth.' Science expresses us when and how the universe came into being; the Bible tells us why. They are two isolated domains of knowledge.

The above interpretations lean towards overlapping in an ambiguous way in the minds of many who have not thought rationally about the consequence of these interpretations on the Gospel.

All such 're-interpretations' stem from an effort to harmonize the Bible's Creation/Fall/Flood account (Genesis 1–11) with the claim of modern historical science that the universe is billions of years old. In this interpretation, rocks containing fossils on Earth fashioned over eons of time, mostly before people came to being.

The fossil record, so *interpreted*, expresses of death and suffering on a massive scale—which mostly occurred before people were created, or evolved. Nevertheless, this interpretation has grave consequences for the rest of the Bible, because it:

## 1. Undermines the Goodness of God

Non-Christians object, 'How can you have faith in in a loving God when there is so much suffering in the world?' They cite animal suffering as part of the problem. Conferring to the history in Genesis, God created everything and He pronounced it as 'very good' after he completed creating the first people, Adam and Eve (Genesis 1:31). It was so good that the people and animals were vegetarian (Genesis 1:29–30)—it is difficult to visualize a world like that. It was human sin, insurgence against the Maker and Sustainer of the universe, that carried death and suffering into God's good creation (Genesis 3).

Romans 8:18–25 upholds that the whole creation, not just people, has been *"subjected to futility"* and is now *"groaning"* and *"in bondage to decay,"* waiting for its redemption. Foremost commentators on Romans such as F.F. Bruce, C.E.B. Cranfield, and James Dunn agree that Paul here refers to the Fall. This is in harmony with the real history of Genesis 3, where the creation, not just the people, was cursed on account of the man's sin. For example, the ground would now bring forth thorns and thistles (Genesis 3:18). There are thorns well-preserved in the fossil record, allegedly some 300 million years before man came on the scene. If this is actually so, as the above 're-interpretations' maintain, then the Bible deceives.

In actual fact, we live in a corrupt creation because of man's sin; it was not created this way. Christians have had this interpretation from the beginning. John Milton's classic poems *Paradise Lost* and *Paradise*

*Regained* reflect this Christian worldview that was once acknowledged almost without question. But if God created over billions of years, He is most definitely *not* 'good.' In such an interpretation, He would have sanctioned and overseen death, disease, cruelty, and suffering for billions of years—before sin entered the universe—and entitled his death-ridden creation all 'very good.'

## 2. Undermines the Gospel

The New Testament undoubtedly explains that the reason for Jesus' death and Resurrection depends on the real historical events of Genesis 1–3, that death came in the creation through the sin of the first man:

*"For since by a man came death, by a man also came the resurrection of the dead. For as in Adam all die, so also in Christ all will be made alive."* (1 Corinthians 15:21–22; see also Romans 5:12–21).

Jesus is called *"the last Adam"* (1 Corinthians 15:45) because He came to undo the work of the first Adam. He took upon Himself, in His body on the Cross, the Curse of death for the lost race of Adam (Galatians 3:13; Colossians 1:22).

Plainly, the teaching about the cause for Jesus' death is contingent upon the proceedings in Genesis being factual: that physical death originated with Adam's sin and that it was not already a part of the created order. Those who undervalue the history of Genesis often claim that Adam's death was only 'spiritual,' separation from God. But it was also bodily death: *"from dust you came and to dust you will return"* (Genesis 3:19). Thus, Jesus also died a bodily death on the Cross. He also rose from the dead, bodily, victorious, having dealt with the Curse of death that came through Adam.

If death was permanently a part of 'creation,' how can it be *"the last enemy"* (1 Corinthians 15:26) and why did Jesus have to die?

## 3. Undermines Eschatology – End-Times Doctrines

The Bible voices of a future where the current order will be demolished and God will make a new heavens and earth where there will be no more suffering and pain—the former things will have passed away (2 Peter 3:10–13; Revelation 21:4–5). Nonetheless, if God 'created' things much as we see them, with death and suffering inherent to the created order, which the previously mentioned interpretations of Genesis supposing, why would God want to abolish the current order and create a new one?

Why does Revelation parallel the removal of the Genesis Curse with the removal of death and pain (Revelation 21:4, 22:2), if the Curse did not carry those things into the world in the first place? It does not make sense.

It also weakens the teaching about the future restoration (Romans 8:21, Acts 3:21)—'restoration' means coming back to a former state, so are Christians supposed to be reinvigorated by a return to millions of years of death and suffering?

## 4. Undermines Hermeneutics – How We Understand the Bible

If Genesis cannot be understood as history, as it is meant to be, as we will demonstrate, then how should we comprehend the rest of the Bible? Maybe the account of the Exodus or the Exile in Babylon did not essentially happen, it is the same form of literature; maybe these literatures are just theological arguments, the framework idea? May be the accounts in the New Testament of Jesus' teaching, death and Resurrection are not really history, although it appears like they are?

Additionally, any interpretation that separates Genesis from history:

**a.  Undermines Confidence in the Rest of the Bible**

If Genesis cannot be understood as straight-forward history, where does history begin? Numerous accept that Abraham (Genesis 12) was a real person, but refer to some of his ancestors as metaphors, especially

Adam. But Jesus' genealogy goes back to Adam (Luke 3)—so where do metaphors begin and end?

Jess took Genesis as history. Was the Son of God mistaken? 'Darwin's Bulldog', Thomas Huxley, put his finger on the problem when he remarked long ago,

"I soon lose my way when I try to follow those who walk delicately among 'types' and allegories. A particular passion for unambiguousness forces me to query, bluntly, whether the writer means to say that Jesus did not trust the stories in question, or that he did? When Jesus spoke, as of a matter of fact, that 'the Flood came and destroyed them all', did he trust that the Deluge really took place, or not?"

**b.    Undermines other Doctrines that are Based on Genesis**

For example, doctrines relating to marriage, moral law, the wearing of clothing, and the meaning and purpose of our presence are all based on the history of events in Genesis.

## Why not Believe they are Ordinary days?

Many theologians acknowledge that Genesis appears like straightforward history, but then again, do not believe it. Why? The following characterizes the rational:

"It is apparent that the most straightforward interpretation of the Genesis record, *without regard to all of the hermeneutical considerations suggested by science,* is that God created heaven and earth in six solar days, that "man" was created in the sixth day, that death and chaos came in the world after the Fall of Adam and Eve, that all of the fossils were the result of the disastrous universal flood which secured only Noah's family and the animals therewith."

"Without regard to all the hermeneutical considerations suggested by science," Genesis is a straightforward historical account of real events.

In other words, for many theologians, 'science' is the expert witness, not the Word of God. We submit that this assurance in 'science' to be able to command a 're-interpreting' of Genesis is inappropriate. The speculations of 'historical science', or origins science, provide no stable foundation for anything, let alone meddlesome with the Word of the infinite eternal God who knows everything, see 'Is it science?'

Indeed, the broadly valued systematic theologian Louis Berkhof attested that, conflicting to historical science interpreting Genesis, we require the Bible to understand natural history:

"Originally God revealed Himself in creation, but through the disfigurement of sin that original revelation was masked. Moreover, it was wholly inadequate in the condition of things that obtained after the fall. Only God's self-revelation in the Bible can now be well thought-out adequate. It only conveys a knowledge of God that is pure, that is, free from error and superstition, and that answers to the spiritual needs of fallen man … Some are inclined to speak of God's general revelation as a subsequent source; but this is barely correct in view of the fact that nature can come into consideration here only as interpreted in the light of Scripture."

This appropriately states a major opposition to those who argue that nature is the 67th book of the Bible and who utilize that 'book', as interpreted by the majority of scientists, to, in turn, interpret the days of creation as long periods of time.

## How has Genesis been Understood in the Past?

There are two explanations for looking at the history of how Genesis has been interpreted:

1.    Generally: If long-age interpretations had continuously been prevalent, then a case could be made for assuming that the Bible hints at this. But if they were absent until they became popular in 'science,' it's more likely that such interpretations were driven by trying to reconcile the Bible with 'science.' The author believes that science is degenerate if it is not in harmony with the Bible!

2. Unambiguously, for those who promote 'deep time' within the church: In order to overcome the charge that they are driven by 'science' and not the biblical text, they often *claim that interpreters throughout history have allowed for long creation days.* Therefore, it's important to examine the evidence for this claim.

## The Church Ancestors

Basil the Great (AD 329–379), in a series of sermons on the six days of creation, the *Hexaemeron*, argued that the pure meaning was envisioned: the days were ordinary days; God's commands instantaneously filled Earth with shrubbery, caused trees to shoot up and swiftly filled rivers with fish; that animals did not originally eat each other; that the sun was created after the earth; etc. He also spoke in contradiction of evolutionary ideas of humans springing from animals. Observe that Darwin did not discover evolution; such concepts go back to anti-theistic philosophers before Christ—such as Anaximander, Epermeniids and Lucretius. It has been a pagan, anti-God idea from its earliest origins.

Some have misinterpreted the Church Fathers' positions because they have not read them wisely. It was normal in the Eastern Orthodox Church (EO) to understand the Creation Week as factual, but they often, in parallel, saw it as typologically pointing to a total Earth history of seven thousand years until the end. They most certainly did not favor the days of Creation Week as long periods of time.

The late Seraphim Rose, an EO priest, methodically documented the interpretations of the Church Fathers of the EO church, illustrating that they regarded Genesis the way modern creationists do. Terry Mortenson, who earned a Ph.D. in the history of geology, studied the book:

"His [Rose's] primary sources are first 'Fathers' who wrote explanations on Genesis: John Chrysostom (344–407), Ephraim the Syrian (306–372), Basil the Great (329–379) and Ambrose of Milan (339–397). But he also employed many other 'Fathers' of that and later centuries who wrote on some aspect of Genesis 1–11."

Rose presented how the EO Church Fathers were undivided in their interpretation of the historicity of Creation Week, the Fall and the global Flood. They also trusted that God's creative acts were *instantaneous.* They saw the pre-Fall world as essentially and overwhelmingly dissimilar from the post-Fall one of today.

Some quote "Augustine and Origen" to validate the smuggling of 'deep time' into the Bible. These two gentlemen, being of the Alexandrian School, inclined to allegorize many passages of Scripture. Their illegalization of the days of creation did not arise from within the text, but from external influences, namely their adherence to neo-Platonic philosophy, whereby

*Fig.3.26: John Calvin*

they 'reasoned' that God would not smear himself with being bound by time constraints, etc.,. Nonetheless, conflicting to the positions of those who would utilize "Augustine and Origen" to prop up their own 'deep time' accommodation, both said that God created everything in an instantaneous, *not* over long epochs of time. And they *explicitly* argued for the biblical timeframe of thousands of years, as well as the global Flood of Noah.

Now, some may argue that the Church Fathers blundered in their interpretation, that we now have greater knowledge. But modern academics are not the first who have recognized about the original languages and cultures of the Bible. The burden is on those proposing a new interpretation to evidence their case.

## The Reformers

Calvin said, Figure 3.26; "The day-night cycle was established from Day 1, before the sun was created, [commenting on "*let there be light*"]" and "Here the error of those is clearly contested, who uphold that the world was made in a moment, [almost certainly referring to Augustine and Origen]. For it is too violent a cavil to contend that Moses distributes the work which God perfected at once into six days, for the mere purpose of conveying instruction, [foreshadowing the framework idea?].

Let us rather settle that God himself took the space of six days, for the purpose of accommodating his works to the capacity of men." And, "They will not refrain from guffaws when they are informed that but little more than five thousand years have passed since the creation of the universe." And, "And the flood was forty days, &c. Moses copiously insists on this fact, in order to show that the whole world was immersed in the waters."

Luther, Figure 3.27 wrote even more plainly of these subjects, undoubtedly affirming his reception of the historicity of Genesis. He also dealt with skeptics' claims of hypothetical inconsistencies between Genesis 1 and 2.

**Fig.3.27**: *Martin Luther*

Challengers of the historicity of Genesis love to refer to Ronald Numbers' book *The Creationists*. Numbers apparently illustrated that young-earth 'creationism' was invented by a Seventh-day Adventist, George McCready Price, in the 1920s. This has to be one of the most farfetched examples of historical revisionism, on par with the myth that the ancients in general, and the church in particular, held to a flat Earth, which was totally demolished by historian Jeffrey Burton Russell. It is as if Numbers, a historian, knows nothing of history before Price. The above material on the Church Fathers and reformers is adequate to illustrate the mistake of Numbers' work. But there is much more that disproves it. See the research of the earth science historian Terry Mortenson on the geologists of the early 1800s who safeguarded the biblical age of Earth and the global Flood of Genesis.

Why *must* they be Ordinary Days?

## 1. Genesis was Written as History, not Poetry

Hebrew has distinct grammatical forms for recording history, and Genesis 1–11 utilizes those. It has the equivalent structure as Genesis 12 forwards and most of Exodus, Joshua, Judges, etc. It is *not* poetry or allegory.

Genesis is sprinkled with "*And … and … and …* ", which typifies historical writing, this is technically called the *vav* (ו), often rendered as *waw*, consecutive.

The Hebrew verb forms of Genesis 1 have a precise feature that fits precisely what the Hebrews utilized for recording history or a sequence of past events. That is, only the first verb is perfect (*qatal*), while the verbs that continue the narrative are recognized as *vav* (or *waw*) consecutives (*vayyiqtols*). In Genesis 1, the first verb, *bara* (create), is perfect, while the following verbs that move the narrative forward are imperfect. An appropriate translation in English distinguishes this Hebrew form and translates all the verbs as past tense.

Genesis 1–11 also has more than a few other hallmarks of historical narrative, such as 'accusative particles' that mark the objects of verbs. Terms are often prudently well-defined. Also, parallelisms, a feature of Hebrew poetry, e.g., in many Psalms, are almost absent in Genesis.

The rare pieces of poetry, e.g., Genesis 1:27 and 2:23 remark on actual events nonetheless, as do many of the Psalms, e.g., Psalm 78. Even if Genesis were poetic, it would not essentially make it non-historical.

The sturdiest *structural* parallel of Genesis 1 is Numbers 7:10–84. Both are arranged accounts, both comprise the Hebrew word for day (יוֹם *yôm*) with a numeric—indeed both are numbered *sequences* of days. In Numbers 7, each of the 12 tribes carried an offering on the different days:

"The one who brought his offering on the first day was Nahshon, son of Amminadab of the tribe of Judah. …

On the second day Nethanel son of Zuar, the leader of Issachar, brought his offering …

On the third day, Eliab son of Helon, the leader of the people of Zebulun, brought his offering. …

On the twelfth day Ahira son of Enan, the leader of the people of Naphtali, brought his offering. … "

The parallel is even stronger when we note that Numbers 7 not only has each day (יוֹם *yôm*) numbered, but also opens and closes with "*in the day that*" to refer collectively to all the ordinary days of the sequence. Notwithstanding the utilization of "*in the day that*" in verses 10 and 84, no-one doubts that the numbered day sequence in Numbers 7 (verses 12, 18, 24, 30, 36, 42, 48, 54, 60, 66, 72, 78) contains anything but ordinary-length days, because these days lack a preposition like 'in'. This refutes the claim that "*in the day that*" (יוֹם *yôm*) in Genesis 2:4, summarizing Creation Week, illustrates that the Genesis 1 days are *not* normal-length. This is simply a Hebrew idiom for 'when', see NASB, NIV Genesis 2:4 cf. Numbers 7:10, 84.

In this organized narrative, Numbers 7, with a sequence of numbered days, no one claims that it is simply a poetic framework for teaching something theological and that it is not history. No-one doubts that the days in Numbers 7 are ordinary days, so there is no grammatical basis for denying the same for the Genesis 1 days. That is, Genesis 1 is straightforward history.

Hebrew academics harmonize that Genesis was written as history. For example, the Oxford Hebrew scholar James Barr wrote:

" … probably, so far as I know, there is no professor of Hebrew or Old Testament at any world-class university who does not believe that the writer(s) of Gen. 1–11 envisioned to convey to their readers the ideas that is creation took place in a sequence of six days which were the same as the days of 24 hours we now experience the figures contained in the Genesis genealogies given by

i.  Simple addition a chronology from the beginning of the world up to later stages in the biblical story
ii. Noah's flood was understood to be world-wide and extinguish all human and animal life except for those in the ark."

Barr, consistent with his neo-orthodox views, does not *believe* Genesis, but he understands what the Hebrew writer clearly envisioned to convey. Some criticize our use of the Barr quote, because he does not believe in the historicity of Genesis. But that is precisely why we use his statement: he is a *hostile witness*. With no need to try to harmonize Genesis with anything, because he does not see it as carrying any authority, Barr is free to state the clear intention of the author. This differences with some 'evangelical' theologians who attempt to hold some sense of authority without essentially believing it says much, if anything, about history—'wrestling with the text', we've heard it called.

Hebrew scholar Dr Steven Boyd has illustrated, employing a statistical comparison of verb type frequencies of historical and poetic Hebrew texts, that Genesis 1 is clearly historical narrative, not 'poetry'. He concluded, "There is only one acceptable view of its plain sense: God created everything in six literal days."

Some other Hebrew scholars who support literal creation days include:

- Dr Andrew Steinmann, Associate Professor of Theology and Hebrew at Concordia University in Illinois.
- Dr Robert McCabe, Professor of Old Testament at Detroit Baptist Theological Seminary in Allen Park, MI.
- Dr Ting Wang, lecturer in biblical Hebrew at Stanford University.

## 2. **The Use of Day' in** Genesis 1 **Compared to another Hebrew Scripture**

A straightforward value of understanding a Bible context is the comparability to the use of words and phrases with other parts of the Bible.

How is the word 'day' used in Genesis 1? This is the passage of utilization of 'day', as accurately as possible, as per the *New American Standard Bible* here:

"And God called the light day and the darkness he called night. And there was evening and there was morning, one day … and there was evening and there was morning, a second day … a third day … a fourth day … a fifth day … the sixth day."

It is substantial that the standard Hebrew lexicon designates 'day' in Gen. 1:5 as a 'day of twenty-four hours'. This 'day' is distinct by an evening and a morning cycle; night and day, as well as a number. There should be no necessity to go furthermore—it is as pure as day what 'day' means in Genesis 1! As the nineteenth-century liberal, Professor Marcus Dods, New College, Edinburgh, said:

"… if, for example, the word 'day' in these chapters does not mean a period of twenty-four hours, the interpretation of Scripture is hopeless."

Note that *'day'* is used utilized a number in Genesis 1. It is used as a singular or plural with a number 410 times outside of Genesis 1 and it continuously means an normal day.

*'Evening'* and *'morning'* are used together without *'day'* 38 times outside Genesis 1 and it continuously designates an ordinary day. *'Evening'* or *'morning'* are used 23 times each with *'day'* outside Genesis 1 and it always means an regular day. And *'night'* is used with *'day'* 52 times and it always specifies an regular day.

Scripture and logic dictate that we have no choice but to comprehend *'day'* in Genesis 1 as an 'ordinary' day.

**The Hebrew Word for 'Day', Yôm, is Used in Several Ways in Genesis 1 that Show that the Days were Ordinary Days.**

## 3. **Creation Week is the Basis of the 7-Day Week**

Exodus 20:11 recapitulates the Creation Week. It removes any likelihood of an lengthy timescale by *any* explanatory scheme, outline hypothesis, day-age impression, all gap theories, God's-days-not-our-days, days of revelation, etc., since it is specified as the basis for our seven-day week with a day of rest (v. 10):

"For in six days the Lord made heaven and earth, the sea, and all that is in them, and rested the seventh day. Therefore, the Lord blessed the Sabbath day and made it holy."

Exodus 20:1: *"And God spoke all these words, saying, …."* These are the very words of God himself, not the thoughts of Moses, or some reviser or even one of the fantasy scribes, who supposedly lived a millennium after the event, long discredited nonsense taught, sadly, at many theological institutions.

God took six days to make *everything*—there is nothing other than the "*heavens and earth, the seas and all that is in them*". This is an all-inclusive statement that highlights wholeness. 'God made the universe' would be an suitable paraphrase. Then God ceased from his work on the seventh day, the day of 'rest'. God did not need six days to make everything and He did not need to rest (Isaiah 40:28), nonetheless, He did it in this method and timeframe as a pattern for our week. That's where our 7-day week came from.

This Scripture alone counters all efforts to stretch the timeframe for the universe's existence.

Other arguments used against six days:

## 1. Sometimes 'day' can Mean other than an Ordinary Day

No one denies that 'day' can have several meanings, as it does in English, but the context of a numbered day with an evening and a morning defines the days in Genesis 1 as normal days. *"In the day that …"* in Genesis 2:4 is a Hebrew expression for 'when', as clarified earlier, and it does not have a number *or* evening *or* morning to describe it as an normal day.

Consider the following English sentence:

In my Father's Day, he would go to bed early Sunday evening and rise early in the morning of the next day, and devote the next six days travelling, during the day, to cross the whole country. The numerous meanings of 'day/days' are quite clear from the context and other languages are similar, including Hebrew. Indeed, all the meanings of day/days in this sentence are found in Genesis 1, 2 and Exodus 20:11.

Some cite *"with the Lord, a day is as a thousand years"* (2 Peter 3:8) to make each of the days of creation a thousand years long (or longer). This is a misapplication of Scripture. Observe that the Bible **compares** the thousand years with a day, it is *as* or *like* a day, not that it *is* a day.

The Bible teaches us here merely that what might appear like a long time to us waiting for the second coming of Christ is nothing to the eternal God—He is patient, waiting for people to repent of their sin. This has nothing to do with the connotation of 'day' in Genesis 1. Actually, the figure of speech is so effective precisely because the day is *literal* and contrasts so intensely with 1,000 years—to the eternal Creator of time, a short period of time and a long period of time may as well be the same.

A similar passage in Psalm 90:4 equivalences a thousand years to a watch in the night, three or four hours, in God's sight, yet no one claims that the night watch could last a thousand years! This passage again underscores that Scripture here contrasts God's eternal view with our time-based one. As the valued commentator "John Gill" said, "the words aptly express the disproportion there is between the eternal God and mortal man." They have nothing to do with the meaning of 'day' in Genesis 1.

## 2. Genesis 1 and 2 are Contradictory Accounts of Creation, So, Why Should We Believe Genesis 1 as History?

Genesis Chapters 1 and 2 are not dissimilar accounts of creation and they are not inconsistent. Genesis 1 in agreements with the creation of everything, the universe, the 'big picture' (see Gen. 1:31–2:4a). Genesis 2 recaps the creation of the man and woman, providing details not tailored in the first chapter and mainly their condition in the distinct garden God equipped for them. Chapter 2 is *not* alternative creation account: there is no mention of the creation of the earth, sun, moon, stars, seas, land, sky, sea beings, creeping things, etc.

Some refer to an apparent difference in order of creation between Chapters 1 and 2, claiming a problem with the plants and herbs in Genesis 2:5 and the trees in Genesis 2:9, which in some English translations appear as though they came into being *after* Adam, allegedly opposing the order in Genesis 1, plants on Day 3, people on Day 6.

Nonetheless, Genesis 2 emphases on subjects of direct importance to Adam and Eve and the garden, not creation over-all. Observe that the plants and herbs are designated as 'of the field' in Chapter 2, compare 1:12, and they required a man to tend them, (2:5). These are evidently cultivated plants, not plants in general. Also, the trees (2:9) are only the trees planted in the garden, not trees in general. These proceedings relay to God creating the garden, not creation in general.

The reference of the forming of the 'beasts of the field' and 'birds of the air' in Genesis 2:19, before the creation of Eve, is also allegedly a problem.

The supposed inconsistencies fall away when we understand that Hebrew has no precise verb form to indicate the pluperfect, 'had formed', 'having formed'. A number of Hebrew academics and analysts, such as Keil & Delitzsch and Leupold, have known that the context of **Genesis 2** suggests the pluperfect tense for these proceedings—they are being narrated for the purposes of Chapter 2. For example:

"Now out of the ground the Lord God had formed every beast of the field" (2:19, ESV). Such a translation, which is lawful, eliminates any clue of inconsistency.

There is no requirement to conclude that Genesis 2 denies Genesis 1 and so this is not a lawful argument against taking **Genesis 1** as straightforward history.

**Genesis 2** is not a ***different*** Account of Creation—it is a ***more detailed*** Account of the Sixth Day of Creation.

## 3. Adam could not have named all the animals in one day (Day 6)

Adam did not name every *species* of living thing on Earth today, which *would* be rather hard—he only had to name the animals that God brought to him. The animals named were "*the cattle, the birds of the sky, and every beast of the field*" (**Genesis 2:20**)—the creatures pertinent to man's macro-environment. The sea creatures and "*everything that creeps upon the earth*" were not encompassed. Similarly, even within the named set, there would not have been hundreds of *species* of parrots to name, but maybe only a single parrot *kind*, or a few, for example. God seemingly gave Adam the naming exercise as an act of sovereignty, Adam was to rule—**Genesis 1:28**—and naming something is an exercise of sovereignty. The naming also accentuated to Adam that he was missing something: a mate. Eve was then created, with Adam being most grateful!

We require to remember that Adam was created perfect, with language, and would have had no trouble in his unfallen state in naming this subcategory of creatures in a few hours.

## 4. The Sun was not Created Until "Day 4," So, how could the First Three Days have been Ordinary Days?

The creation of light before the sun was renowned by early Church Fathers and the later Reformers deprived of any problem, nonetheless, some advance it today as if creationists had never thought of it. E.g., in AD 180, Theophilus of Antioch noted that it made nonsense of sun-worship since God made the plants before the sun, and Basil said the same.

The most straightforward characterization of a day is 'the time for Earth to make a complete rotation on its axis'. All we require for a day is Earth to be rotating. To establish the day with evening and morning, we then require a directional source of light so that the rotating Earth causes the night and day cycle that is pronounced for each day in Genesis 1. The Bible says that in the latter part of the first day, subsequent the period of darkness (Genesis 1:1–2) "*God said, 'Let there be light' and there was light*" (v. 3). So, we have a source of light and a rotating Earth and we have days happening. The source of light was the illuminating light of the Glory of the Infinite God:

"*and there was evening and there was morning, one day.*"

Those who would claim that the first days had to be a different length have to presume that God altered the speed of rotation of Earth on its axis, when he created the *greater light* as the light bearer (Genesis 1:14), which is hardly likely.

Scripture gives no hint that the days were any different: the same formula applies for Days 2 and 3 as for Days 4 and 5 (*there was evening and there was morning, a second/third/fourth/fifth day*).

## 5.  The Seventh Day has not Finished, so the other Days Could be Long Periods of Time

Some claim that because the seventh day (Genesis 2:2–3) did not have the 'evening' and 'morning' demarcation, it must still be continuing; it is a long period of time, so we can regard the other days as long periods also.

Meanwhile, there was no eighth day of creation, there was no need for an evening and morning to mark off the seventh day from the eighth. Also, evening and morning manifested the beginning and end of a day, so, if their absence means that the seventh day has not finished, then it has not begun either.

This hollow argument is frequently tied to the claim that Hebrews 4 says that the seventh day of creation is a long period of time, so the other days could be also. Here is the argument:

"According to this passage [Hebrews 4:4–11], the seventh day of the creation week carries on through the centuries … the seventh day of Genesis 1 and 2 represents a minimum of several thousand years and a maximum that is open ended (but finite). It seems rational to settle then, specified the parallelism of the Genesis creation account, that the first six days may also have been long time periods."

Nonetheless, Hebrews 4 does *not* state that the seventh day of creation is continuing to the present; it only says God's *rest* is continuing. If someone says on Monday that he rested on Saturday and he is still resting, it would not mean that Saturday has continued through to Monday.

Furthermore, the rest is for those who are in Christ (Hebrews 4:9–11), those who are in the Kingdom of God. In other words, it is a spiritual rest. If the rest being alluded to were a continuation of the seventh day of Creation Week, then *everyone* would be in this rest.

This argument also founders on the rock of Exodus 20:10–11, written by God Himself, where God's seventh day of rest is specified as the foundation for the Sabbath rest commandment, making it plane that God's Day of rest, the seventh day, was a day like the other six days of the Creation Week. It would be a odd *week* where the seventh day had not finished yet.

## 6.  Genesis is Poetry/Figurative, a Theological Argument – (Polemic) and so is not History (the Framework Hypothesis)

This is the foundation of the 'framework hypothesis,' perhaps the favorite view among seminaries that say they receive biblical authority but not six normal days of creation.

It is eccentric, if the literary framework were the true sense of the text, that no-one interpreted Genesis this way until "Arie Noordtzij" in 1924. Actually, it's not so odd, because the leading framework advocates, "Meredith Kline" and "Henri Blocher," self-confessed that their foundation for a bizarre, novel interpretation was a desperation to fit the Bible into the unproven 'facts' of science, which no Bible scholar had believed of until the 20th century.

For example, Kline acknowledged in his main framework article, "To refute the literalist interpretation of the Genesis creation 'week' advocated by the young-earth theorists is a central concern of this article." And Blocher said, "This hypothesis disables a number of problems that inundated the commentators [including] the confrontation with the scientific vision of the most distant past." And he additionally acknowledges that he discards the plain teaching of Scripture because, "The rejection of all the theories accepted by the scientists necessitates substantial bravado." Obviously, the framework awareness did not come from attempting to understand Genesis, but from attempting to counter the view, held by scholar and layman alike for 2,000 years, that Genesis records real events in real space and time.

**i.    Are the Genesis 1 Days Real History?**

However, as revealed above, Genesis is, without any doubt at all, most certainly written as historical narrative. Supporters argue that because Genesis 2 is (they say) organized topically rather than chronologically, so is Genesis 1. Therefore, the days are 'figurative' rather than real days. But this is like arguing that because the Gospel of Matthew is arranged topically, then the Gospel of Luke is not arranged chronologically. And, as we have pointed out above (point 2), it is rational, and in agreement with ancient near eastern literary practice, to have a historical overview (Genesis Chapter 1) preceding a recap of the details (Genesis Chapter 2) about certain proceedings previously stated. Chapter 2 does not have the numbered sequence of days that Chapter 1 has, so how can it determine how we view Chapter 1?

**ii.   Are There Triads of Days?**

One of the supposed main 'evidences' for a poetic structure is an unproven two triads of days. In this opinion, Moses organized the days in a very stylized framework with "Days 4–6" paralleling "Days 1–3." Kline suggests that "Days 1–3" refer to the Kingdom, and "Days 4–6" to the Rulers, as per the following table 3.1:

**Table 3.1 :** A Framework Idea, which Fails Scrutiny

| Days of Kingdom | | Days of Rulers | |
|---|---|---|---|
| Day 1 | Light of God and darkness | Day 4 | Sun, moon, and stars (luminaries) |
| Day 2 | Sky and waters separated | Day 5 | Fish and birds |
| Day 3 | Dry land and seas separated, plants and trees | Day 6 | Animals and man |

However, even if this is factual, it would not rule out a historical sequence—confidently God is proficient of creating in a firm order to teach certain truths. Similarly, other theologians argue that the 'literary devices' are more in the fancy of the advocates than the text. For example, the counterparts of these two triads of days are massively overdrawn. Methodical theologian Dr Wayne Grudem summarizes:

"First, the wished-for correspondence between the days of creation is not nearly as meticulous as its supporters have hypothesized. The sun, moon, and stars created on the fourth day as 'lights in the firmament of the heavens' (Gen.1:14) are located not in any space created on Day 1 but in the 'firmament'… that was created on the second day. In fact, the correspondence in language is quite explicit: this 'firmament' is not mentioned at all on Day 1 but five times on Day 2 (Gen.1:6–8) and three times on Day 4 (Gen.1:14–19). Obviously, Day 4 also has correspondences with Day 1, in terms of day and night, light and darkness, but if we say that the second three days illustrate the creation of things to fill the forms or spaces created on the first three days, or to rule the kingdoms as Kline says, then Day 4 overlaps at the minimum as much with Day 2 as it does with Day 1.

"Furthermore, the equivalent between Days 2 and 5 is not particular, because in some ways the groundwork of a space for the fish and birds of Day 5 does not come in Day 2 but in Day 3. It is not until Day 3 that God gathers the waters together and calls them 'seas' (Gen.1:10), and on Day 5 the fish are commanded to 'fill the waters in the seas' (Gen.1:22). Again, in verses 26 and 28 the fish are called 'fish of the sea', giving repetitive emphasis to the fact that the sphere the fish inhabit was specifically formed on Day 3.

Thus, the fish formed on Day 5 appear to fit in much more to the place prepared for them on Day 3 than to the broadly dispersed waters below the firmament on Day 2. Founding a equivalent between Day 2 and Day 5 faces additional difficulties in that nothing is created on Day 5 to inhabit the 'waters above the firmament', and the flying things created on this day (the Hebrew word would include flying insects as well as birds) not only fly in the sky created on Day 2, but also live and multiply on the 'earth' or 'dry land' created on Day 3. (Note God's command on Day 5: 'Let birds multiply on the earth' [Gen.1:22].)

"Lastly, the corresponding between Days 3 and 6 is not precise, for nothing is created on Day 6 to fill the seas that were gathered together on Day 3. With all of these points of imprecise correspondence and overlapping between places and things created to fill them, the hypothetical literary 'framework,' while having an initial advent of neatness, turns out to be less and less convincing upon closer reading of the text."

iii. **Genesis 2:5 Teaches that Normal Providence was Used?**

Another key disagreement by framework advocates is founded on Genesis 2:5. Kline rightly states that God did not make plants before Earth had rain or a man, although this is talking about *cultivated* plants not all plants. So, Kline asks, what's to stop God making them anyway because He could miraculously sustain them? The answer, according to Kline, is that God was working by ordinary providence:

"The unargued assumption of Gen. 2:5 is clearly that the heavenly wisdom was operating during the creation period through processes which any reader would identify as ordinary in the natural world of his day."

Observe that Kline acknowledges that this supposed assumption is *not* argued in the text. This would clarify why no exegete saw this for thousands of years. Then he brands another astonishing jump to say that there was normal providence operating throughout Creation Week:

"Embedded in Genesis 2:5 is the principle that the *modus operandi* of the divine providence was the same during the creation period as that of normal providence at the present time."

But this is desperation. Even if normal providence were functioning, it would not trail that miracles were *not*. Actually, there is no miracle in the Bible that does *not* operate in the midst of normal providence. "Michael Horton" points out that those who reject God acting in the normal course of events do it from an *a priori* philosophical supposition and not from anything in the text.

A miracle is properly understood not as a 'violation' of providence but an *addition*. Therefore, when Jesus turned water into wine (John 2), the other features of 'providence' were still functioning. Perhaps Jesus created the stunning variety of organic compounds in the water to make the wine, but gravity still held the liquid in the barrels, taste buds were still working in the guests, their hearts pumped blood without skipping a beat, etc.

Ironically, if we shoulder the evolutionary durations that Kline's notion is meant to accommodate, Genesis 2:5 actually contends *against* normal providence. In the evolutionary scenario, there are billions of years between the arrival of the oceans and the first plants on land.

Observe that the verse designates that the reason why *"no plant of the field had yet sprung up"* was that *"God had not caused it to rain upon the earth,"* i.e., there had not been any rain prior to the creation of land plants. Provided the normal providential operation of evaporation and precipitation, etc., how could there have been no rainfall on the earth in all that vast stretch of time? Such would have been hugely miraculous!

So, in conclusion, Kline incorrectly presupposes normal providence as God's sole *modus operandi* for Genesis 2:5, wildly extrapolates it to the entire Creation Week, and further assumes that normal providence eliminates miracles. This error is intensified by failing to note the thin focus of Genesis 2 on man in the Garden.

**iv.    Is Genesis Merely a Theological Argument - Polemic?**

Though Genesis 1 surely disproves many errant concepts about God, it disproves those concepts precisely because of the real events. For example, it has an implied argument against sun worship *because* God essentially created light (Day 1), before He created the sun (Day 4). The contention depends on the historicity of the events.  God did not create, He said "Let there be light.  God is light.

Is Genesis 1 an argument for the Sabbath? Exodus 20:10–11, which clearly teaches the Sabbath commandment, cites the historical events of Genesis 1 as the basis for the commandment. That is, the works of God recorded in Genesis presage the commandment. The *history* forms the basis of the commandment.

The writings of the framework supporters are marked by lack of clarity. Take a statement by Blocher, for example: 'It [the framework idea] identifies normal days but takes them in the context of one large figurative whole. Nonetheless, cutting through the verbal fog, what they really mean is that they contradict that the days happened in real space-time history.

About the only thing that provides any logical consistency to their views is a clear opposition to the calendar-day understanding of Genesis.

## 7.  God's Days Not Our Days?

A few have argued that the days of Genesis 1 are 'God's days' and so we should not worry about taking it literally (i.e., as history).

This idea, which sounds sketchily pious, if applied consistently, would make comprehending any of the Bible an intolerable mission. The Holy Spirit of God moved the Bible's words so that we offspring of Adam could comprehend the things that God would have us know, about salvation, etc.,. That means that the words carry God's thoughts *to us*. If any words have connotations that only God comprehends, then what is the idea of having them in the Bible? Perhaps 'murder' or 'adultery' are 'God words' that do not mean what we comprehend them to mean—clearly a outrageous idea.

In any case, meanwhile God is eternal and is outside of time, as we have discussed earlier, what would 'God's Day' be; what would it mean? God does not have days and years (see the earlier discussion of 2 Peter 3:8).

## 8. Days of Revelation

Yet another effort to get away from the basic, envisioned meaning of Genesis 1 is to claim that the days stated were the days during which God revealed the creation account to Moses, or someone else. Nonetheless, nowhere does the text give any hint that God is *revealing* things on the days. Advocates of this view attempt to argue that the Hebrew interpreted as 'made' (*asah*) can mean 'revealed' or 'showed'. The Hebrew obviously says that God created (Hebrew *bara*) or made (*asah*) things, not that He revealed them. *Asah* has a broader meaning than *bara*, covering 'to make, manufacture, produce, do' etc., but not 'to show' in the sense of reveal. Where *asah* is translated as 'show'—for example, *"show kindness"* (Gen. 24:12), it is in the sense of 'to do', or 'make', kindness.

Again, Exodus 20:11 highlights that the *whole creation process* happened in the timeframe of an 'ordinary' week, Figure 3.28.

***Fig.3.28**: Creation Week – 6-Days Creation*

# PROBLEMS WITH LONG-AGE INTERPRETATIONS

## 1. The Order of Events

Efforts to stretch the timeframe of Genesis 1 by making the days into eras of Earth history fail to house the millions of years anyway—the order of creation opposes the order appealed by the very same secular historical 'science' that is being accommodated, **Table 3.2.**

**Table 3.2 :** Contradictions between the Order of Creation in the Bible and Evolution/Long Ages

| Bible account of creation | Evolution/long-age speculation |
| --- | --- |
| Earth before the sun and stars | Stars and sun before Earth |
| Earth covered in water initially | Earth a molten blob initially |
| Oceans first, then dry land | Dry land, then the oceans |
| Life first created on the land | Life started in the oceans |
| Plants created before the sun | Plants came long after the sun |
| Fish and birds created together | Fish formed long before birds |
| Land animals created after birds | Land animals before whales |
| Man, and dinosaurs lived together | Dinosaurs died out long before man appeared |

## 2. What Pollinated the Plants?

The plants were created on "Day 3," but the pollinators were not created until "Day 5" or "Day 6." If these days were epochs of hundreds of millions of years or more, what pollinated the plants to ensure their endurance? Some plants have complicated symbiotic relationships with their pollinators—for example, the yucca plant and its moth pollinator.

## 3. Adam's Age

God created Adam on "Day 6." Adam lived through "Day 7" and died at an age of 930 years (Genesis 5:5). If each day were an epoch of time, even, only thousands of years, or the seventh day was still continuing, it would make no sense of Adam's age at death.

This is a Question of Authority:

Is historical 'science' or Scripture the authority?

For those who:

a.  regard Scripture, the Word of God, as the ultimate authority, and

b.  take the historical roots of the Gospel seriously, with the reality of Adam and the Fall affecting the created order, belief in six 'ordinary' days is the only logically consistent position to take.

Efforts to separate Genesis from the real history of the universe end up making Christianity into an 'upper story' insignificance, where 'faith' is realized as little more than a virus of the mind, or an exercise in wishful thinking, like have faith in fairies at the lowermost of the garden.

## YOUNG UNIVERSE

Some scientists and academic institutions believe that frogs have turned into princes—if the process is stretched over hundreds of millions of years.

As far back as documented history goes, there have been those who have worked diligently to shed light on this intricately designed world without a Designer. To escape accountability to a powerful Creator, several types of evolutionary concepts have arisen. These are stories of how the world made itself, of how design, complexity and order have come about without any supernatural cause.

Paul elaborated in Romans 1:20 that a falling man has to work hard to influence himself that there is no Creator. The design in nature speaks volume of the Creator's existence. Mamy people in their heart of hearts, had littel concwpt that God is responsible for the world.

How does one explain a highly intricate and complex universe without a Designer? One way would be to claim that it has at all times existed. However, the earthly laws formalized by the Laws of Thermodynamics, as nothing stays the same forever; everything has to have had a beginning.

Another method a falling man may claim that the universe has just burst into existence by itself, wholly formed, with no cause. Nevertheless, this would be a difficult idea to be recognized, as again both science and common logic illustrate that nature of itself achieves no such miracles.

The understandable clarification is that it was made as an entirely functional universe by an "Uncaused Cause" outside and much greater than the physical universe. — merely as stated in Genesis. Nevertheless, that would be equivalent to accepting the existence of the God of the Bible and His right claim over one's life.

If pne disregards this clarification, one is left with only one more choice — some kind of evolutionary theory which speaks that today's intricate complexity came about progressively from something simple developed to a little less complex, which in turn came developed something further complex, and so on forward. This breaks the Poof of miracle, which nature is being asked to perform, into much more complex steps. People *will* believe that frogs have turned into princes — if the process is stretched over hundreds of millions of years.

## AVOIDING THE TRUE GOD

Needless to say, simple-to-complex theories are the only theories imaginable if you are attempting to clarify the origin of the universe deprived of God. Countless pagan cosmologies have some type of primeval ooze or chaos impulsively giving upsurge to the universe — or to the deities who then shaped the universe. Similarly, modern evolutionary cosmologies have virtually always had the universe beginning off with something categorically simple, comparable to a cloud of hydrogen, or a 'primeval atom' which burst.

This unquestionably begs the question of where this original constituent came from, since it's difficult to comprehend of anything simpler. Once again there are only a limited options. For instance, the primordial atom could have been sitting there for an eternity before it burst. The idea, which is shared among theistic evolutionists, that it was created misses the point.

In the end, the cause why this primeval blip of decisive simplicity has arisen in human thought is because of the longing to disregard a Creator. A Designer capable of creating only an extremely simple preliminary object would hardly seem worth mentioning, anyway.

Nonetheless, in the event you still sought to cling to a Creator for even this simple beginning, a number of theories suggest that this first object just materialized, uncreated. The primeval atom is said to have spontaneously came to the scene out of the vacuum via a "quantum fluctuation." For such very small objects, "quantum mechanics" advocates that typical "logic" may not always apply, so the "Poof" explanation will not appear as preposterous as in the case of the whole universe.

One might reflect that evolutionary story-tellers would sense they have now lastly flourished in removing the Infinite Creator God. The universe has fundamentally been made to appear out of nothing, by itself, and to have created its own order and complex intricacy.

Currently, though, an uncooperative problem has instigated to stand in the way of hard work to eradicate design. It is being more and more acknowledged that the universe is far too unconvincing to have come about in the way earlier believed. If the forces tying subatomic particles collectively were just a fraction different from what they actually are, we would not even have stars. Also, it is conceded that today's universe is in a 'very peculiar state' — it is not 'characteristic' of the way a universe might be predictable to emerge from a "big bang." More and more, we read that it seems to have been designed with just the right characteristics to have "evolved" in the means it supposedly has.

## THE ANTHROPIC – MAN – PRINCIPLE

Consequently, to evade even this hint of design, we now have the "anthropic principle" in cosmology, (anthropos = man). Many have misinterpreted this to mean that cosmologists are now acknowledging that the universe was "made for man." Rather, they are saying "if the characteristics of this universe had not been just precise for man to evolve, there would be no observer to notice this. The fact that we are here to observe it means that it *must* appear to us as if it was designed just right for us to evolve.

### Evading Design

While not typically specified, this position usually encompasses a belief that our universe is only one of numerous universes, each with possibility dissimilar properties. Out of these, the universe in which the properties were just right to evolve planets and people is the universe we happen to be observing.

This is only a minor step from conceivably the definitive in evolutionary fantasy, which is nowadays being seriously proposed by some cosmologists. That is, that our universe is in a way "living" and evolving by natural selection! By devising figures in mathematical abstractions, one can "form" objects dropping into a theoretical black hole thrust through some equally theoretical "space-time warp" and appear as an expanding universe in "some other set of dimensions."

In this manner, the universe is specified to be making "offspring" — heaps of other universes. Professor of physics Lee Smolin, of Pennsylvania State University, contends that every time a novel universe is conceived, the laws of physics in it are a slight different. This is imaginary to be like mutations in biological evolution theory, permitting natural selection to function to select the "fittest" universe.

This is, undeniably, virtuously a system of belief, beyond any conceivable test or trial, and consequently outside of any actual science. In a way, we can feel heartened about the truth of Psalm 19, "The heavens declare the glory of God," when we observe great minds engaging in such strange conjecture to evade the understandable conclusion stated in this Godly verse.

## The Virtues of True Science

*"This most beautiful system of the sun, planets, and comets, could only proceed from the counsel and dominion of an intelligent Being. Figure 3.29"*—Sir Isaac Newton

**Fig.3.29:** *Sir Isaac Newton*

## Faith and Science

Many anti-Christians proclaim that Faith and science have been opponents for centuries. This is the opposite of the truth. Knowledgeable historians of science, counting non-Believers, have pointed out that contemporary science first flourished under a Christian worldview while it was *stillborn* in other cultures such as ancient Greece, China and Arabia.

This should be no wonder when we enquire why science works at all. There are particular indispensable characteristics that make science possible, and they purely did not exist in non-Christian cultures.

1. There is such a thing as *objective truth*. Jesus said, "I am the way, and the truth, and the life. No one comes to the Father except through me" (John 14:6). Nonetheless, postmodernism, for example, repudiates objective truth. One example is, "What's true for you is not true for me." So perhaps they should try jumping off a cliff to verify if the Law of Gravity is true for them. Another postmodern claim is, "There is no truth"—so is *that* statement true? or "We can't know truth"—so how do they *know* that?

2. The universe is *real*, since God created the heavens and the earth (Genesis 1). This sounds understandable, nevertheless, many eastern philosophies trust that everything is an illusion, accordingly, is *that* belief an illusion as well?). There is no point in attempting to explore an illusion by investigating on it.

3. The universe is *orderly*, as God is a God of order, not of confusion— (1 Corinthians 14:33). Nonetheless, if there is no creator, or if Zeus and his gang were in custody, why should there be any order at all? If some Eastern religions were true that the universe is a great "Thought," then it could change its mind any instant.

4. It is not feasible to demonstrate from nature that it is orderly, because the evidences would have to *presuppose* this very order to attempt to prove it. Similarly, in this *fallen* world with natural tragedies and thunderstorms and over-all chaos, it is not so apparent that it was made by an orderly Creator. This is a main message of the book of Ecclesiastes—if we attempt to live our lives only according to what is "under the sun", the result is pointlessness. Henceforth, our principal end is to "Fear God and keep his commandments" (Ecclesiastes 12:13).

   A essential side of science is springing laws that deliver foreseeable results. This is likely only because the universe is orderly.

5. Meanwhile, God is sovereign, He was free to create as He pleased. Therefore, the only means to find out how His creation works is to *investigate* and *experiment*, not depend on man-made philosophies as did the ancient Greeks.

   This is exemplified with Galileo Galilei (1564–1642). He illustrated by experiment that weights fall at the same speed (apart from air resistance), which contested the Greek philosophy that heavy objects fall faster. He also illustrated by *observation* that the sun had spots, disproving the Greek notion that the heavenly bodies are "perfect."

   Another instance is Johannes Kepler (1571–1630), who revealed that planets moved in ellipses around the sun. This disproved the Greek philosophies that contended on circles because they are the most "perfect" shapes; this didn't tie up with the observations, so they added more and more awkward system of circles upon circles called *epicycles*.

*Fig.3.30: Galileo, Johannes Kepler and Isaac Newton – Scientists of Faith*

However, when it comes to *origins* of the universe as opposed to today's processes, God has declared that He created it about 6,000 years ago over six regular-length days – 24 hours a day, and judged the corrupted earth with a globe-covering flood about 4,500 years ago. It's therefore, no accident that Kepler computed a Creation date of 3992 BC, and Isaac Newton (1643–1727), undoubtedly, the supreme scientist of all time, also strongly safeguarded biblical chronology.

6.  Man *can and should investigate* the world, since God provided us *dominion* over His creation (Genesis 1:28); creation is not itself divine. So, we do not need to sacrifice to the forest-god to cut down a tree, or soothe the water spirits to measure its boiling point. Somewhat, as Kepler held, his scientific thoughts were "thinking God's thoughts after Him."

    Numerous other founders of modern science also discovered their scientific research as delivering glory to God. Newton said, Figure 3.29:

    "This most beautiful system of the sun, planets, and comets, could only proceed from the counsel and dominion of an intelligent Being. … This Being governs all things, not as the soul of the world, but as Lord over all; and on account of his dominion, he is wont to be called 'Lord God' Παντοκράτωρ [*Pantokratōr* cf. 2 Corinthians 6:18] or 'Universal Ruler'. … The Supreme God is a Being eternal, infinite, absolutely perfect … ".

    "Opposite to godliness is atheism in profession and idolatry in practice. Atheism is so senseless and odious to mankind that it never had many professors."

7.  Man can *initiate* thoughts and activities; they are not merely the results of deterministic laws of brain chemistry. This is a inference from the biblical instruction that man has both a material and immaterial aspect (e.g. Genesis 35:18, 1 Kings 17:21–22, Matthew 10:28). This immaterial feature of man means that he is above matter, so his thoughts are likewise not bound by the makeup of his brain.

8.  Nevertheless, if materialism were factual, then "thought" is just an epiphenomenon of the brain, and the results of the laws of chemistry. Therefore, *given their own presuppositions*, materialists have not spontaneously arrived at their conclusion that materialism is true, because their conclusion was *predetermined by brain chemistry*. But then, why should their brain chemistry be trusted over mine, since both obey the same infallible laws of chemistry? Consequently, in truth, if materialists were correct, then they cannot even aid what they believe (counting also on their belief in materialism!). Yet they frequently call themselves "freethinkers", overlooking the blatant irony. Genuine initiation of thought is an insurmountable problem for materialism, as is consciousness itself.

9.  Man can reason rationally and logically, and that logic itself is objective. This is a deduction from the fact that he was created in God's image (Genesis 1:26–27), and from the fact that Jesus, the Second Person of the Trinity, is the *logos* (John 1:1–3). This aptitude to think logically has been weakened *but not eliminated* by the Fall of man into sinful rebellion against his Creator. (The Fall means that sometimes the reasoning is imperfect, and sometimes the reasoning is effective but from the wrong premises. Consequently, it is idiocy to elevate man's reasoning above what God has revealed in Scripture.) But then again, if evolution were true, then there would be selection only for *survival advantage*, not essentially for rationality.

10. Outcomes should be described *honestly*, since God has prohibited false witness (Exodus 20:16). Nonetheless, if evolution were true, then why not lie? It is not that astounding that scientific fraud is a cumulative problem.

11. Observe, it's important to comprehend the point here—*not* that atheists cannot be moral but that they have *no objective basis for this morality from within their own system*. The fanatical atheistic evolutionist "Dawkins" admits that our "best impulses have no basis in nature," and another atheistic evolutionist, "William Provine," said: "Naturalistic evolution has indistinct consequences that "Charles Darwin" understood perfectly.

    i.   No gods (worth having) exist;

    ii.    no life after death exists;

   iii.    no ultimate foundation for ethics exists;

   iv.    no ultimate meaning in life exists; and

    v.    human free will is nonexistent."

It is therefore, no accident that science has blossomed since the Reformation, wherever the Bible's authority was re-experienced. And it is no coincidence that the nation with the robust remnants of Bible-based Christian faith, the USA, leads the world by a mile in the output of valuable science.

Though, the Western world is basically living on the capital of its Christian inheritance. But the push to indoctrinate students into evolution, and therefore atheism, at least, for all practical purposes, destabilizes these Christian fundamentals of science (cf. Psalm 11:3). Thus, evolutionary education will not advance science, but extinguish it.

# THE HYPOTHESIS OF MULTIVERSE

## SCIENCE FICTION OR SCIENCE MADNESS

### Non-Existence of Multiverse

Even if certain characteristics of the universe appear to necessitate the presence of a multiverse, absolutely nothing has been directly or indirectly observed or seen to advocate it actually happens. This far, the indication advocating the idea of a multiverse is virtuously theoretical, and in many cases, philosophical.

Multiverse, a hypothetical assortment of potentially varied observable universes, each of which would encompass everything that is empirically available by a connected group of observers. The observable recognized present created universe, which is reachable to telescopes, is about 90 billion light-years across, Figure 4.1.

*Fig.4.1: Hypothetical Multiverse*

## HYPOTHESIS OF MULTIVERSE

Multiverse hypothesis proposes that our present universe, with all its 2-3 trillion of galaxies and almost countless stars, straddling tens of billions of light-years, may not be the solitary one. Instead, there may be a completely dissimilar universe, indistinctly disconnected from ours — and additional, and additional ..

### Still a Hypothesis

As we comprehend the hypothesis thus far, no, which is the main reason that technical arguments of the multiverse grow scorned by numerous scientists. Nonetheless, it is possible one day to be scientifically elevated by fraud scientists. Scientists believed it cannot be tested.

### Travel to Multiverse

*Fig.4.2: Travel between Universes*

If you are mistakenly becoming a believer of a multiple "big bang multiverse," then that would desire you to leave our present universe to travel to another would be absolutely as impossible as travelling backward to the time prior the "big bang" that caused our universe even occurred. According to NASA, time travel may be possible, however, it may not be the same traditional way we know, Figure 4.2. According to Einstein's theory of relativity, which states time and motion are relative to each other, as well as nothing can go faster than the speed of light, which is 186,282 miles per second. Time travel may occur through a term coined "time dilation."

We may not utilize a time machine to travel hundreds of years into the past or future. That type of time travel only occurs in books and movies. Nonetheless, the math of time travel does disturb the things we utilize every day. For example, we use GPS satellites to aid us figure out in what way to travel to new places. However, near a black hole, the deceleration of time is extreme. From the viewpoint of an observer outside the black hole, time stops. For example, an object falling into the hole would seem frozen in time at the edge of the hole.

## The Multiverse Concept

The Multiverse is the assortment of alternative universes that share a universal hierarchy. A great variety of these universes were initiated from another as a result of a main choice on the part of a character. Some multiverse can appear to be taking place in the past or future because of variances in how time passes in each universe, Figure 4.3.

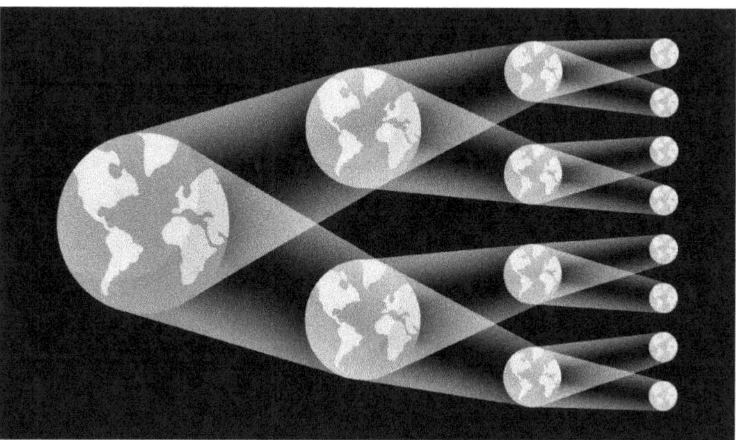

**Fig.4.3**: *Parallel World*

One obvious query that ascends, then, is accurately how many of these parallel universes might there be. In a new study, scientists have computed the number of all conceivable universes, pending with an answer of $10\wedge10\wedge16$. Obviously, this is unbelievable! Scientists claim it takes almost infinite number of parallel Universes to interpretation for all the potentials, but this interpretation is just as binding as any other. There are no experimentations or observations that decree it out. The Many Worlds Interpretation of quantum mechanics embraces that there are a number beyond human perception. Beyond the visible 3D, some scientists accept as true that there may be many more. In actual fact, the theoretical framework of "Superstring Theory" postulates that the universe happens in ten dissimilar dimensions. The world as we recognize it has three dimensions of space—length, width and depth—and one dimension of time. But there's the mind-bending likelihood that many more dimensions exist out there. Consistent with string theory, a model was developed by physicists in the last half century, stated that the universe operates with 10 dimensions, Figure 4.4.

## Levels of Hypothetical Multiverse

Scientists hypothetically assume four levels of the multiverse:

| level I – | Beyond our Cosmic Universe, |
| level II – | Multiverse – Other Post-Inflation Bubbles, |
| level III – | Quantum Many Worlds and |
| level IV – | Other Mathematical Structures |

*Fig.4.4: Parallel Worl in Multi-Dimensions*

Stephen Hawking had never believed in in multiverse. He said "I have never been a fan of the multiverse. If the scale of different universes in the multiverse is large or infinite the theory can't be tested." Hawking and his life-long colleague Hertog stated in their paper, "this account of eternal inflation (multiverse) as a theory of the big bang is wrong."

## Beyond The Universe

The only entity we believe, beyond the universe is the Infinite God who holds it together. by definition, a finite expanding universe raises up the idea that it would have a boundary or edge, separating it from something beyond, Figure 4.5. Thus far we have observed only a speck of the universe through the observable universe. We have not detected yet an edge or a center, as we are unable to observe the end of the universe, yet.

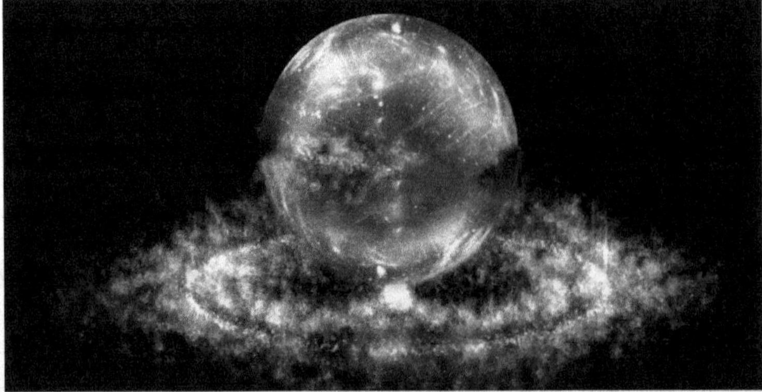

*Fig.4.5: What is Beyond the Universe?*

## Past – Present – Future

Scientists hypothetically believe that the space/time would embrace the whole history of reality, with each past, present or future event occupying an evidently resolute place in it, from the very beginning and forever. The past would consequently, still exist, just as the future already exists, but somewhere other than where we are now present. Contrary to the Christians believe, the entire universe does not have a living soul to contain past, present or future events. However, they are all contained in the infinite entity of God's dwelling, who is outside our present universe. He holds his creation together. past, present, future and eternally.

## God was not Created by the Universe

According to the Book of Genesis, God created the universe - and all the heavenly bodies, the sun, the moon, and the stars, and Earth and man - in six days. But according to contemporary cosmologists the universe hypothetically began with a great explosion known as the "Big Bang," after which the stars and galaxies slowly formed over billions of years. The idea of the multiverse stems from the "big bang" hypothesis — Albert Einstein's once controversial, as he did not accept the hypothesis of the "Big Bang." Einstein had faith that, "God is subtle, but he is not malevolent." By this Einstein intended that even though the universe did not disclose its innermost workings easily, it would not torment us with hopelessly scheming puzzles. Einstein was enthralled with the comprehensibility of the universe, Figure 4.6.

*Fig.4.6*: *God Created the Universe*

## No Multiverses in Real Life

At present, there is no evidence that multiverse exists, and everything we can see suggests there is just one created universe — our own. We can't travel out of our universe, any more than we can travel through time. Both are taking place in a "direction" that physics does not permit us to move.

## The Cosmos

Universe is a designation given to all the matter surrounding us. Also, our universe is named the cosmos. It was formerly a Greek word. In ancient days it was believed that our Galaxy established the entire universe, Figure 4.7.

*Fig.4.7: Contemplating the Cosmos*

## The Quantum Immortality

Quantum Immortality typically denotes to, in a traditional sense, a person who is sufficiently "fortunate" to endure in any episode in the world. Such a quantity of fortunateness is even sufficiently big to preserve that person away from the aging of the human body. That is, actually, when we say that individual turn out to be "immortal," Figure 4.8.

*Fig.4.8: The Quantum Immortality Theory*

## The Twin Paradox

The twin paradox undertakes those one of the two twins travels to space at 99.9% the speed of light and comes back to Earth. It suggests that when the traveling twin returns, she is younger than her twin sister who remained on Earth, Figure 4.9.

**Fig.4.9**: *The Reality of Twin Paradox*

## Multiverse Myths or Facts

Countless physicists believe that multiverses could exist, extending from universes prowling behind the event horizons of black holes, to growing universes expanding like bubbles ocean surface.

A multiverse is something which is actually not that bizarre if you think of it historically, from the point of view of science, theoretical physicist the horizons have unceasingly been expanding. At some time, we supposed that Earth was the only planet and considered the whole universe. We now realized there a much wider universe jam-packed with billions other planets. It is normal to venture within reason that there may be another universe out there. Multiverse is not a theory yet, Figure 4.10. It is a wild hypothesis.

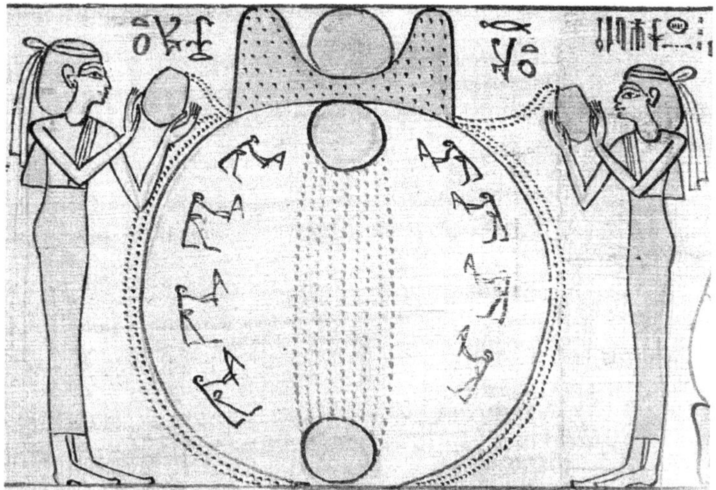

**Fig.4.10**: *Ancient Egyptians Multiverse Myths*

There are some leading multiverse hypotheses, and which of them may harbor a mischief or possibly some goodness.

## Blackholes – Harbor Multiverse

Research scientist suggests a minuscule black holes from the early universe may contain infant universes.

A hypothesis states that blackholes, which are formed in the initial instants of the universe may lead to a better understanding of explain Dark matter/Dark energy. Physicists assume that few of tiny blackholes may contain "infant universes," Figure 4.11.

In the instant, the universe hypothetically came to existence was extremely hot and extremely dense. The hypothesis continued to express that that a random local "fluctuation" caused the density to increase. Once the area is dense to 50% of it size, it would be sufficient to ignite gravitational collapse into a black hole.

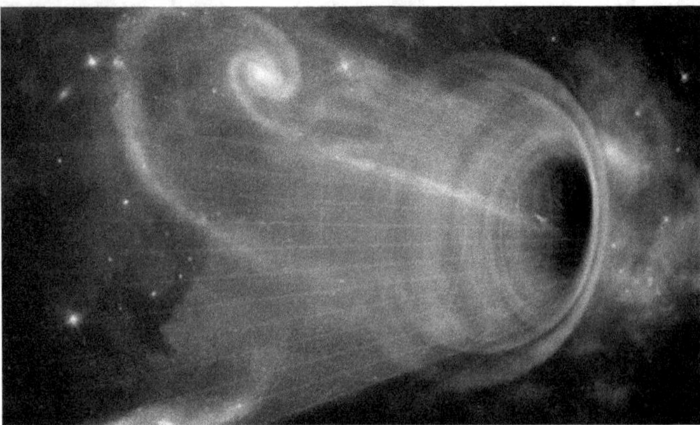

*Fig.4.11: Black Hole Harbor Multiverse*

Dissimilar to the black holes formed through our modern universe, the hypothetical "primordial" blackholes formed at this time did not form from a crumpling star. Instead, a wide range of small masses, were formed.

Since they these tiny masses hypothetically radiate no light. The primordial blackholes are considered a hypothetical potential contender for Dark matter/Dark energy. Such mysterious Dark material captures up about 85 per cent of all visible matter in the universe. Dark matter/Dark energy cannot be seen through traditional or advanced telescopes.

## The Cosmological Inflating Multiverse

This is a hypothesis grown out of imaginative cosmology, predominantly from the detection that our universe is expanding, Figure 4.12. This hypothesis of a multiverse formation is contributed to the early rapid inflation that the present universe experienced some 13.8 billion years ago. It may have taken place in aloof regions of space-time detached from our present universe.

The hypothesis claims that our present universe is a piece of a specific patch of space/time that is evolving. This specific time/space region which contained the initial ingredient to form tiny are to cause "inflation" is homogeneous, isotropic (the same in all directions) and naturally expanding in a well-described method.

The particular regions of the multiverse may hypothetically experience their own "Big Bangs," and consequently, they conduct their own expansions. However, their "fluctuation" as well as their expansion would not affect our universe. They constitute the other universes. The assemblage of all such universes is the multiverse.

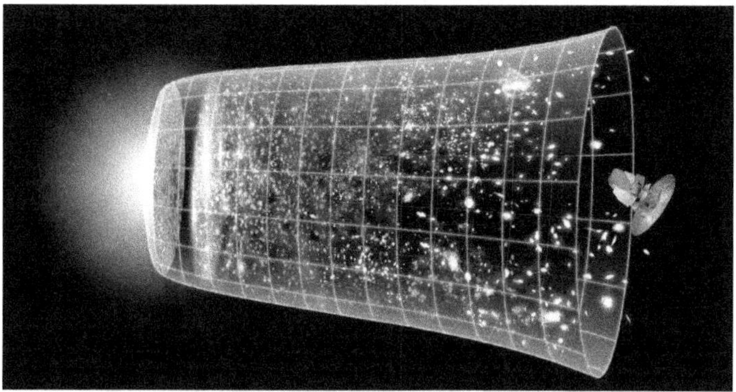

**Fig.4.12**: *Cosmic Inflation*

This hypothetical multiverse imaginary concept caught on in-fiction as it is an exceptional storytelling material, Consequently, it developed into popular subject in cosmology as it could discourse the haunting mysteries of physics, science and cosmology.

The hypothesis of "Inflation" gave rise to the hypothetical idea of the multiverse.

Few research work demonstrated that once inflation occurred, it may form significant disconnected regions.

While, inflation took place hypothetically 13.8 billion years ago in the present universe it may bring inflation back in different region of space-time. Cosmologists imagine that inflation is never ending – referred to a blasphemous "eternal inflation."– and the possibility of an infinite number of bizarre "different universes." Figure 4.13.

**Fig.4.13**: *Multiverse Represented As Bubble expanding in Bouts of Eternal and Chaotic Inflation on Cosmic Canvas – Curtsy Science Photo Library*

American/Russian - theoretical physicist derived a hypothesis for arranging multiverse, Figure 4.1. They imagine the universes as "bubbles" similar to a cosmic canvas, enfolding away from each other.

Cosmologists' opinion differs regarding these universes within a multiverse. Multiverse is currently the topic of speculation among cosmologists and Hollywood. Physicists, speculate that the laws of physics would be the same in each separate regions, where the multiverse is formed. Additionally, other universes may have other versions of the laws of physics. A region of expansion may have variation that could apply across a range of physical parameters, such as gravity and the rate of universe expansion. Some of these universes may have laws of physics that may not fit for the development of large-scale bodies as galaxies or stars. However, they may not even have the same fundamental particles.

Accordingly, these universes may not be variations of our present universe and therefore, may not host any life at all.

## String Theory Multiverse

String theory is a hypothesis derived by physicists to link "quantum mechanics" and "Einstein's General Relativity," which are the finest hypothetical imageries to describe infinitesimally small objects and incomprehensibly large. The essential concept of "string theory" is that vital particles like quarks and electrons are assumed a single point in one-dimensional strings, which vibrating at different frequencies, Figure 4.14.

*Fig.4.14*: String Theory Multiverse

This 'string- scene" gives a prevalent setting for the multiverse, contributed to one of the main elements upon which string theory is hinged on. With the purpose of being mathematically correct, string theory requires "additional dimensions" to hypothetically occur.

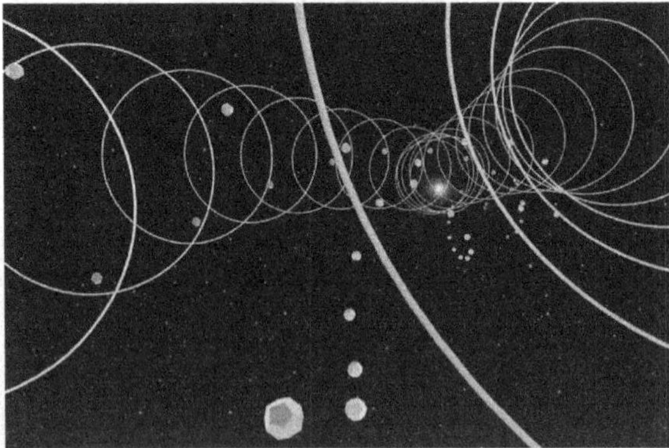

*Fig.4.15*: String Hypothesis Attempts Explaining Fundamental Particles in Nature –
Curtsy – Science Photo Library

These dimensions are not parallel dimensions as we observe in science fiction. As a replacement, string thinkers made themselves to trust these extra dimensions, which are curled up within the traditional 3D of space. These extra dimensions keep on being unseen to us. However, we have hypothetically evolved only to see in 3D. Also, these extra dimensions could bid a "way-in" to the string hypothesis multiverse, Figure 4.15.

## Modelling The Strings Hypothesis

One requires 11 dimensions in total. Also, it was suggested that the extra dimension will hypothetically enable us to get to these other universes. Although, the extra dimensions are the means to connect us to the other universes, they may still remain an obstacle, which is the "cosmic horizon" beyond which we cannot see.

Nonetheless, if there is the "connectivity" between universes within a multiverse is not made, it makes the cosmological hypothesis of a multiverse impossible to test empirically.

The evidence to-date is hypothetical, not experimental. Scientists believe they cannot conduct any direct experiments to authenticate or falsify what goes on in other universes.

Scientists' inability to verify these hypothetical concepts may not be able to prove the existence of a multiverse. Scientists may not be able to disprove it either.

## The Blackhole Multiverse

*Fig.4.16: Event Horizon at Blackhole*

The death of a massive as it dissipates its own fuel for nuclear fusion, it collapses into a black hole. Blackhole is a region of space/time constrained by a surface so-called an "event horizon," at which nothing can escape, not even light, Figure 4.16.

Einstein's General Theory of Relativity explains that a large mass generates large gravitational force, which curves space-time. Einstein's theory also states that at the core of the Blackhole the mass of a star is enormously compressed to a singularity, where the space-time curvature becomes unlimited. Consequently, the laws of physics break down.

This is an idea that plights physicists. Sadly, many scientists, who wish to do away with God as the infinite creator, used the "Singularity" either as means where the "Big Bang" started the creation of the universe, or used the "Singularity itself to hypothesize multiverse. In this case, the singularity is meant to replace an entire universe and in turn, a multiverse.

"Singularity" is non-physical as they cannot be measured. Their existence shows that the hypothesis is imperfect. There is also a related hypothesis, stating that every Blackhole harvests a new, baby universe inside. The other side of the "Event Horizon" –– A hypothetical "wormhole" comes out acting as a bridge linking between this infant universe to the parent universe in which the Blackhole originally exists, Figure 4.17.

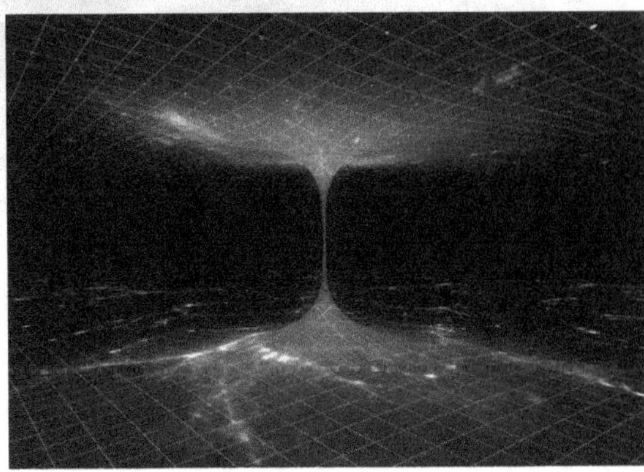

*Fig.4.17: A Linking Bridge of a Wormhole Connecting the Infant Universe to the Original Blackhole Parent - Curtsy Science Photo Library*

Is it possible a black hole broods a new baby universe? This illustration, Figure 4.18, is of a wormhole, a hypothetical crosscut joining two separate points in space/time.

This hypothesis of infant universe, when observed from the side of the new universe, the parent universe seems as the other side of a "White-hole." It is a region of space/time that cannot be crossed from the outside, which can be considered as the reverse of a Blackhole.

It can be represented as the matter going to a Blackhole and come out in a new universe. It is similar to puffing a soap bubble through a circular wand. The wand is the event horizon – albeit, in one dimension less – the soap liquid is the matter crossing the event horizon, and the surface of the bubble is the new universe.

In the hypothesis scientists suggest a universe may produce billions of Blackholes and each of them could produce a bizarre baby universe. In January 2023, researchers at the International School of Advanced Studies (SISSA) in Italy projected that there could be as many as 40 trillion –Blackholes in our Universe alone. That is much too much of baby universes. I wish there is a psychiatrist to every weird scientist!

To add insult to injury, scientists suggest that these infant universes would be concealed from the view of the occupants of their parent universe by the light-trapping surface of the "event horizon." Once that "event horizon" is crossed there is no return. Nothing can enter a white-hole. There is no interaction between parent and infant universe.

*Fig.4.18: Einstein's General Theory of Relativity, Large Objects Cause Space/Time to curve Curtsy - Science Photo Library*

However, if two Blackholes came to being in the same universe, and each of these Blackholes developed a new universe, then there is a likelihood that these two infants' universes merge, similar to two Blackholes merge to produce one Blackhole.

As for the hypothetical possibility of a substitute version of you existing beyond the "event horizon" of a Blackhole, scientists conclude that possibility does not exist. There would be no alternate you. At any time, an object can only exist in one universe.

Christianity is not a religion, it is a philosophy of Eternal Life, Table 4.1.

**Table 4.1 :** World Religion Vs Christian Faith

| Principal World Religion | Christian Faith and Principals |
|---|---|
| Atheism: there is no God. | God exists, and was present "at the beginning." God created the universe; the universe neither spontaneously appeared nor has it existed forever. |
| Agnosticism: it is impossible to know whether God exists. | God has revealed Himself in Scripture as the Creator. |
| Dualism: Good and Evil are eternally co-existent (as Zoroastrians believe). | God was alone when He created. God is perfectly good. Beings who became evil are part of the created order. |
| Finite-god views (e.g., Open Theism and Process Theology). | God created the space/time universe. Thus, He is not limited by anything in the universe, including the future, since God created time itself. |
| Evolutionism: that goo became you via the zoo. | God created all things. |
| Humanism: man is the measure of all things. | God is the ultimate reality. Man is part of the created order. God created us so He is the measure of all things. |
| Materialism: Matter (or mass-energy) is the only reality. This is a synonym of: Naturalism: natural laws describe all things. | God created matter (and mass-energy); or, God created nature. God is thus sovereign over the natural world. Thus matter (mass-energy) is not eternal or self-existent. |
| Pantheism: all is god; god and creation are the same thing. | God created the universe. Thus, God is distinct from His creation. |
| Panentheism: "all is in god." | God *transcends* what He created. |
| **God Created or Quantum Fluctuation?** | |
| Polytheism: there is more than one god. | Only one God created all things. |
| Unitarianism (that God is an absolute unity, e.g., Islam, modern Judaism, Jehovah's Witness doctrine, classical Unitarianism). | Elohim is a plural noun with a singular verb, teaching a plurality in the Godhead. The New Testament reveals this further as the Trinity. |

## God Created or Quantum Fluctuation?

**Fig.4.19:** *Quantum Mechanic or God Created*

In one aspect, *Genesis 1:1* is the most significant verse in the Bible: "In the beginning, God created the heavens and the earth." If we can trust this verse, no other verse in the Bible should be a problem. For example, if God can create the entire universe, then raising people from the dead and causing a virgin to conceive would be simple beyond words, Figure 4.19. Similarly, in this one verse, all other untrue religions are excluded:

On the other hand, if we cannot trust this verse, then nothing else in the Bible makes sense. In the meantime, this verse is so foundational, it is not astonishing that atheists have feverishly criticized this concept. Some of the attacks are childish, while others have the surface of philosophy or advanced science.

## Eternal God

The Bible does not try to demonstrate that God exists—it proclaims this truth as obvious. Nonetheless, a common question from little children, and not-so-little atheists. is: "If God created the universe, then who created God?" Or, "If everything has a cause, then who caused God?" But then again, no thoughtful apologist ever argued that way. As we have illustrated in numerous writings and books, one of the main *real* arguments is:

1.   Everything which has a beginning has a cause.
2.   The universe has a beginning.
3.   Therefore, the universe has a cause.

It is not everything that has a cause, but only everything which begins to exist. The universe necessitates a cause since it had a beginning. This can be illustrated by the *Laws of Thermodynamics*. The *First Law of Thermodynamics* declares that natural processes can neither create nor destroy mass-energy

mass-energy interchange can occur according to $E = mc^2$,

however, the total stays the same. But the *Second Law* states that the quantity of energy *available for work* is running out, or "entropy" is increasing to a maximum. If the total quantity of mass-energy is limited, and the amount of usable energy is decreasing, then the universe cannot have existed forever. Or else, it would *by now* have exhausted all usable energy—the 'heat death' of the universe. For example, all radioactive atoms would have decayed, every part of the universe would be the same temperature, and no further work would be possible. Consequently, the apparent corollary is that the universe began a finite time ago with a lot of usable energy, and is now running down.

Furthermore, Einstein's general relativity, which has much experimental backing, demonstrates that time is related to matter and space, Figure 4.20. Subsequently, time itself would have begun along with matter and space; an insight first explained by Augustine in the fourth century. Since God, by characterization, is the Creator of the entire universe, he is the Creator of time. Consequently, He is not restricted by the time dimension He created, so has no beginning in time—God is "the One who is high and lifted up, who *inhabits eternity*, whose name is Holy" (*Isaiah 57:15*). Therefore, He doesn't have a cause.

It is a metaphysical principle that things which begin have a cause, but it is also self-evident—no-one actually repudiates it in his heart.

All science and history would breakdown if this law of cause and effect were denied. So would all law enforcement, if the police did not believe they wanted to find a cause for a stabbed body or a burgled house. Also, the universe cannot be self-caused—nothing can create itself, because that would mean that it existed before it came into existence, which is a logical ludicrousness.

Notwithstanding this, the favorite philosopher of modern atheists, the Scotsman David Hume (1711–1776), disagreed. He taught that one might conceive of something coming into being without a cause.

However, British analytic philosopher, and conservative Roman Catholic, G.E.M. (Elizabeth) Anscombe (1919–2001) argued convincingly that no one actually perceives of any such thing. To rewording one of her points, what if a banana suddenly appeared on your plate. You would not think, "Hume was right after all—this banana actually did come into being without a cause." No, you would think, "How did that banana get there?" and look for the likely cause. Perhaps there was a hole in the ceiling above it, or in the plate below it. If that were ruled out, then maybe you were temporarily unaware of your surroundings, and in that

*Fig.4.20:* Einstein Theory of Relativity – Mass/ Gravity Bend Space

time, somebody located the banana there without your noticing. Failing that, maybe a magician's trick, or even a miracle, was the cause. Irrespective, even an *unknown cause* would be more likely than *no cause.*

Further, Anscombe pointed out, we would be less likely to think that this banana came into being than that it *already existed* and was somehow *moved* to the place, i.e., the cause was in *transportation* not in creation out of nothing.

However, even though Hume claimed that one could easily conceive of something coming into being without a cause, in truth, he likely never actually conceived any such thing. Indeed, it appears impossible to conceive. Hume himself, in more articulate moments, even acknowledged as much:

Nonetheless, permit me to express you that I never asserted so ridiculous a Proposition as that anything might arise without a cause: I only maintained, that our Certainty of the Falsehood of that Proposition advanced neither from Intuition nor Demonstration; but from another Source.

## Universe Out of Nothing

In spite of the above, a number of atheists have proposed that the universe actually came from 'nothing'. For example, an article about Alan Guth (1947–), the pioneer of the inflationary universe (see ch. 6), stated:

The universe eruption into something from absolutely nothing—zero, nada. And as it got larger, it became full with even more stuff that came from absolutely nowhere. How is that likely? Ask Alan Guth. His theory of inflation helps explain everything, Figure 4.21.

More recently, physicist and atheistic propagandist Lawrence Krauss (1954–) has promoted this conception, and even wrote a book, *A Universe from Nothing*, which had a radiant review by a famous atheist "Richard Dawkins." Though, "Luke Barnes," a non-creationist astrophysicist who is a Postdoctoral Researcher at the Sydney Institute for Astronomy, University of Sydney, Australia, is scornful about Krauss and those who argue like him:

First and foremost, *"I'm getting really rather sick of cosmologists talking about universes being created out of nothing."* Krauss frequently talked about universes coming out of nothing, particles coming out of nothing, diverse kinds of nothing, nothing being unbalanced. This is garbage. The word nothing is often used lightly—I have nothing in my hand, there's nothing in the fridge etc. However, the appropriate meaning of nothing is "not anything". Nothing is not a category of something, not a type of thing. It is the nonappearance of anything.

Some of the best examples of the misconception of evasion encompass giving the word nothing as if it were a kind of something:

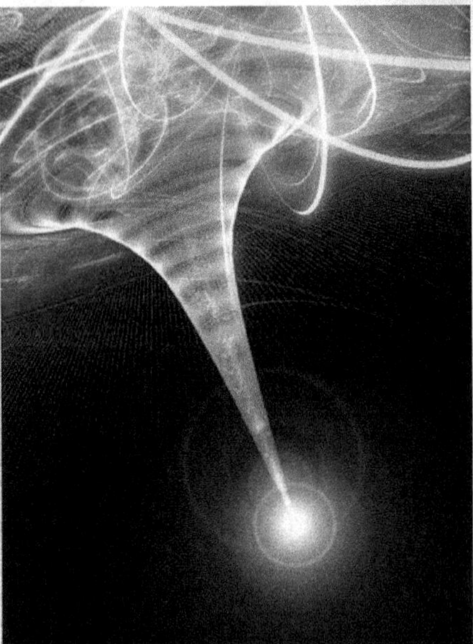

**Fig.4.21:** *Universe Out of Nothing!*

- Margarine is better than nothing.
- Nothing is better than butter.
- Thus, margarine is better than butter.

We can reveal the delusion by merely rewording the premises, evading the word nothing:

- It is better to have margarine than to not have anything.
- There does not exist anything that is better than butter.

The conclusion - margarine is better than butter- does not follow from these premises.

## The Fallacy with Quantum Fluctuation

Some physicists proclaim that quantum mechanics violates this cause/effect principle and can yield something from nothing. For instance, Paul Davies writes:

… space/time could appear out of nothingness as a consequence of a quantum transition. … Particles can suddenly appear out of nowhere deprived of specific causation … the world of quantum mechanics characteristically produces something out of nothing.

However, this is a great misemployment of quantum mechanics. Quantum mechanics never yields something out of nothing, Figure 4.22. Davies himself acknowledged that his setting *should not be taken too seriously.* Also, concepts that the universe is a "quantum fluctuation" must presume that there was *"something"* to fluctuate—their "quantum vacuum" is a lot of matter-antimatter possibility—not "nothing." Consequently, this is another evasion.

Nevertheless, Krauss is still kept to these misconceptions, as "Luke Barnes" points out, clarifying in more explicitly how the term 'nothing' is distorted:

**Fig.4.22**: *Quantum Fluctuation and Object Movement*

Once again, let us examine at Krauss' propagation once more. Does it make sense to say that there are diverse kinds of not anything? That not anything is not stable? This is bollocks. What Krauss is actually speaking about is the "quantum vacuum." The quantum vacuum is a type of something. It has characteristics. It has energy, it fluctuates, it can cause the expansion of the universe to accelerate, it submits the -highly non-trivial- equations of quantum field theory. We can define it. We can compute, predict and misrepresent its characteristics. The "quantum vacuum" is not nothing.

This proposes a very simple test for those who request to talk about nothing: if what you are talking about has properties, then it is not nothing. It is pure prevarication to relate to the "quantum vacuum" as nothing when a philosopher begins posing the question *"why is there something rather than nothing?"* She is not enquiring "why are there particles rather than just a quantum vacuum?" She is asking "why does anything exist at all?" As Stephen Hawking once questioned, *"why does the universe go to all the bother of existing?"*

We can now understand that this question cannot be answered if we followed any of the traditional scientific methods. Employing scientific theories are necessary to ensure safe and logical application to serve their intended purpose in physical reality. Equations are developed to describe properties, and therefore, define the purpose of application and its consequences. There cannot be equations that describe "not-anything."

Scientists of imaginative unrealistic science may derive any equation they desire—they will not be able to create from that equation the unreality imagined to form a nonexistence reality.  The derived equations, no matter how complex or simple, they gave to describe the existing object in the real world.

Barnes' objections to Krauss's evasions are shared by my colleague and philosopher "David Albert," professor of philosophy at Columbia University, NY, who also has a doctorate in theoretical physics. David reviewed Krauss's book unsympathetically in the *New York Times*, David known for unfriendliness to orthodox Christianity:

*"Where, for starters, are the laws of quantum mechanics themselves supposed to have come from?* **Krauss is more or less upfront, as it turns out, about not having a clue about that**. *He acknowledges (albeit in a parenthesis, and just a few pages before the end of the book) that everything he has been talking about simply takes the basic principles of quantum mechanics for granted. …"* [Emphasis Added].

Krauss appears to contemplate that these vacuum states amount to the "relativistic-quantum-field-theoretical" type of there not being any physical stuff at all. And Krauss has an argument—or somewhat believes he does—that the laws of "relativistic quantum field" theories entail that vacuum states are unstable. And that, in a nutshell, is the account Krauss propositions of the reason there should be something rather than nothing.

However, David said, *"that is just not right. "Relativistic-quantum-field-theoretical" vacuum states—no less than giraffes or refrigerators or solar systems—are particular arrangements of elementary physical stuff. The true relativistic-quantum-field-theoretical equivalent to there not being any physical stuff at all isn't this or that particular arrangement of the fields—what it is (obviously, and ineluctably, and on the contrary) is the simple absence of the fields! The fact that some arrangements of fields happen to correspond to the existence of particles and some don't is not a whit more mysterious than the fact that some of the possible arrangements of my fingers happen to correspond to the existence of a fist and some don't. And the fact that particles can pop in and out of existence, over time, as those fields rearrange themselves, is not a whit more mysterious than the fact that fists can pop in and out of existence, over time, as my fingers rearrange themselves. And none of these popping—if you look at them aright—amount to anything even remotely in the neighborhood of a creation from nothing."*

Krauss's is just the up-to-date in a sequence of philosophically incompetent books by the *soi-disant* 'new atheists.' It is difficult to disagree with the Thomist philosopher Edward Feser, Associate Professor of Philosophy at Pasadena City College:

"Krauss spate of bad books on philosophy and religion by prominent scientists—

- Dawkins' The God Delusion,
- Hawking and Mlodinow's" The Grand Design," and
- "Atkins' On Being,"

*among others—is distinguished not only for the sophomoric philosophical and theological errors they contain but also for their sheer repetitiveness. Krauss' fallacious account of how* **"something can come from nothing,"** *though presented as a great breakthrough, and praised as such by Dawkins in his afterword, is largely a rehash of ideas already put forward by Hawking, Mlodinow, and some less eminent physics popularizers. Dawkins has been peddling the "Who created the creator?" meme since the eighties."*

Honest critics have uncovered their blunders and fallacies again and again. Yet these novelists keep reiterating them anyway, for the most part simply disregarding the critics. What accounts for this? To rewording a well-known remark of Ludwig Wittgenstein's, *"I would suggest that a picture holds these thinkers' captive, a picture of the quantitative methods of modern science that have made possible breathtaking predictive and technological successes."*

## God Began the Universe

The Word of the Infinite God declares that God began the universe. The truth of the universe's foundation points strongly to the Creator reliable on the biblical God. Some skeptics, following "Hume," have declared that something can begin without a cause, however, this is not only irrational, it is arguably unimaginable. The 'New Atheists' have resorted to "quantum bluffing" to entitlement that something really can come from nothing. Nonetheless, they must fudge about the word 'nothing.' This certainly should mean *nothing—no* properties. However, their proposed "quantum vacuum" is not nothing; it must be *something, with* properties—e.g., the "quantum vacuum," which is being bound by the laws of "quantum physics," so that it can 'fluctuate.' "In the beginning God created the heavens and the earth." It stands to reason.

## MULTIVERSE – UNKNOWN SCIENCE OR

### Illogical Reasons

A published opinion by some scientists demonstrated how "unscientific" and anti-God some of their articles have become.

Scientists expressed their believe in multiverse theory, or the concept that our universe may be only one of countless universes that currently exist. Such conjectures try to explain away the advent of design in the universe. This has substantial spiritual implications. In an article called "*What's God got to do with it,*" the article states:

"WHAT would you rather believe in, God or the multiverse?"

It implies as an occasion of cosmic apples and oranges, but then again, more and more we are being told it is a choice we must make.

Consider the interchange earlier this year between Richard Dawkins and physicist Steven Weinberg in Austin, Texas. Deliberating the details that the universe came to be fine-tuned for our existence, Weinberg told Dawkins:

"*If you discovered a really impressive fine-tuning … I think you'd really be left with only two explanations: a benevolent designer or a multiverse.*" *(Emphasis in original).*

"Weinberg went on to simplify that appealing a benevolent designer does not count as a sincere explanation, however, I was fascinated by Weinberg's scenario. Is that actually our only choice? Supernatural creator or parallel worlds?"

## Multiverse Ignoring Evidence

Why doesn't it count as a genuine explanation?

If the "Ockham's Razor principle is followed, (see below) then the most straightforward explanation would be that there had to be a designer, Figure 4.23.

Weinberg correctly explained that if one was to see a finely-tuned universe then it would imply design, but then effectually ruled out that designer. God is omitted from the question on philosophical bases. Why philosophical bases? Well, the universe does look as if it is designed and very finely-tuned for life, so if it is not God then what is left?

"Short of appealing a benevolent creator, many physicists view only one conceivable explanation," "Our universe may be but one of perhaps infinitely many universes in an unimaginably vast multiverse." Cosmologist Bernard Carr, said: "If you don't want God, you'd better have a multiverse."

*Fig.4.23: Ockham's (or Occam's) Razor*

## Ockham's (or Occam's) Razor

[Ockham's (or Occam's) Razor, also known as the Law of Parsimony or the Law of Succinctness, is the name given to a philosophical principle. It was used so frequently and so cuttingly by the English theologian and philosopher, William of Ockham, in the fourteenth century, that it became known as Ockham's Razor.

Today Ockham is a village in Surrey, England, about 40 km (25 miles) south-west of London. William was born, there c. 1285 and died at a convent in Munich, c. 1349.

## A Razor-Sharp Maxim

*He principle that bears William's name, in one of its several Latin forms, is: 'Entia non sunt multiplicanda praeter necessitatum', i.e., 'Entities [of explanation] should not be multiplied beyond necessity'.*[3,4]

When applied to different theories about nature, this does not mean that nature is not complex, but only that, when accounting for any observed phenomenon in nature, the explanation that involves the fewest assumptions is most probably the correct one.

As well as being good philosophy, this method of reasoning is also sound science. For example, the main advantage of the Copernican Theory (that the earth revolves round the sun) over the Ptolemaic Theory (that the sun moves round the earth) was the reduction in the number of separate assumptions from 79 to 34. Later Isaac Newton was able to account for the movements of both the earth and the heavenly bodies by just one assumption—the Law of Gravitation.]

Of course, it is nothing original to comprehend that people do not want there to be a God to be answerable to. Both creation and evolution are faith base systems about past events, and "evidence" or facts are usually understood within the framework of that belief system. Therefore, if the indication for multiverse theory could be understood with the framework for cosmic evolution I could then comprehend where they are coming from.

"The scientist writer who is an proponent of multiverse," goes on to reel out the defense for multiverse theory but it is actually no evidence at all:

"There are sufficient explanations to take the multiverse seriously. Three key theories: —

- quantum mechanics,
- cosmic inflation and
- string theory—all converge on the concept."

Resorting to unsubstantiated hypothesis to sustain unsubstantiated theories!

These three keys hypotheses are just tossed into the blend as if they are substantiated science fact, nonetheless, they are not. Multiverse is a wholly hypothetical paradigm, but what brands the multiverse opponent explanations even worse is that the scientist opponent to multiverse attempts to use even more speculation in an attempt to bolster the former.

Primarily, "quantum mechanics" deals largely the nature and behavior of subatomic particles. "Quantum mechanics" is acceptable in itself, but the scientist opponent to multiverse is speaking about an extremely unconvinced understanding of it, called the "many worlds" interpretation. Quantum theory forecasts particular probabilities of numerous events happening, e.g., the decay of a radioactive atom. Nonetheless, the many worlds opinion declares that there are parallel universes, one for each possibility.

The multi-universe (multiverse) inkling is a subset of this "many worlds" idea. However, it goes further. Let us explain how it works. If there were heaps of other universes, with the laws of physics a little bit different in each one, then it would become likely that at least one would happen to have the properties required for intelligent observers to exist. If it did not, you and I would not be here to ask the question of "Why does our universe look special?" In the first place, this is really a non-answer—envision somebody taking a lethal dose of poison, surviving, and then being asked, "how did you survive?" It would be obtuse to answer, "If I didn't, I wouldn't be here talking to you."

It is a nonsensical effort to explain away design, since even if there were other universes, the laws of physics that command our own mean that it would be impossible for us to detect them anyway. This is science fiction!

Secondly, "cosmic inflation" is rarely understood by most scientists, while it supports the multiverse ideology. However, Inflation in general is yet another hypothesis to rescue a hypothesis—the "big bang." The "big bang" hypothesis is plagues with numbers of problems, one of which is the uniformity of temperature to within 1 part in 100,000.

According to "big bang" time scales, there has not been even a tenth of the amount of time fundamental for heat to have voyaged from the hot parts to the cold parts to amazingly, equalize the temperature. This is a light-travel problematic agonized by "big-bangers," termed the "*horizon problem.*"

The "inflation theory" is a mathematical model attempting to clarify the problem of How can distant starlight reach us in just 6,000 years?

Thirdly, string theory, or the theory that the universe might exist in multiple branes or dimensions, is currently wholly unobservable, unverifiable, and untestable. Though, its supporters would also advocate that it is not falsifiable, and consequently, it might be right. To employ this logic is totally circular in its cognitive and short on substance. Once again, its boundaries have reached the realm of science fiction. String theory is nothing more than elegant philosophical mathematics endeavoring, once again, to resolve some of the observation evidence that is not consistent with a "big bang" model.

## The Worldly Arguments

After utilizing unscientific concepts to support an unscientific multiverse idea, scientists' proponent to multiverse then said:

"But the reason physicists talk about the multiverse as an alternative to God is because it helps explain why the universe is so bio-friendly. From the strength of gravity to the mass of a proton, it's as if the universe were designed just for us. If, however, there are an infinite number of universes—with physical constants that vary from one to the next—our cozy neighborhood isn't only possible, it's inevitable."

This declaration borders on the bizarre! Confidently, it is logical to infer that the cause the universe is very "bio-friendly" is that it was made to be that way (*Isaiah 45:18*). There cannot be a sturdier argument for design from experimental science—i.e., evidence that is observable, repeatable and testable. However, observe how proponents' scientists to multiverse resorts to unknown entities to claim a certainty ("**If ... there are** an infinite number of universes") it's **"inevitable"** (emphasis made). scientists then had the audacity to say that:

" ... if this theory doesn't pan out our only other option is a supernatural one is to abandon science itself."

These scientists then defined as an "unfounded leap of logic." How is summoning a designer to account for design features illogical? Really, it is their statement in this respect that is illogical. This is indescribable elephant-hurling in the extreme, mainly as they went on to say that creationists are erroneous in their claim that multiverse theory was invented to help as science's get-out-of-God-free card.

Well, given their best efforts to provide evidence for a multiverse theory they are playing that card very well deprived of any help from creationists. The scientists are invoking pure philosophy in an effort to disrupt the very logical arguments that creationists have been utilizing. They fail to acceptably answer them, because scientists merely pleas to unsubstantiated and speculative theories. Their motivation to evade a Creator at all costs is exposed later in the writings when they quote famous evolutionary cosmologist "Michio Kaku":

"To make matters worse, physicists are also dragging morality into the picture. In a recent show about the multiverse that aired on the History Channel, physicist Michio Kaku asked: 'Why should I obey the law knowing that in some universe if I commit a crime, I'm going to get away with it?' The ID [intelligent design] community has already tried to draw lines from Darwin to the Holocaust in their attempt to paint rational people as Satan's minions. Are physicists really suggesting that the multiverse gives us license to commit evil? It's an absurd notion, which moral philosophers have already killed off in other guises."

Note how multiverse scientists reproach him for pointing out something that Bible-believers (and Kaku) have long pointed out. Their appeal to unspecified "moral philosophers" did not clarify how this rational inference has been replied. No one from this camp has ever recommended that balanced people are Satan's minions, -Hitler maybe, but that's just a nice bit of hyperbole they tossed in for dramatic effect-, and we have never claimed that non-believers or evolutionists cannot be moral.

What we have reliably clarified is that they have no rational basis for being moral. Indeed, how can they even describe what "evil" is except there is a law giver; an ultimate authority who outlines what is wrong and right? With the moral goalposts being moved so rapidly today one man's "evil" can quickly become another's "If it feels good, do it!" This is what Michio Kaku was logically inferring out, but in their keenness to shape straw man arguments and disrepute those who promote design, they missed it.

Multiverse scientists went on to say:

*"Pitting the multiverse against religion presents a false dichotomy. Science never boils down to a choice between two alternative explanations. It is always plausible that both are wrong and a third or fourth or fifth will turn out to be correct."*

This is true to some degree. One can never fully recognize everything there is to know because tomorrow we might discover or learn something that we didn't recognize today. However, here scientist's proponent to multi verse idea are once again playing their own suitably named "get-out-of-God-free card". Once attempting to make sense of one's world, one can only unfailingly utilize what we comprehend and recognize to be true. They illogically resorts to unknowns -- what we cannot observe and test -- to clarify what we can observe and test. They then throw some more elephants by resorting to yet another "unidentified, untestable piece of philosophy.

"What might a third choice look like here? Physicist John Wheeler once presented a proposal: maybe we should consider cosmic fine-tuning not as a problem but as a hint. Possibly it is evidence that we someway bestow the universe with particular characteristics by the mere act of observation. It's an impression that Stephen Hawking has been considering about, too. Hawking promotes what he calls top-down cosmology, in which observers are making the universe and its entire history right now. If we in some sense create the universe, it is not surprising that the universe is well suited to us."

-- Once again refer to A new age of quantum madness for an answer to Hawking's theory. --Certainly, by now you can understand that scientist's proponent to multiverse have actually confined to religious ideas, which is ironic because the entire point of their writings were to somehow mention that "resorting to a Creator God who made the universe is unscientific and religious." their final determination sounded more like a New Age idea, which is exclusively religious. And lastly, they say:

"That's speculative, but at least it's science."

We set the rules and God's not allowed!

Our goal is always to write factually, but it's hard to keep a straight face when attempting to write a reply to this last sentence. Certainly, were they being thoughtful? Therefore, let us just answer the last sentence by reiterating in point form.

- The universe looks designed, but a designer is not allowed.
- Thus, there must be some other clarification.
- Let us confine to some other religious ideas to explain the appearance of design (the multiverse).
- Then let's use even more religious notions to sustain our religious idea.
- And then let us claim it is science to demonstrate no designer was essential.
- We win- Hallelujah!

This is additional instance of *"whoever defines the terms wins the debate"* (cf. The rules of the game: As the 'rules' of science are now defined; creation is forbidden as a conclusion—even if true).

Our goal here is to illustrate that one should not be scared by the hypothetical claims of evolutionists. When learning such imposing claims, it might be judicious to ask *"What is the evidence?"* and with any luck you will understand, as this writing evidently illustrates, that people -- and indeed, scientists are just normal people-- will confine to whatever to repudiate God. At the end of the day the writer proponent to multiverse falls on his own sword of illogicalness anyway. Multiverse hypothesis includes complex and intricate mathematical extrapolation. Certainly, multiverse *hypothesis* would be more suitable than multiverse *theory*. If it were somehow actual, one would still be left with the problem of who derived the complex equations for the multiverse. As physicist Paul Davies has said *"There is no known law of physics able to create information from nothing."* "If all things being are equal, the simplest solution tends to be the best solution.

# THE 'STRING' HYPOTHESIS

## Cosmic Evolution

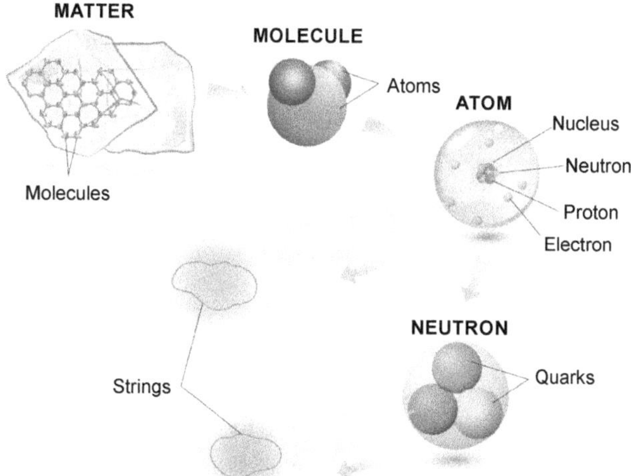


***Fig.4.24:*** *String Hypothesis = Einstein's famous equation, E=MC², demonstrated that energy and matter are essentially related and one can be converted into the other*

Majority of people have learnt of the expression 'the big bang'. Its utilization is so predominant among conventional scientists and the media that it has become the recognized 'fact' for how the universe commenced, Figure 4.24. Though, there are a cumulative number of secular scientists who are doubtful of this hypothesis of cosmic evolution, and much of their doubt has been triggered by cumulative findings that fly in the face of 'big bang' hypothesis. In May 2004 'An Open Letter to the Scientific Community' signed by dozens of secular scientists was promoted in the famous *New Scientist*. At the time of writing this chapter, the total number of scientists signing the letter who are skeptical of the 'big bang' has amplified to over 400.

One of the distinctive complications for those who have faith in that the universe came into existence by itself is that the universe *does not* present itself in such a method. E.g., the complication of the observable universe suits better with the idea that it has been particularly created *ex nihilo*. The 'first cause' problematic is one of the excessive stumbling blocks for evolutionary cosmology. For instance, reflect on the following reasonably effective argument.

1.  Everything which has a beginning has a cause.
2.  The universe has a beginning.
3.  Consequently, the universe has a cause.

Briefly, if the universe had a beginning, then it must have had a cause.

## Looking Like Design

The so-called 'Anthropic Code' originates from the observation that many features of the universe provide at least the *appearance* of having been designed exactly for human life. There is a wealth of design structures that sit awkwardly for those who accept as true that we exist by means of some giant cosmic lottery. Therefore, it is frequently claimed that there must be a countless of other universes, with diverse structures, even diverse laws of physics. And in our own universe the cosmic dice just occurred to drop in such a way that the laws were those that would permit people to evolve. Which is the reason we are here in the first place to enquire about such 'anthropic' questions. In other words, we are not here by design, it just looks that way, maybe by reason of 'evolution by natural selection' amid numerous -unobserved- universes.

Albeit the 'big bang' is fervently seized by some believers of design, it also is employed to challenge as well as to avoid a Designer. The theory claims that all the matter, space and energy of the entire universe pre-existed in a particle no bigger than the head of a pin. Then, for no cause, it abruptly exploded, and the energy became matter that shaped galaxies, stars and ultimately people.

Efforts to prop up this theory in the face of swelling glitches have led to numerous weird hypotheses invoking enigmatic hidden forces to basically hold the universe together or to push it apart. All of these speculations attempt to avoid the obvious spiritual implications of design. *For by him all things were created, in heaven and on earth, visible and invisible, whether thrones or dominions or rulers or authorities—all things were created through him and for him. And he is before all things, and in him all things* **hold** *together. (Colossians 1:16–17)*.

## Theory of Everything

Numbers of scientists and cosmologists today suggest that other universes may exist in parallel with our own. Some have faith in that extra-terrestrials may exist in these other universe(s) and have yoked the technology to fly between their own dimension and ours, and that this could interpret for the countless of UFO sightings and extra-terrestrial happenstances over the years, Figure 4.25.

One must contemplate that countless of these 'cosmic convolutions' lack experimental sustenance and are hypothetical ideas outfitted in longwinded convoluted mathematics. One such impression is 'String Theory.' While vastly controversial, it has been fast in admiration, and research is well-funded, predominantly by those attempting to prop up 'big bang' ideology.

*Fig.4.25: Extraterrestrial Hypothetical Image*

**There is not a single bit of experimental indication to subftantiate the claims.** String theory is currently totally unobservable and untestable. Though, its supporters would also advocate that it is not falsifiable, and therefore, it might be correct. This is the nature of such 'sophisticated' theories. One not uncommonly hears comparable claims about evolution in general.

## Stumbles Hypothesis

The doubt of this hypothesis is that the promoters are just struggling around doing ever more esoteric mathematics, however, not actually making any progress at all. It is a misgiving established by "David Gross," the Nobel Prize-winning - American Theorist and one of the creators of string theory.  At an impressive conference of the best and brightest in physics he admitted: *"We don't know what we are talking about."*

One of the reasons that numerous are enthusiastic about the possibility of string theory is that they wish it might evolve to develop the indefinable 'theory of everything'—the consecrated treasure that has escaped physicists for decades.

The essential powers that govern our universe are gravity, electromagnetism, and the unknown strong and weak nuclear forces -- that hold atoms together. String theory is an effort to unite these forces into one mathematical theory, an effort that numerous physicists now trust to be impossible anyway.

## Christians be Aware

Similar to the expression 'the big bang', 'string theory' is considered another slice of the public language of cosmology, and sadly, numbers of professing evangelical Christians, as the old-earth creationist 'Dr Hugh Ross,' have flown onto the 'string theory' movement. Many, always excited to find a way of not seeming too 'out of kilter' with the secular scientific community, falsely proposition that the other dimensions appealed by 'string theory' could be the 'other' spiritual dimensions stated in Scripture. This is sacrilegious!

Of course, it is rational to shoulder that there is another realm or dimension in which angels and/or the infinite God may exist, and Paul spoke about a 'third heaven' *(2 Corinthians 12:2)*. These may be compartments or areas of a spiritual realm or dimension, though not necessarily extra dimensions in themselves. The Bible does not speak about other dimensions precisely, and venturing on such matters in too much detail can lead to positions neighboring on the sacrilegious. This is one of the problems of taking the presently prevalent views of secular evolutionary scientists and attempting to acclimatize them to Scripture.

## Here's the Kicker

Evolutionary notions as 'string theory' begin from a worldview framework that there is no God. Consequently, it is perilous practice to intermarry secular conjecture with the Bible. For example, similar to the 'big bang,' there are numbers of varieties of 'string theory.' Thus, which one should we suspend our theology on? If the secular theory deviates tomorrow, do we then revise our theology? Lastly, when the Bible cannot be warped appropriately to include man-made ideologies that we undertake to be truthful, what transpires to one's faith?

It's a much more consistent practice to start with Scripture when attempting to comprehend our world and universe. The realism is, there is still considerable knowledge we don't comprehend. Nonetheless, we can out faith in God. On balance, He is the One who created it all at the outset. He inspires us to *'Trust in the Lord with all your heart, And lean not on your own understanding; In all your ways acknowledge Him, And He shall direct your paths'* (Proverbs 3:5–6).

## Distant Starlight Reaches Earth in 6,000 Years

How could starlight arrive to earth in 6000 years?

"God is not a man that He should lie, nor a son of man that He should change His mind." *(Numbers 23:19, NIV)*

The above Scripture positions the indispensable trustworthiness of God. In view of this, if the whole universe -- and not just the planet earth was created nearly 6,000 years ago, how do you comprehend such measures as, for example, *"supernova 1987A,* Figure 4.26" in which a star burst at a distance of approximately 170,000 light years from the solar system?

$$[170,000 \times 6,000,000,000,000 \text{ miles}]$$
$$= [1,020,000,000,000,000,000 \text{ miles}]$$

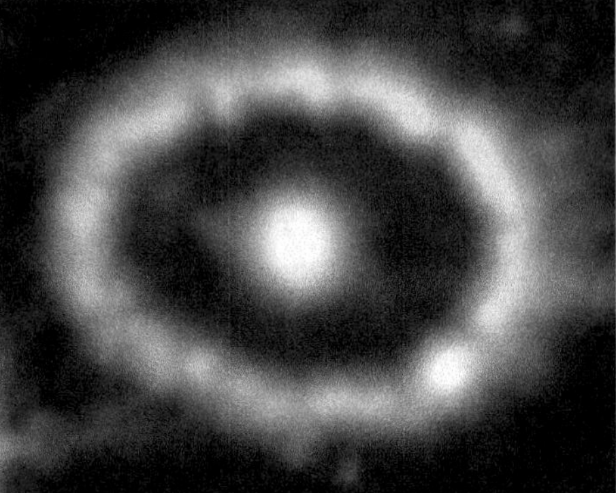

*Fig.4.26: Supernova*

The astronomical distance scale is obviously not absolute, however, there is little foundation for believing it to be overvalued by a factor of about 30 at such a relatively close distance. Other astronomical observations seem to signify vastly more distant objects, seen therefore at a much earlier epoch.

To advocate that the whole observable universe was created with a history of actions --that never really happened-- appears to be erratic with the character of the God defined in the Bible.

"God is not a man that He should lie, nor a son of man that He should change His mind." *(Numbers 23:19, NIV)*

Certainly, and it is God's dependability that gave us the foundation for the learning and development of science originally and is the reasons science developed in cultures that detained a Judeo-Christian worldview. Likewise, God's trustworthiness delivers the foundation for our considerations of the articulateness of Scripture, that is, the confidence that it was written to be realized and that God would not deceive or mislead us. The pure meaning of the opening chapters of Genesis can, consequently, be believed as an historic account of what really occurred during the Creation Week, the Fall, the Flood and the dispersal at the Tower of Babel.

Unquestionably, it does not shadow that declarations made by fallible *people* grounded on their observations of the physical world are essentially effective or trustworthy.

It is interesting to note that 'big bangers' have *exactly the same problem.* That is, the background radiation temperature is almost uniform, to one part in 100,000, at about 2.725 K, even when we look in the opposite directions of the cosmos.

Since the 'big bang' would predict enormously different temperatures, how did they become so even? Only if energy was moved from hot parts to cold parts. Though, there has not been nearly enough time for this to happen even in the expected time since the unproven 'big bang.'

The misotheistic publication *New Scientist* admitted in 13 things that do not make sense (19 March 2005, updated 14 April 2009):

This "horizon problem" is an immense annoyance for cosmologists, so immense that they have crop up with some fantastic wild answers. "Inflation", for example. You can crack the horizon problem by having the universe expand super-fast for a time, just after the 'big bang,' gusting up by a factor of $10^{50}$ in $10^{-33}$ seconds. However, is that just hopeful thinking?

Other 'big bangers' have attempted to remedy this annoyance by suggesting that the speed of light was much faster in the past, e.g., João Magueijo and John Barrow. Yet when some creationists proposed something similar a few decades ago, it was a heresy! i.e., anything goes when it comes to saving the 'big bang' dogma, but saving Genesis by precisely the same means is forbidden. Though, the observations of uniformity in the cosmic background radiation, which confront reasonable evolutionary clarifications, are consistent with a single Creator of space/time who holds the universe together *(Colossians 1:17)*.

## GOD STRETCHED OUT THE HEAVENS

In at least 11 places, the Scriptures speak of God 'stretching out the heavens' (e.g., Job 9:8, Isaiah 40:22 and 42:5, Jeremiah 10:12, Zechariah 12:1)

- Job 9:8 **who** alone stretched out the heavens and trampled the waves of the sea;
- Isaiah 40:22 It is he who sits above the circle of the earth, and its inhabitants are like grasshoppers; who stretches out the heavens like a curtain, and spreads them like a tent to dwell in;
- 42:5  Thus says God, the LORD, who created the heavens and stretched them out, who spread out the earth and what comes from it, who gives breath to the people on it and spirit to those who walk in it:
- Jeremiah 10:12 it is he who made the earth by his power, who established the world by his wisdom, and by his understanding stretched out the heavens.
- Jeremiah 10:12 **the** oracle of the word of the LORD concerning Israel: Thus declares the LORD, who stretched out the heavens and founded the earth and formed the spirit of man within him:
- Genesis 1:15 and let them be lights in the ***expanse of the heavens*** to give light upon the earth." And it was so

and in Genesis 1:15 the words 'And it was so.' are chronicled in linking with the events of "Day 4" of Creation Week, point toward the conclusion of the events described on that "Day." It is a rational conclusion to draw that God ***stretched out the heavens*** to the massive extent of the observable universe in just "one 24-hour day" and then ended the action of '***stretching out***'. This is more rational than the inflation fudge of big bangers. That is, where the universe just occurred to expand much faster than light.

We should also observe that God created the Earth first before the sun, moon and stars, and by implication the planets etc., consequently, it would seem rational to accept the universe was ***stretched out*** with the Earth at or very near its center. Besides, Psalm 147:4 and Isaiah 40:26 imply that there is a finite number of stars in the universe.

- Psalm 147:4 He determines the number of the stars; He gives to all of them their names.
- Isaiah 40:26 Lift up your eyes on high and see: who created these? He who brings out their host by number, calling them all by name; by the greatness of his might and because he is strong in power, not one is missing.

Therefore, the Bible appears to impart that we live in a limited universe that has, at the very least, our Milky Way galaxy at its center.

### Distant Starlight and the Biblical Timescale

According to Einstein's Theory of Relativity, we understand that the Space/Time are inseparable, Figure 4.27. As space stretched light is consequently transported to wherever the space is located.  If the space/time is configured upward, downward, sideway, or stretched to a determined location, light is transported as an inseparable jointed twin.

We therefore, have the explanations to comprehend the methodology of the Infinite all powerful God had allowed the domain of the first sunlight traveling through the initial space to be stretched at an instant of God's choosing.  Starlight can reach Earth from such vast distances at any time the Lord God desired. The days of the Creation Week were recorded from the point of view of an observer on the earth so the time reference in Genesis is Earth time. On "Day 4," as God commenced stretching out the heavens, the mass of the universe, presumably including the 'waters above' which were separated out on "Day 2" would have been confined to a much smaller volume of space than is the case today.

**Fig.4.27:** *Space/Time Inseparable Correlation*

By the end of Day 4, when God completed his work of creating the sun, moon and stars, and had stretched out the heavens to their vast extent. There would have been more than enough time for the light from distant stars to have reached the earth so that when Adam gazed at the night sky on that sixth night, he would have seen much the same as what we see today.

6,000 years have passed since the Creation Week. The outlined explanation above considers the light we see today from any star that is within 6,000 light years away from the earth will have originated on "Day 4" itself. This would include most of the visible stars, all of which are part of the Milky Way galaxy. We are effectively looking at God's creative activity on "Day 4" as we gaze into the universe!

Therefore, the supernova 1987A – At 170,000 light years away, we are in fact looking at an event that occurred on "Day 4," whose light was observed exactly at the right moment.

Is an exploding star consistent with a perfect creation? God said that the stars were created to be for signs and seasons *(Genesis 1:14)*, and God said, "Let there be lights in the expanse of the heavens to separate the day from the night. And let them be for signs and for seasons, and for days and years, and God foreknew all that would happen right from the very beginning.

## Hawking Goes Beyond the Evidence

Stephen Hawking once again spread the headlines with his new book, "*The Grand Design*," which allegedly explains the universe without God.

## A brief history of Hawking

Stephen Hawking (born in 1942, died in March 14, 2018) is one of the well-known scientists in the world today. Thus far his accomplishments are not so broadly known as Sir Isaac Newton or Professor Albert Einstein. His highest influence was perhaps a combination of "quantum mechanics" and relativity: in 1974 he mathematically illustrated that blackholes, a verified prediction of general relativity, will gradually lose mass through a quantum mechanical effect that would result in emission of radiation. It is distinguished that the much overhyped peer-review process originally disallowed this eponymic Hawking Radiation.

In 1979, he was presented the prestigious Lucasian Chair of Mathematics at the University of Cambridge, England, which he detained for 30 years. This chair was once held by the great creationist scientist, Sir Isaac Newton.

Thus far he has never won the Nobel Prize, since the commission insists on experimental evidence—only fairly lately has science fashioned good evidence for blackholes as a whole, so evidence for a very weak radiation from them is outside current detection devices. And in 2004, he retracted one of his major theories.

In 1975, he contended that a blackhole would annihilate all information about the nature of the matter inside, even overwhelming "quantum mechanical" laws that would preserve it. However, he lately believed that some information would escape.

Hawking's renown mostly rests on his popular-level book "*A Brief History of Time: From the Big Bang to Black Holes* (1988)." This was an enormous best-seller, nonetheless, has also been so-called "the most *widely unread book in the history of literature*". Additional key contributor is how he has overcome his extreme disability began by "Amyotrophic Lateral Sclerosis (ALS)," normally called Lou Gehrig's Disease, that has left him contingent on a wheelchair and speech synthesizer.

This book was more famed for its philosophy than for its science, with his well-known rhetorical question, 'What place, then, for a creator?' and the conclusion: If we discover the response to that, i.e., "*why it is that we and the universe exist,*" it would be the eventual triumph of human intellect—for then we would know the mind of God.

**Nonetheless, in Hawking's book, he had to admit:**

"This [big bang] picture of the universe … is in agreement with all the observational evidence that we have today. … Nevertheless, it leaves a number of important questions unanswered … (the origin of the stars and galaxies)."

Actually, the big bang has come under severe attack from other cosmologists. Nevertheless, even granting that Hawking is right for the purpose of discussion, an incapability to explain such important cosmological things as stars and galaxies is a major inadequacy.

Later, Hawking slowly understood that a 'theory of everything' is a imaginary that founders on "*Gödel's incompleteness proof:*" that in any theoretical system as intricate and difficult as arithmetic or above, there would continuously be factual declarations that cannot be established within the system.

## Scientists and 'God'

1.  When non-Christian scientists express or write of 'God', they do not typically mean the God of the Bible, unless they are mocking what the Bible says. Then, they may mean something similar to 'the substance for existence', 'a profounder level of explanation', or something similarly ambiguous or obscure. For Stephen Hawking, 'God' seems to be 'a whole theory of the universe' that 'breathes fire into the equations.'

2.  Any theory that is in harmony with the true God is itself not true and will not endure the test of time.

3.  While Hawking and Davies say that 'What transpired before the "big bang"?' is a non-question, Christians comprehend that God continuously existed before He created the universe, and He continuously was and still is now transcendent to this universe, since He 'inhabiteth eternity' *(Isaiah 57:15)*. Christians are bright to experience instantaneous spiritual interaction now through prayer with this transcendent holy God.

## Origins and the Bible

1.  Any theory of origins that is conflicting to the early chapters of Genesis, *Exodus 20:11; 31:17,* and the many other references in the Bible, that ascribe creation to the work of God, is not true and will not endure the test of time, no matter how well it is endorsed by human-centered educational institutions, Figure 4.28.

2.  In adopting the above defiance, are Christians in threat of reiterating the blunder of the seventeenth-century Church when it contrasted Galileo for suggesting a heliocentric (sun-centered) mechanism for our planetary system? Answer: No. While the Church authorities in Galileo's day erroneously believed that the Bible supported the Greek idea of a geocentric (earth-centered) system, there was nothing essentially anti-Creator about Galileo's idea that the earth moved.

**Fig.4.28**: *The Bible contains 66 books and was written by more than 40 authors over a period of roughly 1,500 years.*

By dissimilarity, the 'big bang' and all other theories of origins that are founded on human philosophies try to say how the universe made itself by its own progressions and properties, and with no supernatural input.

Hawking has also made some headlines with some way-out statements. In 2000, he declared that genetic engineering of humans is unavoidable, acknowledging it will cause countless social and political complications. One article quotation marks Professor Hawking as saying: *"It may not be in accord with democratic or egalitarian principles, but evolution has never been politically correct."* Unfortunately, evolution-founded eugenics concepts were politically correct for the first few decades of the 20th century. Earlier in the same year, Hawking warned against contact with extra-terrestrials, as the significances would be devastating: "If extra-terrestrials visit us, the consequence would be much as when Columbus landed in America, which didn't pose well for the Native Americans." Thus far he maintained that we should takeover space or perish: *"Our only chance of long-term survival is not to remain inward looking on planet Earth, but to spread out into space."*

## Atheistic Faith Masquerading as Science

As typical with atheistic scientists, Hawking's atheopathy long predated his science. His influential mother Isabel was a "Communist," and in his teen years he venerated the strongly anti-Christian mathematical philosopher Bertrand Russell.

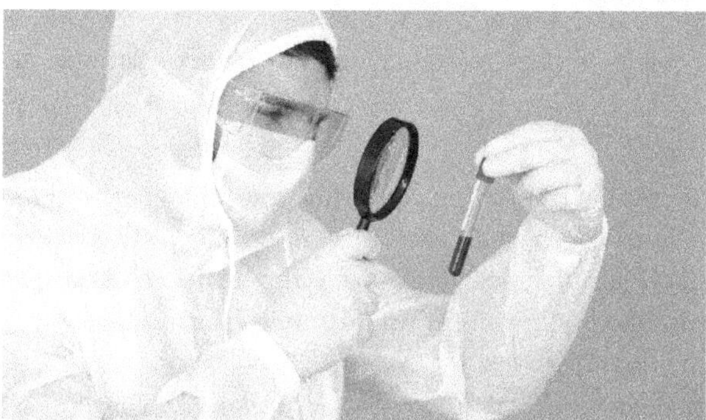

*Fig.4.29: What to say when someone asks for proof of God's existence?*

As with Dawkins, his opinions for atheism were immature. He said, *"We are such insignificant creatures on a minor planet of a very average star in the outer suburb of one of a hundred billion galaxies. So, it is difficult to believe in a God that would care about us or even notice our existence,"* Figure 2.29.

Nevertheless, King David was similarly aware of our tininess equated with the universe's massiveness, and came to a dissimilar deduction in *Psalm 8:3–5*:

"When I look at your heavens, the work of your fingers, the moon and the stars, which you have set in place, what is man that you are mindful of him, and the son of man that you care for him. Yet you have made him a little lower than the heavenly beings and crowned him with glory and honor."

Likewise, as C.S. Lewis pointed out, the medieval theologians were well conscious that compared to the vastness of heavens, the earth was but a point in space. While, somehow modern antitheists think this is news, regard it as a deep disproof of God, as if God desired a small universe to exist. And if the universe were small, then these same atheopaths would perhaps whimper, *"If God is so great, then why didn't He create anything else?"*

## Jane Wilde and Hawking

Hawking met his wife to be Jane (née Wilde) in 1962, a year before he was diagnosed with his degenerative disease. Jane is an intellectual in her own right, with a doctorate in Medieval Portuguese Literature. She is also a Christian, and ill-advisedly violated Paul's instruction against unequally yoked with an unbeliever *(2 Cor. 6:14)*—they married in 1965. Jane did not imagine him to live very long; he had been assumed two years— the time the eponymous Lou Gehrig (1903–1941) endured his diagnosis. However, one way or another, this engagement made an enormous change in his life, as he self-confessed, *"what really made a difference was that I got engaged to a woman named Jane Wilde. This gave me something to live for."* His biographers said:

There is little doubt that Jane Wilde's appearance on the scene was a major turning-point in Stephen Hawking's life. The two of them began to see a lot more of one another and a strong relationship developed. It was finding Jane that enabled him to break out of his depression and regenerate some belief in his life and work. For Hawking, his engagement to Jane was probably the most important thing that ever happened to him. It changed his life, gave him something to live for and made him determined to live. Without the help that Jane gave him, he would almost certainly not have been able to carry on or had the will to do so."

Their marriage soon brought to fruition three children. Thus far although Stephen lived far longer than anyone predicted, his body worsened markedly. Jane said in 1986, *"Without my faith in God, I wouldn't have been able to live in this situation."* Unfortunately, Stephen's antitheism developed more unbending and vicious, which caused foremost conflicts. Nevertheless, she stuck with him, following *1 Corinthians 7:12–17*. However, sooner or later Stephen terminated the marriage after 25 years. Jane later wrote an perceptive autobiography, seeing-through the conflicts between her Christian faith and Stephen's dogmatic nonbelief.

## Atheistic Book

Hawking one more time has made the headlines with his novel book, co-authored with science writer and physicist "Dr Leonard Mlodinow," oddly entitled "*The Grand Design*." This evidently shows that no Creator was required. However, once again, he went far outside the evidence. And there were many people who have not been enthralled with the book who we would supposed to welcome it. The "*ultra-liberal and anti-Christian New York Times published a review:*"

The actual news about "*The Grand Design*," though, is not Mr. Hawking's hypothetical discarding of God, information that would astonish no one who has followed his work closely. The factual news about "*The Grand Design*" is how poorly tinny and tasteless it is. The spare and earnest voice that Mr. Hawking employed with such plea in "*A Brief History of Time*" has been substituted here by one that is alternately condescending, as if he were Mr. Rogers elucidating rain clouds to toddlers, and impenetrable.

"*The Grand Design*" is crammed with grating yuks. "*If you think it is hard to get humans to follow traffic laws,*" we read, "*imagine convincing an asteroid to move along an ellipse.*" (Oh, my.)

And "*The Times*" (UK) published a review:

"*The present book consists of only 208 large-print pages, yet I thought it too long — it reads like a stretched magazine article. Even allowing for the need for the pleasures of digression, there is too much padding and too much recycling of long-stale material. It gives me no pleasure at all to say that I doubt whether 'The Grand Design' would have been published if Hawking's name were not on the cover.*"

## Extrasolar Planets

Hawking cites the detection of extrasolar planets as a rejuvenating point in contradiction of Sir Isaac Newton's faith that the universe must have been planned:

That brands the chances of our planetary circumstances—the solitary Sun, the lucky combination of Earth-Sun distance and solar mass—far less extraordinary, and far less convincing as evidence that the Earth was prudently designed just to satisfy us human beings.

Nevertheless, extrasolar planets have instigated far more difficulties for evolutionary models of astronomical systems, counting the Nebular Hypothesis. For example, to obtain "hot Jupiter's," evolutionists must suggest that they shaped far enough from the star for water vapor to condense, then migrated inwards. Other extrasolar planets have exceedingly slanted or even retrograde orbits, i.e. in the opposite direction to their star's spin. Rather, extrasolar planets suggest our solar system is unique and young.

## Hawking Fails Logic and Meta-Science

Hawking's crucial proclamation is that the 'big bang' followed inescapably from the laws of physics so required no creator: "*because there is a law such as gravity, the universe can and will create itself from nothing.*"

However, logic doesn't appear to be his convincing point; '*self-creation*' is self-contradictory. Something can do something—including create—only if it exists; something not so far existing has no power to do anything, *including create itself.*

And for such a great scientist, he appears rather inexperienced about the *meta-science* problems, i.e., the expectations that *overlie* science and permit it to work. For example, his remark presumes that laws can do anything, but I've pointed out before that this kind of claim:

… deals with natural laws as actual entities. In truth, scientific laws are *descriptive* of what we observe occurring frequently, just as the outline of a map defines the form of a coastline. dealing with scientific laws as prescriptive, i.e., the cause of the observed regularities, is like claiming that the drawing of the map is the cause of the shape of the coastline.

Similarly, "Professor John Lennox," who overwhelmed "Dawkins" in a debate, in his review of Hawking, pointed out:

Nonetheless, opposing to what Hawking claims, physical laws can never deliver a whole clarification of the universe. Laws themselves do not create anything; they are merely a explanation of what ensues under certain circumstances. *The very reason science flourished.*

What Hawking seems to have done is to confuse law with agency. His invitation on us to select between God and physics is a little like someone demanding that we choose between "aeronautical engineer Sir Frank Whittle" and "the laws of physics" to explain the jet engine.

That is a confusion of category. The laws of physics can enlighten how the jet engine works, but somebody had to build the thing, put in the fuel and start it up. The jet could not have been created without the laws of physics on their own—but the mission of development and creation required the genius of "Whittle" as its agent.

Correspondingly, the laws of physics could on no occasion have essentially built the universe. Some agency must have been involved.

To employ a simple equivalence, Sir Isaac Newton's laws of motion in themselves not ever guided a snooker ball running across the green baize. That can only be achieved by people utilizing a billiard ball cue and the movements of their own arms.

Hawking also employed the usual game of "warfare of religion v science" for all its worth. Therefore, his leading actor has long been Galileo, notwithstanding the fact that his dispute was really science v. science.

Hawking also praises the Ionian Greeks with realizing the nature of scientific laws. Nevertheless, he disregards the all-embracing research that illustrates that science itself flourished only under the Christian worldview, and was stillborn in other cultures, *including ancient Greece.* Consequently, it flourished in the European Middle Ages under an over-all Christian worldview, and even more so with the unambiguous biblical worldview of the Reformation. This was as a result of the suppositions obligatory for science to work in the first place, including the realism and shrewdness of both the universe and our own thoughts.

"Lennox," a mathematician who is also very educated in the philosophy of science, raises the similar points:

The very motive science succeeded so energetically in the 16th and 17th centuries was exactly due to the certainty that the laws of nature which were then being discovered and well-defined echoed the influence of a divine law-giver.

One of the essential themes of Christianity is that the universe was built consistent with a coherent, intelligent design. Far from being at odds with science, the Christian belief essentially brands perfect scientific logic.

Many years back, the scientist "Joseph Needham" conducted an heroic investigation of technological development in China. He desired to learn why China, for all its initial gifts of innovation, had tumbled so far behind Europe in the development of science.

He unenthusiastically derived the deduction that European science had been stimulated on by the prevalent faith in a cogent creative force, recognized as God, which enabled all scientific laws understandable.

We acknowledge and deeply regards the devotion and the creativities of Hawking work, while we certainly disagreed with him on several issues, in particularly, his vehement denial to his own caring creator. Below, some of Hawking's achievements.

- Hawking's supreme works were in blackhole physics.
- He has bravely fought against an awful physical disability. His Christian wife Jane was a great sustenance, but eventually he left her after 25 years of marriage.
- His celebrity basically rests on his weak efforts to reject God based on provocative physics. His atheism was displayed early, and was an assumption he brought to his physics; it was not a resultant from his science. It was also a cause of mounting conflict in his marriage, that strained it to breaking point.
- Hawking's up-to-date work contains falsehood in logic and the philosophy of science. E.g., "self-creation" is rationally inconsistent, and the laws of science cause nothing to occur but define what does occur.
- He propositions a theory of multiverses, but this is not scientific as they cannot be observed.
- His M-theory is not reinforced by a shred of experimental evidence.

Thus, a summary of Hawking's approach is:

- The universe appears designed, but a designer is not permitted.
- So, there must be some other explanation.
- Let's resort to some other religious notions to clarify the appearance of design (the multiverse).
- Then let us utilize even more religious concepts to support our religious idea.
- And then let us claim it is science to illustrates no designer was necessary.
- We win!

## Impact of Detecting Gravitational Waves on Biblical Creation

*Fig.4.30: Einstein predicted the existence of gravitational waves in 1915. Now*

- Gravitational waves as predicted by Einstein, Figure 4.30, were observed by the (Laser Interferometer Gravitational-Wave Observatory) LIGO, Figure 4.31, observatories for the first time on September 14, 2015.
- The detection strongly supports Einstein's general theory of relativity published in 1916 where Einstein predicted such a phenomenon. No evidence for violation of general relativity was observed, Figure 4.31.
- A binary pair of black holes were observed to coalesce—the first time their existence confirmed.
- Their distance, determined from luminosity, is about 1.3 billion light-years.
- The black holes had masses of 36 $M_\odot$ (mass of Sun) and ٢٩ $M_\odot$ before coalescence and ٦٢ $M_\odot$ after they combined. An equivalent of ٣ $M_\odot$ was radiated away as gravitational waves.

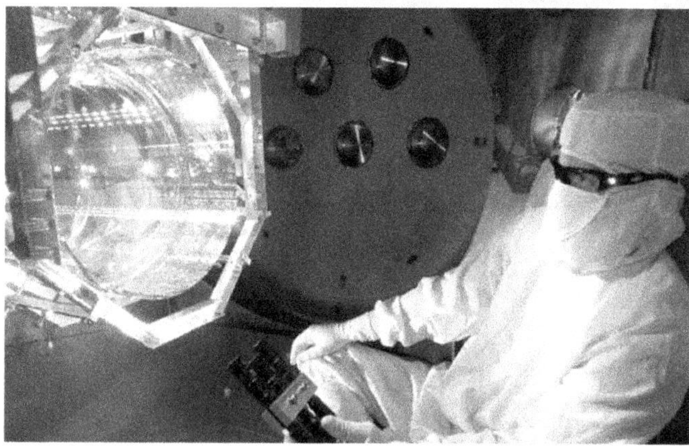

**Fig.4.31**: *Laser Interferometer Gravitational Wave Observatory (LIGO) optics technician inspecting one of LIGO's core optics – Curtsy – (AFP/Caltech/MIT/LIGO Lab)*

- There is very high confidence that the event seen at two widely separated sites must be real. The quality of the detected signals are high and were the same at each site.
- This is, in principle, repeatable (with other binary sources) and therefore is operational science. No fudge factors were invoked.
- The laws of physics used are the created laws of our God.
- The detection provides strong confirmation that the current value of the speed of light has not changed since creation. Therefore, the idea of $c$-decay is ruled out.
- There are other more plausible solutions to the biblical creationist starlight-travel-time problem.
- Big bang cosmology is not operational science. This observation in no way strengthens claims that the alleged big bang happened. The big bang necessarily still needs many unverifiable fudge factors. It is still unreasonable.

Big bang cosmology is not operational science. The hypothetical big bang "origin of the universe" from a universal singularity - not a black hole, which is an extravagant term for nothing, is not repeatable science. Nor are there other universes (multiverse) that we can observe to test how a characteristic universe began in a big bang or otherwise. The failed "BICEP2" claim of detection of 'primordial gravitational waves; and the 'smoking gun' indication of the inflation epoch, exemplifies the problem. The claim at the time was that it was 'smoking gun' evidence. That is an overt declaration that the event itself was not observed, but unobserved forensic or hypothetically circumstantial evidence after the fact.

Then there is the problem of degeneracy. In astrophysics and cosmology this means that there are a overabundance of conceivable theories to clarify the same cosmological observations. Just detection of Cosmic Microwave Background (CMB) radiation, which was a 'big bang' prediction of George Gamow in 1948, is not adequate reason (evidence) to settle that the big bang occurred -- at some instant in the unobserved past. You would have to demonstrate that all other conceivable causes for the CMB radiation are ruled out. In addition, there is inconsistent evidence that supports the concept that the CMB radiation is not even from the background and therefore, it can't be leftover radiation from the big bang fireball, as is alleged.

If conflicting evidence was discovered that ruled out this gravitational wave detection then that should be extremely considered. But you think that that is improbable. Ruling out the very improbable prospect of gross deception, by a numerous of scientists involved in the discovery, it is hard to see that this could be anything else other than an honest detection, subsequently it has all the hallmarks of the laws of physics that we do understand. No unidentified unknowns were invoked to get the observations to fit the theory. No evidence for defilement of general relativity was observed.

Now, these laws of physics, are divinely designed laws from the hand of the Creator of this universe. The detail that the blackhole system is so far away --admittedly there are some expectations to derive that fact-- and the same laws we have discovered on earth apply out there expresses to us of the constancy of those laws. They are the creation of an Intelligence, a Creator, and we are just learning how delightfully He made this universe.

One may be doubtful that there will be a crowd of claims on the internet and in other news media that this discovery someway authenticates the big bang origin of the universe. Nonetheless, it doesn't!

The standard big bang cosmology is founded on the solution of Einstein's field equations found by Friedmann and Lemaître, in the 1920s. Those same field equations were linearized in what is called the post-Newtonian approximation, and from that Einstein advanced the theory for gravity waves spreading through "*spacetime*." But there are many conceivable mathematical solutions of Einstein's field equations for the whole universe, many of which have already been rejected, as not fitting what we observe. The presence of a solution does not mean it has any physical significance. Einstein himself gained the Einstein static universe solution, which he later rejected. Because he had included the "*cosmological constant* ($\Lambda$)" to maintain a *static* universe, when he learnt of Hubble's 1929 discovery of an expanding universe, he exclaimed that its inclusion was the biggest blunder of his career.

Every solution necessitates a set of expectations, which are called boundary conditions. These are expectations about the initial conditions, and in the case of the Friedmann-Lemaître solution it necessitates the "*cosmological principle*," which is an assumption that states that the universe is 'isotropic' and homogeneous, or uniform. That means that the matter density in the universe, on the large scale, is the same all over the universe, and that there is no unique center nor any boundary or edge to the universe. It also adopts the laws of physics are the same everywhere and at every epoch.

Biblical creationists would agree that the laws are the equal at every place in the universe, but not essentially at every epoch, because there was a very special Creation epoch—Creation week. Big bang cosmology also has an exemption, at the big bang itself, which is effectively a miracle without any satisfactory cause (or explanation).

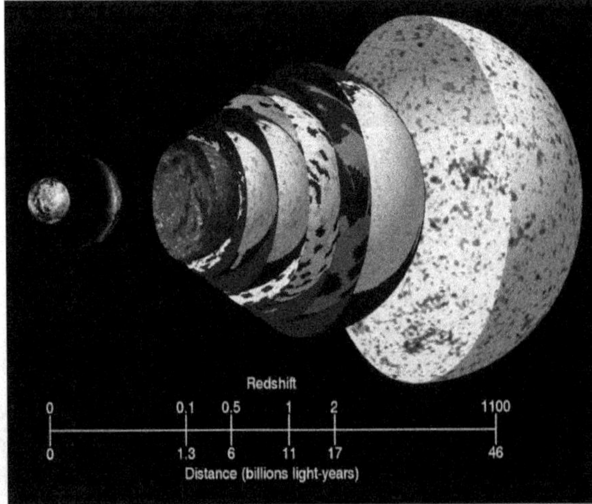


*Fig.4.32*: *An illustration of the cosmic radiation background at various redshifts in the Universe.*
*CMB: CURTSY NASA/WMAP*

In addition, the issue of the topology of the universe—whether it has a unique center and an edge—the "*cosmological principle*" has a few immense problems. One of them is the 'Axis of Evil.' This is the purpose of a peculiar alignment of the temperature fluctuations originate in the Cosmic Microwave Background CMB radiation, Figure 3.32, from both the "Wilkinson Microwave Anisotropy Probe" WMAP and the "Planck" satellites. Those data autonomously determined the identical anomalous axis in the universe, aligned with the plane of our solar system, in the specific direction determined by the two points where the sun's path crosses the earth's equator each year. Nevertheless, such an extraordinary axis in space should not exist. The local physics of our solar system and that of the 'big bang' fireball should have no connection. This disproves the homogeneity and isotropy requirement of the cosmological principle, and because it does so much damage to their theory, the big bang cosmologists have called it the 'Axis of Evil'.

Another immense problematic that has developed as a significance of acceptance of the standard "Lambda Cold Dark Matter" $\Lambda$CDM big bang cosmology for the universe is the acceptance in dark energy and dark matter. Because the observations on the large-scale measurements in the universe do not fit the modern form of the Friedmann-Lemaître model, dark energy and dark matter were invoked to get agreement. Dark energy, a kind of anti-gravity, was derived through the cosmological constant ($\Lambda$) however, dark matter was essential to reinforce the total quantity of matter since the small amount of normal observed matter was inadequate to allow the theory to agree with the observations. Dark energy and dark matter are unknowns to science and hence we call them fudge factors, unknown unknowns, or 'gods of the gaps' for modern cosmology.

Fascinatingly, the calculation employed to find the masses of the merger of blackholes in the analysis utilized the standard acknowledged speed of light, $c$. That is, it utilized the same constant value that we measure today. Does that express to us something? we think it does.

Many biblical creationists prefer a much greater value for the speed of light in the past, from a time immediately after creation of the universe, after which it decreased or decayed down to its present value -- the concept is known as cdk, from $c$-decay. They utilize this supposed much greater value of $c$ in the past as a solution of the biblical creationist light-travel time problem. Nevertheless, now this novel discovery illustrates that, at a time in the past illustrative of a distance in the cosmos of 1.3 billion light-years, the value of the speed of light ($c$) was identical to today's current value. Irrespective of which creationist cosmology you like, the gravity waves observed in September 2015 must have left their source very soon after Creation week. Consequently, the cdk idea is thoroughly rejected.

Un conclusion, Einstein's general relativity is further strengthened as respectable operative science with no fudge or nonsense factors. Any change in the speed of light is rejected. Nevertheless, there exist other much more reasonable answers to the biblical creationist starlight-travel-time problem. With a constant speed of light, general relativity theory provides us the required evidence that time is an absolute in the universe, which means that our Lord the infante God has "stretched" the space as stated in previous sections to allow space/time to reach Earth when Adam opened his eyes for the first time. The age of the Earth is about 6,000 years since creation. There are no other implications that impact on biblical creationist explanations for the origin of the universe.

## THE FALSEHOOD OF BIG BANG AND THE MULTIVERSE

### The Untruth of the Origins and Evolution of the Universe

Dr Paul Steinhardt a distinguished Professor of Physics at Princeton University, raised his scientific opinion on the subject of young-earth vs old-earth creation, and regarding the astronomer Dr Ross, who is well recognized for his compromising extreme position in Figure 4.33, trying to force fit Genesis creation into the falsehood "Big Bang" cosmology and old-earth fossil record/isotope dating.

*Fig.4.33: In The Beginning God Created Genesis 1:1 – Curtsy – NASA*

Among a flood of highly technical evidence and arguments that must surely have gone beyond reasonable understanding of the vast majority of the people, Dr Ross quoted the recent discovery of 'gravity waves' as subsidiary to the 'inflationary Big Bang' cosmology and forcefully declared that "Only an inflationary universe can sustain life!" With this in mind.

Professor Steinhardt carefully explained: "When a team of cosmologists publicized in they had sensed 'gravitational waves' caused in the primary instants after the Big Bang, the "origins of the Universe" were once again major events among scientists. However, the described discovery created also a worldwide awareness in the scientific community, the media, in particularly and the public at large."

"Rendering to the team at the "BICEP2 South Pole telescope," Figure 4.34, there is less than one chance in two million of the 'gravitational waves' being a random incidence. The outcomes were greeted as evidence of the "Big Bang inflationary theory" and its offspring, "the multiverse."

*Fig.4.34*: BICEP2 Telescope (pictured), the "Keck Array and the BICEP Telescope" have squeezed parameter space for models of inflationary cosmology. *Curtsy – BICEP2*

Nobel prizes were foretold and scores of theoretical models reproduced. The proclamation also prejudiced decisions about academic appointments and the refusals of technical papers and grants. The announcement even had played a vital part in governmental planning of large-scale projects."

These all occurred as the consequence of a '*press conference*,' not the publication of a peer-reviewed scientific paper.

Steinhardt continued to raise his deep concerns. After careful re-examination by scientists at Princeton University and the Institute for Advanced Study, also in Princeton, has determined that the "BICEP2 B-mode pattern" could be the consequence typically or completely of foreground effects deprived of any influence from 'gravitational waves.' Other dust models well-thought-out by the "BICEP2 team" do not altera this negative conclusion, the Princeton team illustrated."

All the participants of this re-examinations concluded that "These results suggest that "BICEP1" and "BICEP2" data by itself cannot differentiate between foregrounds and a primordial gravitational wave signal, and that future "Keck Array" ['ground-based telescope'] observations at 100 GHz and "*Planck*" [satellite] observations at higher frequencies will be vital to govern whether the signal is of primordial origin."

Dr. Steinhardt then said: *"The sudden reversal [of the previous announcement] should make the scientific community contemplate the implications for the future of cosmology experimentation and theory. ..."*

"This time, the teams can be guaranteed that the world will be paying close attention. This time, acceptance will necessitate measurements over a variety of frequencies to single out from foreground influences, in addition to tests to dismiss other sources of misperception. And this time, the declarations should be made after submission to journals and vetting by skilled referees. If there must be a "press conference," confidently the scientific community and the media will mandate that it is accompanied by a comprehensive set of documents, as well as particulars of the systematic examinations and satisfactory data to permit objective substantiation."

"The BICEP2 event has also exposed a truth about 'inflationary theory.' The shared view is that it is a extremely predictive theory. If that was the circumstance and the discovery of 'gravitational waves' was the 'smoking gun' evidence of inflation, one would think that non-detection illustrates that the theory fails. Such is the character of typical science. Hitherto, some advocates of inflation who celebrated the "BICEP2" announcement even now maintain that the theory is likewise valid whether or not 'gravitational waves' are detected. How is this conceivable?"

"The response provided by advocates is disturbing: the 'inflationary paradigm' is so flexible that it is resistant to experimental and observational tests.

**First**, inflation is determined by a hypothetical scalar field, the inflation, which has properties that can be altered to yield efficiently any outcome.

**Second**, inflation does not terminate with a universe with uniform properties, but virtually unavoidably leads to a multiverse with an infinite number of bubbles, in which the cosmic and physical properties differ from bubble to bubble.

The portion of the multiverse that we observe resembles a part of just one such bubble. Scanning over all conceivable bubbles in the multiverse, everything that can physically occur does occur an infinite number of times. No experiment can rule out a theory that permits for all conceivable outcomes. Hence, the paradigm of inflation is unfalsifiable."

"This may appear puzzling provided the hundreds of theoretical papers on the predictions of this or that inflationary model. What these papers characteristically fail to recognize is that they disregard the multiverse and that, even with this unfounded choice, there exists a spectrum of other models which yield all means of varied cosmological outcomes. Taking this into account, it is evident that **the inflationary paradigm is basically untestable, and therefore scientifically worthless.**"

Dr. Steinhardt is Director of the Princeton Centre for Theoretical Science, which means he is a powerful voice in the scientific community. He is a well-known critic of 'inflationary Big Bang' cosmology and co-author of several books on cosmology subject. Dr. Steinhardt believes Dr Hugh Ross has placed his belief in a scientific theory that is patently 'meaningless.' The 'inflationary theory' leads to the multiverse concept and "in the multiverse, everything that can physically happen does happen an infinite number of times." How's that for a God-substitute!

# THE SOUTH POLE "BICEP2 TELESCOPE"

## Cosmic Inflation

Further to the previous section, the BICEP2, Figure 4.35, crew of scientists and astronomers conducting their study out of their South Pole telescope falsely issued the spectacular claim of detecting of 'cosmic inflation' through a signal that was mistakenly expected in the Cosmic Microwave Background (CMB) radiation from complementary 'gravitational waves' in the period of time much less than a second after the unproven "big bang."

Some scientists have expressed some doubts in the results. Additionally, other scientists much intimate to the field also doubted the discovery.

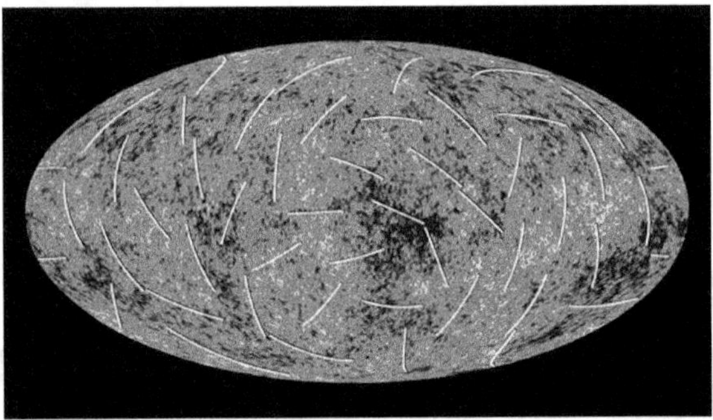

*Fig.4.35: BICEP2 Sought Characteristic Swirls in the Polarization of the Universe's so-called Relic Radiation from the Big Bang – Curtsy NASA / WMAP Science Team*

Immediately after the BICEP2 team's 25-page paper was accepted for publication in the prestigious journal *Physical Review Letters, the team* had added a half-page warning stating that they might be erroneous.

It was afterward established that they were most probably erroneous according to their not properly accounting for the effect and the influence of the foreground contamination of their alleged signal from dust discharged in the Galaxy. That places great interest in one of the substantial dangers of dashing to publish when they have not ruled out all other possible sources. Cosmology is predominantly much more problematic than other branches of science.

## The Clear of Gravitational Wave

The European "Planck" satellite team has afterward investigated at the foreground dust contamination problem:

The Planck team has indeed shadowed up the BICEP2 outcome, examining data from the same area of sky employing a variety of frequencies that variety from 30 giga-hertz to 857 giga-hertz. BICEP2 observed in just a single frequency, 150 giga-hertz. And the news was not good for the BICEP2 crew, as the new study reports.

*"Unfortunately, according to the secured analysis, the effect of contaminants and in particular of gases existing in our galaxy cannot be ruled out,"* the study was co-authored by Carlo Baccigalupi of the International School for Advanced Studies in Trieste, Italy.

Many others had already analyzed the possible results if dust emission, utilizing the "Planck" satellite data, was properly considered for in the BICEP2 examination. No gravitational wave effects of the alleged inflation epoch could be justified.

After this, all accurate data was exposed in the public domain, and many other data also were released were published on the web the reasons why the claims were either premature or erroneous. The "BICEP2" team joined up with the European "Planck Consortium" who had much improved data than they had on the influences of dust in the Galaxy. The outcomes of their cooperative work have now been submitted also to *Physical Review Letters.*

The following was taken out from an online BBC News site:

The "BICEP2" cosmologist team employed every piece of dust data it could source on the portion of the sky it was observing above Antarctica.

What it was short of, though, was entrée to the dust data being amassed by the "Planck space telescope," which had charted the microwave sky at numerous frequencies than "BICEP2."

This permitted "Planck" to more easily describe the dust and distinguish its confusing effects. The "Planck Consortium" arranged to start working with "BICEP2" back in the summer. The European group combined its high frequency data—where dust gleams most brightly—and the US team added extra data composed by its next-generation instrument in Antarctica called the "Keck Array."

Nevertheless, the outcomes of the joint valuation would propose that whatsoever signal "BICEP2" perceived, it cannot be detached at any important level from the spoiling influences. On the other hand, the original observations are likewise well-matched with there being **no primordial gravitational waves.** Thie means no "Big Bang" evidence is exhibited.

"This joint effort has revealed that the detection of primordial B-modes is no longer reliable once the emission from galactic dust is removed," Jean-Loup Puget, primary investigator of "Panck's" High Fidelity instrument, said in the Esa statement.

"So, unfortunately, we have not been able to confirm that the signal is an imprint of cosmic inflation."

Therefore, after all that it is ultimately claimed that they have no confirmed detection.

## Dubious Science

A word of caution for those who would resort the dubious science of 'big bang' cosmology, which is today founded on a disbelieving worldview as contrasting to a biblical worldview, in support of their biblical apologetics.

This image, 4,36, from the European Space Agency's "Planck" satellite illustrates the space observatory's view of the same area observed by the Antarctica-based "BICEP2" project. The "Planck" data proposes that light patterns that established cosmic inflation theory were really caused by space dust. Credit: ESA/Planck Collaboration. Acknowledgment:

- M.-A. Miville-Deschênes,
- CNRS - Institut d'Astrophysique Spatiale,
- Université Paris-XI, Orsay, France

Accordingly, the biblical creationists reject -as fairy tales- any story that does not follow the creation account of 24-hour days and a 6-day creation about 6000 years ago. Nonetheless, quite clearly the *Genesis 1* account does not approve the 'big bang,' which in no way even follows the biblical sequence of events.

If God's Word designated the origin of the universe in a 'big bang' inflation setting 13.8 billion years ago, long before God made the earth and the rest of the solar system, with all the other countless details that disagree with the biblical account, we would believe it. Nevertheless, it does not and that is where we should take our stand.

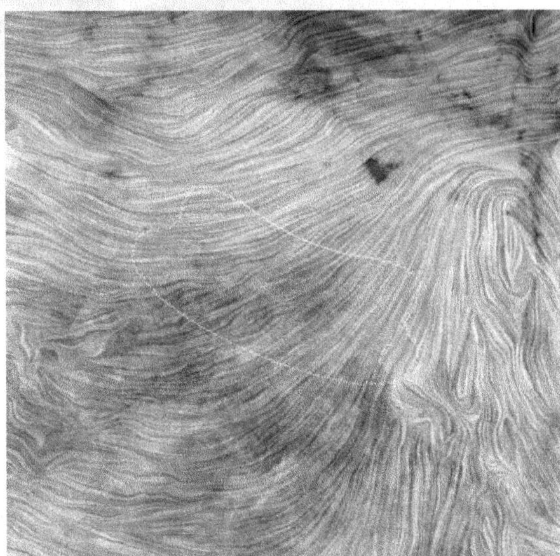

*Fig.4.36: No gravitational wave effects of the alleged inflation epoch could be justified – Curtsy space.com*

Now let us add *one more* word of caution. Even if the "Planck Consortium" scientists had so-called established the original BICEP2 B-mode, Figure 4.36, polarization results, in those swirls in the CMB radiation, it still would not demonstrate the idea of cosmic inflation. It surely would reinforce the case in favor of it nonetheless, you would still have to exclude all other conceivable causes, besides foreground dust emissions in the Galaxy, even those we have not yet considered. This is the specific problem with cosmology, as we cannot interact with the universe, we have no control experiment. Therefore, it is similar to saying we cannot obtain a control signal from a 'typical' universe that did not start off with cosmic inflation. Thus, what are the probabilities of mistaking the real source of those swirls? As it turns out – it is very high.

## Big Bang Inflation Detection was Wrong

1. Usually, *Physical Review Letters* has a stringent page limit of 4, and occasionally permitting up to 6 for enormously significant discoveries. To provide you some sense of the heaviness they sited on this 'discovery' they allowed 25 pages. Hitherto, half a page was dedicated to a face-saving exercise describing the reasons they might be erroneous. This clearly came about as the journal Editor had already accepted the paper for publication prior the counter claims and doubts had been completely declared.

2. Wall, M., Epic Big Bang Discovery Might Just Be Space Dust, 23 September 2014; space.com.

3. Mortonson, M. J., Seljak, U., A joint examination of Planck and BICEP2 B modes as well as dust polarization uncertainty, *JCAP* **10**:035, 2014; available at http://arxiv.org/pdf/1405.5857v2.pdf.

4. Amos, J., Cosmic inflation: new study declares BICEP detection was wrong, 30 January 2015; bbc.com.

## APPENDIX

- The laser interferometers used to detect these signals are about 4 km long. The strain sensitivity refers to the detection sensitivity in terms of the fractional change in length of the arms $(\Delta L1\text{-}\Delta L2)/L$. So, any putative signal can be detected that results in an absolute change in the arm length of as little as a few parts in $10^{-19}$ m. That is about a ten thousandth of the diameter of a hydrogen nucleus, i.e. of a proton.

- The luminosity of the source and its redshift were determined from the brightness of the source signal where standard big bang cosmology was applied. Nevertheless, that choice of cosmology has little impact on the veracity of the detection.

- GW means gravity wave and the date of the detection is included in the nomenclature.

- $M_\odot$ represents the mass of our sun, a solar mass unit.

- 1 ms = 1 millisecond.

- Even though I am calling this operational science, such science done in the cosmos, where the researchers have no local laboratory wherein, they can interact with their experiments, is a weaker form of the science. Therefore, a higher standard of evidence should be required before conclusions may be drawn. And even then the tentative nature of the science needs to be properly understood.

- Degeneracy in this context means there are multiple solutions that cannot be distinguished from observation.

- This is when the sun is seen exactly overhead at the equator. It occurs only twice a year due to the tilt of the earth's axis. As the earth travels around the sun, the sun is seen overhead at lower or higher latitudes. Twice a year at the summer and winter equinox, Earth's equatorial plane passes through the center of the sun. Those two points on the opposite sides of Earth's orbit, in the plane of the orbits of the planets, describes a unique direction in space.

- CDM refers to cold dark matter.

- Dark matter historically was invoked before this. It was found that is needed in spiral galaxies to get the dynamics of the rotation of the galaxies to fit standard theory. This then spread to galaxy clusters and super-clusters also.

- Canonical speed of light is defined as $c = 299{,}792{,}458$ m/s.

- They worked together on what are now called the Penrose–Hawking singularity theorems, and co-authored The Nature of Space and Time, Princeton University Press, 1996. Penrose is now Emeritus Rouse Ball Professor of Mathematics at the University of Oxford, and the brother of Jonathan Penrose, who won the British Chess Championship 10 times.

- Creationists and most evolutionary cosmogonists agree that the universe had a beginning.

## REFERENCES


1. Abbott, B. P., *et al.*, Observation of Gravitational Waves from a Binary Black Hole Merger, *Phys. Rev. Lett.* 116(6), 2016 | doi:http://dx.doi.org/10.1103/PhysRevLett.116.061102.

2. Hartnett, J.G., Has the 'smoking gun' of the 'big bang' been found? March 2014; creation.com/bbgun.

3. Hartnett, J.G., New study confirms BICEP2 detection of cosmic inflation wrong, February 2015; creation.com/inflation-wrong.

4. Hartnett, J.G., and Tobar, M.E., Properties of gravitational waves in Cosmological General Relativity, *Int. J. Theor. Phys.* 45 (11):2213– 2222, 2006.

5. Hartnett, J.G., The singularity—a 'Dark' beginning, July 2014, creation.com

6. Hartnett, J.G., An eternal quantum potential or an eternal Creator God, January 2016, biblescienceforum.com.

7. Hartnett, J.G., 'Light from the big bang' casts no shadows, *Creation* 37(1):50–51, 2015; see also biblescienceforum.com.

8. Hartnett, J.G., CMB Conundrums, *J. Creation* 20(2):10–11, August 2006. Return to text

9. Hartnett, J.G., Development of an 'old' universe in science, July 2015, biblescienceforum.com.

10. Hartnett, J.G., Big bang fudge factors, December 2014, biblescienceforum.com.

11. Hartnett, J.G., Is dark matter the unknown god?, *Creation* 37(2):22–24, 2015, biblescienceforum.com.

12. Hartnett, J.G., Starlight and time: Is it a brick wall for biblical creation? July 2015, biblescienceforum.com.

13. Hartnett, J.G., The Lecture: Starlight and time—Is it a brick wall for biblical creation?, July 2015, biblescienceforum.com.

14. Hartnett, J.G., Solutions to the biblical creationist starlight-travel-time problem, November 2014, biblescienceforum.com.

15. Batten, D., (Ed.), *et al.*, How can we see distant stars in a young universe? *The Creation Answers book*, ch. 5, Creation Book Publishers, Queensland, Australia, 2006.

16. Cho, A., Gravitational waves, Einstein's ripples in spacetime, spotted for first time, *Science*, sciencemag.com, accessed February 2016.

17. Sarfati, J., *Refuting Compromise (updated & expanded)*, Master Books, 2014.

18. Flauger, R., Hill, J. C. and Spergel, D. N., Toward an Understanding of Foreground Emission in the BICEP2 Region.

19. Steinhardt, P.J. and Turok, N., *Endless Universe: Beyond the Big Bang--Rewriting Cosmic History*, Broadway Books, New York, 2007.

20. Coping with peer rejection, Editorial, *Nature* 425:645, 16 October 2003.

21. Christian philosopher William Lane Craig showed up the shortcomings of Hawking's logic in 'What place, then, for a creator?': Hawking on God and Creation, *British J. Philosophy of Science* 41:473–91, 1990.

22. Hawking, S., *A Brief History of Time,* Bantam Doubleday Dell Pub, 10th Ed., 1998.

23. Sample, I., Ultimate equation is pie in the sky, says Hawking, *Guardian,* 23 February 22 ,2004 June 2006.

24. See Bates, G., *Did God create life on other planets? Otherwise why is the universe so big? Creation* 29(2):12–15, 2007 and his book *Alien Intrusion: UFOs and the Evolution Connection*, CBP, 2005, 2010.

25. White, M. and Gribbon, J., *Stephen Hawking: A Life in Science*, Book Club Associates, London, 2002.

26. Hawking, Jane, Music to Move the Stars, McMillan, New York, 2004; see review by Jerry Bergman: "*Stephen Hawking: the closed mind of a dogmatic atheist*", *J. Creation* 19(3):29–33, 2005.

27. Hawking, S. and Mlodinow, L., *The Grand Design*, Bantam Press, 2010.

28. Garner, Dwight, Many Kinds of Universes, and None Require God, *New York Times,* 7 September 2010.

29. Mayor, Michael; Queloz, Didier, A Jupiter-mass companion to a solar-type star, *Nature* 378 (–355:(6555 1995 ,359 P-I-P-E doi:378355/10.1038a0.

30. Spencer, W., The Origin and History of the Solar System, in: Walsh, R.E., ed., *Proceedings of the Third International Conference on Creationism*, pp. 523–513, Creation Science Fellowship, Inc., Pittsburgh, PA, 1994.

31. Henry, J., Solar System formation by accretion has no observational evidence, *J. Creation* 24(2): 87–94, 2010.

32. Oard, M., *The naturalistic formation of planets exceedingly difficult, J. Creation* 16(2):20–21, 2002.

33. Sarfati, J., Solar system origin: Nebular hypothesis, *Creation* 32(3):34–35, 2010.

34. This involves a complex theory where the dust disk as well as the other existing planets change the positions of Uranus and Neptune. See Spencer, W., *Migrating planets and migrating theories, J. Creation*

21(3):12–14, 2007, as well as the *Creation* magazine articles online at creation.com/uranus and creation.com/Neptune.

35. Bernitt, R., *Extrasolar planets suggest our solar system is unique and young*, J. Creation 17(1):11–13, 2003.

36. Sarfati, J., Miracles and science, creation.com/miracles, 1 September 2006.

37. Sir Frank Whittle (1907–9 August 1996), is generally regarded as the father of modern jet propulsion.

38. Lennox, J., As a scientist I'm certain Stephen Hawking is wrong. You can't explain the universe without God, *Mail Online*, 3 September 2010.

39. Sarfati, J., *The Galileo quadricentennial: myth vs fact*, Creation 31(3): 49–51, 2009.

40. Sarfati, J., The biblical roots of modern science, *Creation* 32(4):32–36, 2010.

41. Sarfati, J., *Why does science work at all*, Creation 31(3):12–14, 2009. Both this and the previous article provide the material for *The Greatest Hoax on Earth*? pp. 310–320, 2010.

42. Barrow and Tipler, *The Anthropic Cosmological Principle* (Clarendon, 1986), use this objection to evade the implications of a Designer. However, Christian philosopher and apologist William Lane Craig points out the fallacy in 'Barrow and Tipler on the Anthropic Principle vs. Divine Design', *Brit. J. Phil. Sci.* 38:389–95, 1988 (see also *online version*). Once this fallacy is removed, the book becomes a compendium of data of modern science which point to design in nature inexplicable in natural terms and therefore pointing to a Divine Designer. However, some of the alleged design features presuppose a big bang, so are not considered here.

43. Tegmark, M., *Parallel* universes: Not just a staple of science fiction, other universes are a direct implication of cosmological observations, *Scientific American* 288:30–41, May 2003. Of course, there is no actual observation of these other universes, just observation of fine-tuning in ours that is explained away by multiverses.

44. Penrose, R., review of *The Grand Design, Financial Times* (UK), 4 September 2010.

45. Ideas needed—The hunt for a theory of everything is going nowhere fast, *New Scientist* 188(2529):5, 10 December 2005 (emphasis added).

46. See also Bates, G., *Is 'String' the next big thing*, Creation 30(2):32–34, 2008, and response to critic at creation.com/string-theory-philosophy-challenged.

47. They co-authored The Large Scale Structure of Space-Time, Cambridge University Press, 1973. George Francis Rayner Ellis is now Emeritus Distinguished Professor of Complex Systems in the Department of Mathematics and Applied Mathematics at the University of Cape Town in South Africa.

48. Gibbs, W. Wayt, Profile: George F.R. Ellis; Thinking Globally, Acting Universally, *Scientific American* 273(4):28, 29, 1995.

49. Farmelo, G., Review of *The Grand Design, The Times*, 11 September 2010.

50. A reason for being or an explanation to justify a person or a thing's existence.

51. Why it's not as simple as God vs the multiverse, *New Scientist*, 2685, page 48, 6 December 2008.

52. Hubble, E.P., *The Observational Approach to Cosmology* The Clarendon Press, Oxford, UK, pp. ,59–50 1937.

53. Assis, A.K.T., Neves, M.C.D. and Soares, D.S.L., *Hubble's Cosmology: from a finite expanding universe to a static endless universe*, Second Crisis in Cosmology Conference, 7–11 September 2008 | arxiv.org/abs/0806.4481v1, 27 June 2008.

54. Humphreys, D.R., *New time dilation helps creation cosmology*, Journal of Creation 22(3):84–92, 2008.

55. See Sarfati, J,. *The Fall: a cosmic catastrophe: Hugh Ross s blunders on plant death in the Bible, Journal of Creation* 19(3):60–64, 2005. Return to text.

56. An Open Letter to the Scientific Community (published in *New Scientist* 22 May 21 ,(2004 February 2007; see CMI commentary, Wieland, C., Secular scientists blast the big bang: What now for naïve apologetics? *Creation* 27(2):23–25, 2005; creation.com/bigbangblast.

57. Sarfati, J., *Refuting Compromise*, Master Books, Green Forest, USA, p. 179, 2004; If God created the universe, then who created God? *Journal of Creation* 12(1):20–22, 1998, creation.com/whomadeGod.

58. See Hartnett, J., Has 'dark matter' really been proven? Clarifying the clamor of claims from colliding clusters, creation.com/collide, 8 September 2006.

59. The last word, 'Strings and M-theory are based on little more than fancy math's and a grab-bag of ideas', *BBC Focus*, p. 98, May 2006.

60. For a thorough refutation of Hugh Ross's theological and scientific claims, read *Refuting Compromise* by Jonathan Sarfati, and for a refutation of big bang teaching, see *Dismantling the Big Bang* by Alexander Williams and John Hartnett (both available from CMI).

61. See also Grigg, R., The Gospel in time and space, *Creation* 21(2):50–53, 1999. Also creation.com/time/space.

62. Logically, God must exist outside of our spacetime because He created it in the first place. However, He also interacts in our own spacetime dimension. Return to text.

63. Even self-described Ross supporter, the philosopher/apologist William Lane Craig, has severely criticized Ross's teachings on this: ' … I find his attempt to construe God as existing in hyper dimensions of time and space and to interpret Christian doctrines in that light to be both philosophically and theologically unacceptable', Hugh Ross' extra-dimensional deity: a review article, *J. Evang. Theol. Soc.* 42(2):293–304, 1999.

# THE MYSTERY OF GRAVITY

---

Gravity grasps us definitely on the ground and retains the earth rotating the sun. This unseen force too induces down rain from the sky and allows the daily ocean tides.

In the 19th century, Sir John Herschel called this enormous and stunning circular shape, Keyhole Nebula, inside the Carina Nebula. Its distance is about 8,000 light-years from Earth. This Keyhole Nebula encompasses some of the greatest immense and hottest stars recognized, Figure 5.1, with masses approximately 100 times that of our sun, and 10 times as hot. Nebulae such as these may comprise within them 'black holes', caused as a result of the collapse of a star leftover under its own gravity. Such domain would have gravity exceptionally powerful that even light rays could not escape.

Gravity maintains our earth in a spherical form, and averts our precious life-giving atmosphere from escaping into space.

It would appear that this ordinary gravity force should be one of the greatest comprehended perceptions in science. Nevertheless, just the opposite is true. In countless ways, gravity remains a deep mystery. Gravity delivers a spectacular example of the restrictions of our present scientific knowledge.

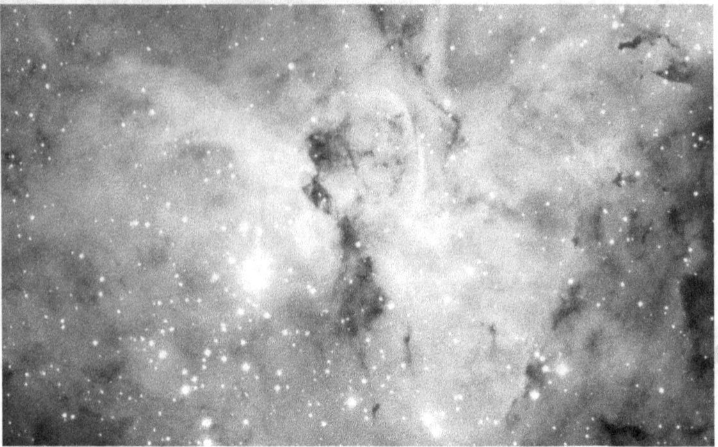

*Fig.5.1: Keyhole Nebula*

# CHARACTERISTICS OF GRAVITY

In 1686 Sir Isaac Newton questioned the nature of gravity, and determined that gravity was an attractive force between all objects. He understood that the equivalent force that causes an apple to fall to the ground also grips the moon in its orbit. Earth's gravity essentially causes the moon to divert about one millimeter away from a straight-line path, each second, as it orbits the earth, Figure 5.2.

*Fig.5.2: The Moon Inches away from the Earth Every Year*

Newton's universal Law of Gravity is one of the great science discoveries of all time. Gravity is one of four recognized essential forces of nature, Table 1.  Observe that gravity is without a doubt the weakest of the four, up till now, it controls on the scale of large space objects. As Newton illustrated, the attractive gravity between any two masses becomes smaller and smaller as the distance between them becomes bigger, but it never quite reaches zero.

**Table 5.1:** The Four Fundamental Forces of Nature

| Force name | Relative strength | Responsible for |
|---|---|---|
| Strong Forces of Nature | 1 | Stability of the atomic nucleus. |
| Electromagnetic | $10^{-2}$ | Atomic, molecular bonding. |
| Weak Forces of Nature | $10^{-6}$ | Radioactive decay processes. |
| Gravity | $10^{-43}$ | Stability of space objects. |

Therefore, every particle in the entire universe essentially attracts every other particle. Gravity is a long-range force dissimilar to the strong and weak forces, Table 1. Magnetic and electric forces are likewise long-range, but gravity is exclusive in being both long-range, and always attractive, therefore, never abandoning out, dissimilar to electromagnetism, where the forces may either attract or repel.

In 1849, starting with the great creationist physicist "Michael Faraday," physicists have examined repeatedly for a concealed correlation between gravity and the electromagnetic force. There is an continuing research to unite all four essential forces into a single equation or *'theory of everything'*, with failure, up until now. Gravity remains the slightest comprehended force.

Gravity cannot be isolated in any way. Prevailing objects, whatever their make-up, have no influence of any kind on the attraction between two detached objects. This point to no antigravity compartment can ever be built in the laboratory. Neither does gravity rely on the chemical structure of objects, but only on their mass, which we distinguish as weight, the force of gravity on something is its weight — the greater the mass, the greater the force or weight. Slabs composed of glass, lead, ice or even Styrofoam, if they all have equal mass, will experience, and exert, equal gravitational forces. These are empirical findings, with no fundamental theoretical explanation.

## GRAVITY THROUGH DESIGN

The force "F" between two masses "$m_1$"and "$m_2$," when detached by a distance "r," can be written as  where "G" is the gravitational constant, first measured by "Henry Cavendish" in 1798.

This equation illustrates that gravity reduces as the separation distance, "r," between two objects becomes big but *never* fairly reaches zero.

$$F_9 = \frac{G_{m_1 m_2}}{r^2}$$

The inverse-square nature of this equation is captivating. On balance, there is no indispensable reason that gravity should conduct itself in this manner. In a chance, evolving universe, some random exponent like "$r^{1.97}$ or $r^{2.3}$" would appear much more likely. Though, precise measurements have illustrated a precise exponent out to at least 5 decimal places, "2.00000." As one researcher put it, "*this result seems 'just a little too neat.*'" We may conclude that the gravity force illustrates exact, created design. Essentially, if the exponent deviated just slightly from exactly 2, planet orbits and the whole universe would become unstable.

For the technically minded, [$G = 6.672 \times 10^{-11}$ Nm$^2$kg$^{-2}$]

## The Truth about Gravity

How is this force capable to perform across the massiveness of empty space?

And why does this force exist in the first place?

Science has on no occasion been very fruitful in responding to these most straightforward questions about nature. Gravity cannot on one way or another gradually ascend by mutation or natural selection. It was current from the very start of the universe. Accompanied by every other physical law, gravity is confidently a testament to a deliberate creation.

Efforts to clarify gravity have encompassed unseen particles, called "gravitons," that travel between objects. "Cosmic strings and gravity waves" have also been supposed, but none have been inveterate. We merely do not recognize how objects physically interrelate with each other over massive distances.

## The Creator and Gravity

Two Word of the Creator references are accommodating in seeing the nature of gravity and physical science in over-all. First, *Colossians 1:17* explains that the infinite eternal creator – Jesus Christ is before all things, and by Him all things consist, and hold all creation together. The Greek verb for consist "συνιστάω *sunistaō*" means to cohere, preserve, or hold together. Extrabiblical Greek utilize of this word images a container holding water within itself. The word is utilized in Colossians in the perfect tense, which usually indicates a present continuing state arising from a completed past action.

One physical mechanism employed is obviously gravity, constructed by the Creator and still upheld without flaw today. Consider the alternative; if gravity stopped for one moment, instantaneous disorder assuredly would result. All heavenly substances, counting the earth, moon and stars, would no more hold together. All would instantly crumble into small rubbles.

Another reference, *Hebrews 1:3*, affirms that Christ sustains all things by the word of His power. Uphold "φέρω *pherō*" again designates the sustaining or maintaining of all things, counting gravity. The word upholds in this verse illustrates much more than merely supportive a weight. It embraces control of all the continuing motions and fluctuations within the universe. This infinite chore is achieved by the Lord's almighty Word, whereby the universe itself was primary called into existence. Gravity, the 'mystery force', which is barely understood after nearly 400 years of investigations, is one of the indicators of this overwhelming divine upholding.

**Table 5.2:** Some values of the gravitational force of attraction between various objects.

| Pairs of Objects | Gravity force (kg at sea level) |
|---|---|
| You and a magazine you are reading | $4.5 \times 10^{-10}$ ($10^{-9}$ lbs) |
| You and the moon | .00045 |
| Two adjacent locomotives | .0022 |
| You and the earth (at sea level) | your weight |
| Moon and Earth | $1.8 \times 1019$ |
| Earth and Sun | $3.6 \times 1021$ |

NB: kg are actually units of mass (the units of force are N = Newtons, where the force on a 1 kg weight at sea level = 9.8 N). The weight (or force) exerted by a given mass depends on how close it is to the earth's gravity, so here sea level is assumed.

## Observe

1. Einstein said: "*it is not a force in the way others are, but an effect of the curvature of space-time.*"
2. According to general relativity, the influence of gravity are not instantaneous, but transmitted at the speed of light.

## Space/time Warps – Blackholes

Einstein' theory of general relativity regards gravity not as a force, but a curvature of space itself near a massive object. Even light, which traditionally follows straight lines, was predicted to bend while travelling through curved space, Figure 5.3. This was first illustrated as the astronomer Sir Arthur Eddington discovered a alteration of a star's seeming position during a total eclipse in 1919, in harmony with the light-rays' being bent by the sun's gravity.

**Fig.5.3:** *Alain r, creative commons Simulated view of a black hole in front of the Large Magellanic Cloud. Note the "gravitational lensing effect," which produces **two enlarged** but highly distorted views of the Cloud. Across the top, the Milky Way disk appears distorted into an arc.*

General relativity also forecasts that if a body were dense enough, its gravity would curve space so powerfully that light could not escape at all. Such a body would absorb light and anything else trapped by its powerful gravity, and so is named a black hole. Such a body could be distinguished only by its gravitational influences on other substances, robust bending of light around it, and by intense radiation emitted by matter falling in.

All the matter captured inside a black hole is compressed into a singularity of near infinite density. So instead, its 'size' is distinct by its "event horizon," a boundary surrounding the singularity such that nothing, not even light, can escape from within the boundary. Its radius is called the Schwarzschild radius, after the German astronomer Karl Schwarzschild (1873–1916), and is specified by the equation

$$R_S = 2GM/c^2,$$

where c is the speed of light in a vacuum.

If the sun were to collapse into a black hole, its Schwarzschild radius would be only 3 km (2 miles).

There is good indication that a very gigantic star, after utmost of its nuclear fuel runs out, would have nothing to counteract a collapse under its own huge weight into a black hole. Black holes with the mass of a billion suns are believed to exist in the centers of galaxies, counting our own, the Milky Way. Numerous scientists trust that the super-bright and enormous distant objects called quasars are power-driven by the energy freed as matter falls into a black hole.

General relativity forecasts that gravity also distorts time. This has also been long-established by very precise atomic clocks ticking a few microseconds per year slower at sea level than at high altitudes anywhere Earth's gravity is slightly weaker. Close to an event horizon, the influence is far more noticeable. If we checked the watch of an astronaut approaching an event horizon, we would see the watch tick ever more slowly. At the event horizon, it would come to a standstill, but we would never see that occur. On the other hand, the astronaut wouldn't see anything different about his own watch, but he would observe our clocks ticking faster and faster.

The major peril to an astronaut close to a black hole is tidal forces, due to the fact that gravity is much stronger on the parts of the body closer to the black hole than on the parts away from it. The tidal forces close to a black hole with a mass of a star are much stronger than a hurricane, powerful enough to stretch an onlooker into tiny pieces. Though, whereas gravitational attraction declines with the square of the distance

(1/r²), the tidal influence declines with the cube of the distance (1/r³). Accordingly, opposing to popular visualizations, large black holes have weaker gravitational (including tidal) forces at their event horizons than small black holes. Consequently, the tidal forces at the event horizon of a black hole the mass of the observable cosmos, for example, would be less noticeable than the gentlest breeze.

Gravitational time dilation at and close to an event horizon is the basis of the new cosmological model of the creationist physicist "Dr Russell Humphreys," in his book *"Starlight and Time."* This model appears to deliver one possible solution to the problem of observing distant starlight in a young universe, and is presently a scientific alternative to the unbiblical 'big bang' theory which trusts on philosophical assumptions outside science.

Apart from Stephen Hawking's theoretical discovery in 1974 that black holes would ultimately evaporate via *Hawking radiation* due to a quantum mechanical effect.

Gravity, the 'mystery force', … is poorly understood after nearly four centuries of research …!

## Isaac Newton (1642–1727)

Isaac Newton published his discoveries about gravity and motion in 1687, Figure 5.4, in his magnum opus "Principia." Some readers rapidly clinched that Newton's universe left no room for God, since everything now could be described with equations. But this was not Newton's view at all, as he illuminated the public in the "Principia's second edition":

**Fig.5.4**: *Image Isaac Newton (1642–1727)*

'Our most beautiful system of the sun, planets, and comets could only proceed from the counsel and dominion of an intelligent and powerful being.'

Isaac Newton was not solely a scientist but also a life-long biblical intellectual. His favorite Bible books were "Daniel and Revelation," in which God's plans for the future are pronounced. Newton essentially wrote more about theology than science.

Newton offered recognition to others such as Galileo. Newton essentially was born the same year that Galileo died, in 1642. Newton wrote in a letter, *'If I have seen further, it is by standing on the shoulders of giants.'* Shortly before his death, perhaps thinking of the mystery of gravity, Newton humbly wrote, *'I do not know what I may seem to the world, but as to myself, I seem to have been only like a boy playing on the seashore … now and then finding a smoother pebble or a prettier shell than ordinary, while the great ocean of truth lay all undiscovered before me.'*

Newton is buried in Westminster Abbey. His Latin epitaph ends with the sentence *'Let mortals rejoice that there existed such and so great an ornament to the human race.'*

## Universal Gravity

The well-known story that Newton hit upon the idea of universal gravitation after observing an apple fall to the ground in his garden is not known for certain to be true. The anti-religious French philosopher and skeptic Voltaire first circulated the story after reputedly hearing it from Newton's grandniece.

## A Scientific Masterpiece

The year 1987 marked the 300th anniversary of the publication of one of the world's masterpieces in scientific literature.

Sir Isaac Newton, English mathematician, physicist and astronomer, published his colossal work "Philosophiae Naturalis Principia Mathematica" (mathematical principles of natural philosophy), Figure 5.5. In this phenomenon work, Newton offered his universal three greatest laws of motion:

- force-free motion is uniform; Universal Gravitation
- accelerated motion is proportional to the impressed force;
- and for every action there is an equal and opposite reaction.

From these, together with the law of universal gravitation, Sir Isaac Newton derived the whole science of matter in motion.

Sir Isaac Newton is considered by many academia, intellectuals and historians as the greatest scientist who ever lived. Nonetheless, he resolutely believed that the Bible was God's Word. He wrote extensively on biblical subjects, and even wrote a book defending Archbishop Ussher's chronology of the world, -- Ussher set the date of Creation as 4004 BC. Newton had faith in a literal six-day creation, and that the worldwide flood of Noah's time accounted for most geological phenomena. He also decisively believed in Christ as his Savior.

In considering on the publication of Newton's Principia, it would be well to reflect on this great scientist's view of Scripture:

"I find more sure marks of authenticity in the Bible than in any profane history whatsoever."

## ISAAC NEWTON AND BIBLICAL CREATION?

Isaac Newton has frequently been quoted as a six-day creationist or registered favorably in support of biblical creation. Furthermore, his chronological writings do appear to permit for a world not much older than 6,000 years. However, his less-than-literal treatment of *Genesis 1–11* should make creationists cautious to appropriate Newton to this end. There is also motive to believe that Newton was perhaps the first to progress the day-age theory, and enthused many of the earliest naturalistic reinterpretations of *Genesis 1–11*.

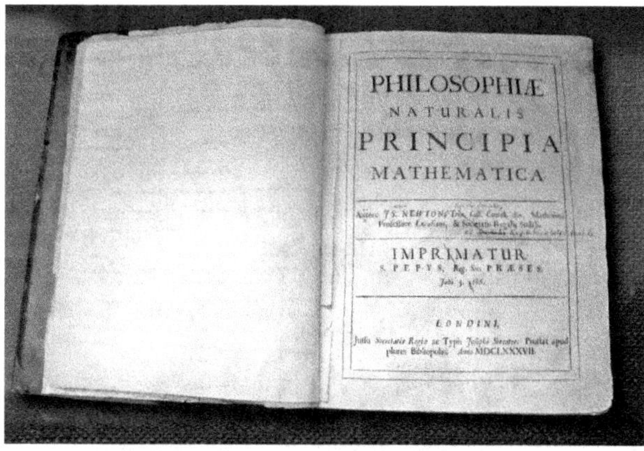

**Fig.5.5:** *Isaac Newton's own first edition copy of Philosophiæ Naturalis Principia Mathematica*

Newton has been entitled "one of the greatest creative men of genius who ever existed," the "high priest of science," and the "last of the magicians." In 1728, one academic reviewer even said that Newton was "the greatest man in the world, not only in this age, but in any age, since the world began". Newton's intellectual impact in the world was such that, "it was not till a century after his death that men liberated themselves from his authority adequately to do important original work in the subjects of which he had preserved."

With a such a inheritance it is only normal for creationists to pursue to utilize Newton in support of biblical creation. However, what were his actual politics and how did he treat Genesis 1–11?

Sometimes Newton's writings on creation, chronology, and the Christian faith were puzzling and less than traditional.

## Newton's Principia Mathematica and Opticks

Majority of Newton's influential work was completed "between the ages of 21 and 23." In these two years, he "expressed his essential laws of mechanics, his optical observations on the character of light, the calculus, and the law of universal gravitation." However, Newton's utmost literary masterpiece, published several years later in 1687, was Philosophiae Naturalis Principia Mathematica, Figure 5.5. This book has been well acclaimed as the "greatest renowned scientific work of all time." It is consequently, both astonishing and momentous to realize that Newton wrote more on theology than he ever did on science.

Accordingly, to what extent did Newton permit his theology to affect his science? Storr, next Keynes, believes that Newton "considered the puzzle of the universe in theological terms." Even if this valuation is defensible, Newton seemingly condemned of the only direct reference to God in the primary publication of "Principia": *"God placed the planets at different distances from the sun"*; because in all subsequent editions, he changed it to *"the planets were to be placed at different distances from the sun"*. This alteration was alleviated by the fact that, to all subsequent editions of Principia, he also added a short theologically unambiguous addendum entitled "General Scholium". In the General Scholium, Newton writes:

*"This most beautiful system of the sun, planets, and comets, could only proceed from the counsel and dominion of an intelligent and powerful Being … . This Being governs all things, not as the soul of the world, but as Lord over all; and on account of his dominion, he is wont to be called* Lord God *[emphasis in original]."*

This is shadowed by two additional pages of theological philosophizing, after which Newton concludes, *"thus much concerning God; to discourse of whom from the appearances of things,* does certainly belong to natural philosophy *[emphasis added]"* So although Newton removed any explicit theologizing from Principia, *these latter remarks appear to support the idea in principle.*

NEWTON'S authorization of intelligent design can also be found in the second edition of Opticks, where Newton wrote, *"And tho' every true step made in this philosophy brings us not immediately to the knowledge of the first cause, yet it brings us nearer to it, and on that account is to be highly valued."* In the fourth edition, he took this even further, saying:

*"… the main business of natural philosophy is to argue from phenomena without feigning hypotheses, and to deduce causes from effects, till we come to the very first cause, which certainly is not mechanical; and not only to unfold the mechanism of the world, but chiefly to resolve these and such like questions."*

Newton had no time for atheism which he regards as *"so senseless & odious to mankind that it never had many professors."*

## Newton's Separation of Science and Theology

Outside of the Royal Society, Newton was also "unmistakably favorable" toward those who utilized his system in their own apologetics. For example, in 1692, Newton wrote to "Richard Bentley" (1662–1742): *"When I wrote my treatise about our Systeme I had an eye upon such Principles as might work with considering men for the beleife of a Deity & nothing can rejoyce me more then to find it usefull for that purpose."* However, in additional short paper, "Seven Statements on Religion," Newton declared as his first principle, *"religion & philosophy are to be preserved distinct. We are not to introduce divine revelations into philosophy, nor philosophical opinions into religion."* Furthermore, as Manuel narrates, "When Newton was President of the [Royal] Society, the journal-books record, he banned anything remotely touching on religion, even apologetics."

This makes it hard to find out, therefore, to what extent Newton permitted theology to permeate his science, assumed that he favored—if not in principle, at least in practice—to keep the two separate. As Snobelen has experienced, there is *"more explicit theology in Charles Darwin's* Origin of Species *(1859) than in the first edition of Newton's great work."* That said, Snobelen still contends:

*"Recent work on early modern science has demonstrated a direct (and positive) relationship between the resurgence of the Hebraic, literal exegesis of the Bible in the Protestant Reformation, and the rise of the empirical method in modern science … . In this, Newton also played a pivotal role. As strange as it may sound, science will forever be in the debt of millenarians and biblical literalists [emphasis added]".*

This may be factual, but Newton does not fit the cataloguing of a traditional 'biblical literalist' simply.

## Newton's Deviation

At the same time as it is not problematic to find indication of Newton's enthusiasm for the Bible, his theological opinions were frequently heterodox. As Thomas Hearne (1678–1735) once wrote of him:

"Sir Isaac Newton, tho' a great mathematician, was a man of very slight religion, in so much that he is hierarchical with the heterodox men of the age."

Likewise, Snobelen acknowledges that though Newton was a *"devoted believer"*, in other ways he was a *"damnable heretic."* It is tough to dispute the fact that Newton was an Arian. In his treatment on the religion of Newton, Manuel refrains from *"pigeonholing"* Newton like this, but still acknowledges that he was anti-trinitarian. This kind of deviation can be seen in 12 proposals he made on the nature of Christ, and in Irenicum, or Ecclesiastical Polyty tending to Peace, where he identifies Jesus as the archangel Michael. Newton also interrogated the reality of a literal devil and evil spirits. Yet, Snobelen maintains that "Newton's demonology was an exegetical option, not a sign of the encroaching Enlightenment."  However, this is hard to accept, given that, as Snobelen himself admits:

"Newton's denial of evil spirits was well outside the theological mainstream in his own day and for a long time afterward. His position would have been viewed as a runway to infidelity, a capitulation to cold, dark atheism, a disturbing disenchantment of the world or even a delusion inspired by Beelzebub himself. If only his witching-hunting colleagues at the Royal Society had known."

## Newton's Radical Friends

Newton's silence in these matters is telling. Furthermore, as King and Popkin have argued, Newton's opinions were not uninjured by the writings of seminal Enlightenment thinkers like Baruch Spinoza (1632–1677) and Richard Simon (1638–1712). As his interpretation of *Genesis 36:31* shows, Newton laboring the same exegetical argument employed by Simon to discharge the Mosaic authorship of certain sections of Genesis. He was also good friends of "John Locke" (1632– 1704), widely viewed as "England's foremost Enlightenment thinker" and "father of liberalism". Newton asked Locke *"in the strictest confidence"* to help him translate a controversial treatise on *1 John 5:7* and *1 Timothy 3:6* into French and have it published anonymously in the Netherlands. The treatise was entitled an historical account of two notable corruptions of Scripture and was addressed to Jean Le Clerc (1657–1736), "Europe's most stubborn central character of rationalist Christian theology". Thus, Popkin writes:

*"Newton, like Spinoza and Simon, took seriously the problems that had arisen in the collection, editing, and transmission of Scripture that made it difficult if not impossible to find the pure original text. Newton, unlike the fundamentalists of the past century and a half, was not committed to claiming the inerrancy of the biblical text, but was committed to finding its message for mankind [emphasis added]".*

Whilst Popkin may be overstressing the case, the fact remains that Newton was not uninjured by the kind of thinking that fueled the Enlightenment. Put more strongly, Newton's influence to the sciences and theology did little to deter its development. As Israel observes:

"Although down to 1750, in Europe as a whole, the fight for the middle ground remained questionable, much of the European mainstream had, by the 1730s and 1740s, firmly adopted the thoughts of "Locke and Newton" which indeed appeared exclusively in agreement and well-matched to the moderate Enlightenment purpose."

Consequently, although Newton *"devoted close to six decades to a fervent study of the Bible, theology, prophecy, church history and natural theology"*, his heterodox reading of Scripture left room for a less-than-literal understanding of Genesis.

## Newton's Earlier Concepts of Genesis

In 1680, "Thomas Burnet (1635–1715)" mailed Newton a pre-publication copy of "Telluris theoria sacra" (1681) for appraisal. Burnet sought to create scientific explanations to "defend the doctrines of the "Universal Deluge," and of a "Paradisiacal" State, and guard them from the quibbles of those that are no well-wishers to sacred history." In the mail which followed, Newton proposes that the *"heat of the sun"* might clarify how the oceans were shaped and dry land appeared on the earth. This imaginative re-interpretation of *Genesis 1:9* necessitates the "diurnal revolutions of the Earth" to be *"very slow"* at the beginning of creation so that "the first 6 revolutions or days might continue time enough for the whole Creation, & ye Sun in that time might convert & shrink the parts of the Earth about the Æquator". In his reply to Newton, Burnet points out that the sun was only made later in the week:

"… methinkes you forget Moses (whom in another place you will not suffer us to recede from) in this account of the formation of the Earth; for hee makes the seas & dry land to bee diuided & the Earth wholly formd before the Sun or Moon existed. These were made the fourth day according to Moses, & the Earth was finisht the 3ᵈ day"

This gives Newton cause to explain, in the letter which follows, that *Genesis 1–2* was written phenomenologically to describe what Moses would have seen, had he been there to witness it himself. The point is not to read Genesis as science or as fiction, says Newton. Instead, Genesis is a *"true description"* of creation accommodated to the *"vulgar"* understanding of Moses' first readers. Taken on its own, Newton's overall conclusion is both conservative and orthodox:

"… me thinks one of the tenn commandments given by God in mount Sina, prest by divers of the prophets, observed by our Saviour, his Apostles & first Christians for 300 years & with a day's alteration by all Christians to this day, should not be grounded on a fiction."

But this stance does not preclude him from asserting that sun, moon, and stars were not created on *"the fourth day nor in any one day of the creation"*; that Moses might not mention their creation at all; that the duration of the first and second days might be *"as long as you please"*; or that Burnet's theory could allow *"a year for each days work"* without misinterpreting the text.

## Newton's Later Concept of Genesis

Overall, it is likely that Newton, at this stage in his career, upheld a more exact approach to Genesis, albeit vaguely. From the mid-to-late 1680s, though, his thoughts on Genesis encouraged in a more radical direction. In a dissertation on Revelation, Newton calls the story of the fall of man a *"parable"*; the trees in Eden, *"mystical"*; and the serpent, *"only a symbol of the spirit of delusion"*. With respects to the six days of creation he writes:

"And so the six days of the Creation may signify not only six years but even six thousand years … or any other six long times. ffor the history of the creation is not in all things litteral. In that Paradise the flaming sword & trees of life & knowledge may be as much figurative descriptions of something we now understand not as the tree of life is in the Paradise to come, & in a parabolical description of the creation a day may be used figuratively as well as other things are especially since there was no light till the end of the first day nor sun till the fourth—to make natural days. The evenings & mornings of Moses respect all parts of the Earth alike so that it was evening all over the Earth in the beginning of each day of Moses & morning all over it in the end of each day: & therefore his evenings & mornings were not natural ones. ffor had they been natural ones it would have been morning in one part of the Earth when it was evening in another."

For this reason, it is obvious that Newton did not believe a literal six-day creation, nor was he *"committed to the literal truth of Holy Scripture"*. Whether or not it was Burnet's writings that lastly persuaded Newton to abandon the literal historicity of the hexameron is not easy to determine. What is recognized, nevertheless, is that in Burnet's succeeding cosmological dissertation, "Archaeologiae philosophicae sive doctrina antiqua de rerum originibus (1692)," he calls the story of Adam and Eve a *parable; rejects the creation of Eve from Adam's rib; considers a speaking serpent to be utterly nonsensical; does not believe that Adam and Eve were capable of sewing their own clothes; denies that the Garden of Eden was guarded by real Cherubim; disbelieves that Adam had named all the animals in a single day or that the universe is less than 6,000 years old.*

## Newton's Chronology

That said, Newton's consecutive writings do treat features of *Genesis 1–11* as literal history. He sustains the repopulation of the world from Noah's sons and traces the origin of nations back to Babel. He also proclaims, in the conclusion to the first draft of his chronology of ancient kingdoms, that "mankind could not be older then [sic] is represented in scripture" which he later revised to "mankind could not be much older than is represented in Scripture [emphasis added]".

Therefore, it is likely that Newton still believed in an earth not much older than 6,000 years. Some scholars even assert that Newton's chronology depended upon or defended James Ussher's "Annales veteris testamenti, a prima mundi origine deducti (1650)." Nonetheless, this claim is uncertain for the following reasons:

- firstly, Newton's chronology does not begin with Adam, it begins with Noah;
- secondly, the manner in which he places Scripture alongside many other historical sources suggests, as Westfall rightly points out, that "Newton's view of human history did not center on the Bible", but that he "treated the historical books of the Old Testament as human documents to be used in concert with other human documents;"
- thirdly, Arthur Bedford (1668–1745), a contemporary of Isaac Newton, wrote one of the earliest critiques of The Chronology of Ancient Kingdoms Amended (1728) in which he demonstrates how Newton's chronology "differs toto cælo from all the learned men in the world" including, most notably, James Ussher.
- In fact, I have also not found a single reference, positive or negative, to Ussher within the Newtonian corpus.
- Finally, in all his chronological writings, Newton never provides an age for the earth or a date for creation. The omission of such an obvious chronological detail is telling.

Newton (2006:191, 376), nevertheless, is helpfully critical of the Egyptian, Persian, and Syrian records, which "out of vanity" have exaggerated the antiquity of their kingdoms by "some thousands of years older than the world". Consequently, Newton's stated motive is to state accurately these erroneous chronologies by referring to the more reliable records well-maintained by the Greeks and Hebrews. However, like Bedford observes: *"As to what [Newton] saith, that he hath made it agreeable with sacred history; it is hard to know, whether he was in earnest or in jest."* This is because, in Bedford's valuation, Newton's chronology disregards, misquotes, and contradicts the biblical record in several places. Newton's conclusion is forceful: *"such poison ought not to go abroad into the world"* for it undermines the integrity of the sacred text.

For these reasons, we should be cautious to endorse Newton's chronology uncritically.

## Newton's Biblical Interpretation Legacy

Burnet was the first to attempt an explanation of:

*Genesis 1–11* : *"And God said, 'Let the earth sprout vegetation, plants yielding seed, and fruit trees bearing fruit in which is their seed, each according to its kind, on the earth.' And it was so."*

In partnership with Newton himself and in relationship of Newtonian physics. And what was the result? A cosmology that had very little to do with Genesis at all. In Burnet's approximation, *"the very letter of the* Hexaemeron *[is] most absolutely contradictory to the nature of things, as well as to all philosophical reasons"* and, *"people could neither understand nor bear a plain and philosophical explication of it".*

What this called for, in practice, was a scientific hermeneutic, whereby "philosophy is the interpreter of Scripture in natural things". It was a principle that resonated strongly with "Charles Blount (1654–1693)," the *"chief deist of his age"*, who eagerly plagiarized sections of Burnet's writings to further his radical agenda in England.

Like Burnet, "William Whiston (1667–1752)" wrote his own philosophical version of *Genesis 1–11,* "A new theory of the Earth (1696)," which he dedicated to Newton. In it he calls literal six-day creation a *"vulgar hypothesis"*, arguing instead that the days in Genesis should be understood as years.  He also argues that *Genesis 1* does not tell us how matter came into being or how the universe was created, being restricted exclusively to the origin of the earth.  Far from critical of Whiston's theory of the earth, in 1702, Newton chosen him as his successor to the Lucasian chair of mathematics at Cambridge.

"Edmund Halley (1656–1742)," a close friend and admirer of Newton, also rejected to accept that the days in Genesis should be taken as *"natural days."*  He upheld that the Scriptures could not provide a dependable account of the age of the earth. In its place, Halley proposed that the salt of the oceans could give a better estimate

In France, "Bernard Le Bovier de Fontenelle (1657–1757)," the first to publish a biography of Newton, also banned *Genesis 1–11* as literal history. He utilized his editorial effect at the French "Académie des sciences" to vigorously indorse geological opinions that prohibited the biblical Flood, at the same time censoring any scientific interpretations that presumed or declared it.  Similarly, "Voltaire (1694–1778)" quotes Newton in backing of his claim that Genesis was not written by Moses, and calls the Pentateuch a *"stupid falsehood"* and *"absurd fable"*, *"written by fools, commented upon by simpletons, taught by knaves"*, and filled with *"innumerable geographical and chronological errors and contradictions"*. :Comte de Buffon (1708–1788)" also utilized Newtonian physics in his meticulously naturalistic reinterpretation of *Genesis 1–11*. This is significant for the modest fact that Buffon, "more than anyone else, was responsible for a new chronology of the earth, that is, for the acceptance of a enormous time scale."

In Germany, one of the most persuasive philosophers of the last three centuries, "Immanuel Kant (1724–1804)," hurled his academic career with a scientific dissertation entitled, "General natural history and theory of the heavens," "or an essay on the constitution and mechanical origin of the whole universe, treated in accordance with Newtonian principles (1755)." In this contentious book, he compels God to a first cause while attempting to clarify, on Newtonian grounds, how matter could arrange itself into the present universe over time

Newton might never have expected or wanted such a legacy, but his influence directly and indirectly affected how *Genesis 1–11* would be read by future generations.

For the 1680s, Newton's dealing with Genesis was far from orthodox, and perhaps the first pronunciation of the day-age hypothesis in history. If accurate, this makes Newton an important figure in the succeeding hermeneutical revolution which formed how *Genesis 1–11* would be read for the next three centuries. This harmonizes well with "Israel's analysis" for the onset of the Enlightenment, which he places within the same time period (i.e. from 1650–1680). Therefore, the timing of Newton's commentaries on Genesis, dated to the late 1680s, occur to associate strongly with the inauguration of the Enlightenment period. Prophetically maybe, in the front matter of the first edition of "Principia," Halley respects Newton as superior to Moses, *"Who opens the treasure chest of hidden truth … . No closer to the gods can any mortal rise."* The inquisitive consequence to all this, is that Newton acknowledged such praise in print as the foreword to his "magnum opus." Did Newton contemplate of himself as Moses' scientific successor? Perhaps, perhaps not. Either way, Moses did not fare well in the next century. As Manuel has observed, for the Enlightenment to flourish in the 18th century, "certain basic intellectual needs" had to be met, the first being: "a replacement of Genesis." Newton did little to deter such a endeavor. It is probable that several theologians from the next generation took their lead from him.

For reasons such as these, Newton's legacy is a greater interruption than help to biblical creationists. Although his science was inspired by Scripture, his view of Scripture was increasingly shaped by his science.

## THE NATURAL LAW – SUPERNATURAL GRAVITY

We all recognize the law of gravity. It's what service the moon ripple tides around the earth, stops galaxies from wandering apart, and holds your soup in the bowl.

We can describe it in wide-ranging terms as the force of attraction that exists between any two bodies, or we may be more exact and express it as an equation that contains the masses of the bodies, their distance apart, gravitational constants and such comparable, and then utilize that equation to aid launch satellites and construct bridges.

The law of gravity is called the 'natural law', or a 'law of nature', as it is not derived from any theoretical proofs, but is merely the result of immeasurable observations of what essentially happens constantly around us. That is, laws are evocative, not prescriptive—they don't cause whatever to occur but define what happens, just as a map doesn't cause the outline of a coastline, but defines what exists.

Nevertheless, we employ the 'law of gravity' unaware every time we serve a tennis ball or hang some washing on the line. No one has ever seen an exception to it—it is a law, continuous and unrestricted at all times, in all places, and under all circumstances, far out in space or in our own physiques.

Nonetheless, it is not the only natural law. Majority of us have at the minimum heard of the laws of thermodynamics, the gas laws, the laws of electromagnetic induction, laws of chemical reaction, the law of biogenesis and many more. These are constructs into the fabric and operation of the universe by its astounding Designer.

Some are articulated scientifically as theorems, such as Bernoulli's theorem that helps us comprehend how carburetors work. These laws cannot be avoided, and they are not postponed or altered in living things.

No one would ever dare to bypass natural laws. Amazingly, as it turns out, anybody wishes to explain the origin of the universe and of life deprived of God must attempt to express a story that avoids at least one, and frequently several, natural laws. It is thought-provoking that such a person would assert that all must be explained in terms of natural laws deprived of supernatural involvement, when essentially this is self-refuting.

### General Events

"The origin of the universe" is readily whispered to have occurred in 'the big bang.' Nevertheless, it is stimulating to learn that few serious cosmologists will essentially talk about the moment of the 'big bang,' but may only talk about the moment "after" the 'big bang.' The cause for this is that the theory in fact calls for us to believe that the whole thing in the universe abruptly appeared, not from a point, but from nothing. Or else, it is not a clarification for the origin of whatsoever at all. Numerous people don't understand that the point that everything allegedly expanded from in this theory has, itself, no explanation, as the concept disobeys the greatest elementary natural law—the law of conservation of matter/energy: "matter/energy cannot be created nor destroyed." Furthermost other theories of the 'origin' of the universe propose something pre-existing. Therefore, only supernatural creation can justify for the rapid appearance of this universe from 'nothing', and only supernatural creation can justify for the fact that this natural law exists at all. Or else, universes could just pop up at random any old time.

Atheistic efforts to describe the origin of life from lifeless chemicals breakdown on natural law also. This is since the laws of chemical response oversee the way substances combine chemically. Even if the often-quoted experiments by Miller, Urey and others established anything realistic about 'the early earth', the idea that the resulting amino acids would formulate proteins in the ocean or lakes and ponds disobeys these chemical laws. This is for the reason that the reaction establishing a protein from amino acids provides water as a by-product, but a chemical law dealing with concentrations of reactants provides the protein break down again if there is excess water around. Proteins consequently, could never accrue in these watery environments, definitely, not long enough to accidentally produce a living structure—no 'primordial soup' could give rise to life 'naturally', Figure 5.6.

The laws of probability joined with these chemical laws also work in contradiction of the concept that even one functioning protein could formulate by accident. Each protein is a very precise arrangement of numbers amino acids not only in their order, but they must also be all 'left-handed'.

No chemical law attaches specific amino acids in any particular order, so their arrangement in any protein in a living thing is determined exclusively by precisely measured processes, like threading beads on a string to produce an anticipated pattern rather than a random, rambling jumble, Figure 5.6. Consequently, only a Creator could have placed amino acids into the order and configuration mandatory for the proteins for the first living things. Natural law demands that it could not occur 'naturally'.

Not only this, but there is the 'law of biogenesis'. This postulate that life breeds only from life.

Have you ever found any living thing that didn't have a 'mother'?

You came from a mother, the vegetables you have had for dinner produced from seeds, and the bacteria in your large bowel came from the division of other bacteria. There has never been any exclusion discovered. The assumption that the law of biogenesis can be disobeyed has no scientific sustenance. Yet naturalists contend that life began by itself 'naturally', deprived of a 'mother', and without God.

Midst many other natural laws we could observe, one of the greatest effective areas is that of information. Information theorems - effortlessly ascertainable natural laws- express us unambiguously that information cannot ascend from the action of time and chance on matter. Information necessitates a mental source with not only intelligence, but also volition. Nonetheless, every cell of our bodies encompasses the corresponding to 1,000 books of information, crowded in a concentration that cannot be exceeded because it is already at the molecular level. Where did it come from? It could only have been created supernaturally.

But then, does the supernaturalist – creationist - have similar problem, since miracles also 'violate natural law'? Not at all. Miracles are actually "additions" to natural law.

*Fig.5.6*: No 'Primordial Soup' could Give Rise to Life 'Naturally'

For example, Archimedes' law of buoyancy declares that the buoyant force of a fluid (i.e., liquid or gas) is equivalent and opposite to the weight of the displaced volume of this fluid. Therefore, if the object is denser than the fluid, the downward force of its weight will overbalance the upward force of its buoyancy, and it will drown. Some disbelievers subsequently, declare that Jesus could not have walked on water because of this law.

However, does this mean that a helicopter rescuing someone from the sea also violates this law?

No, the helicopter provides an "additional force" to the system of the person's weight and the sea's buoyancy. Jesus as the Creator *(John 1:1–3)* *In the beginning was the Word, and the Word was with God, and the Word was God. He was in the beginning with God. All things were made through him, and without him was not anything made that was made.* Likewise, could have exerted an additional force.

Materialistic oppositions to miracles are therefore immaterial, as, once God is acknowledged, the universe is not a closed system. His supernatural involvement is clearly essential for creation and other 'miracles'. Usually, though, God sustains His universe by natural laws like those we have discussed. Actually, it was the idea of a lawmaking God that led to the concept of natural laws in the first place—the birth of modern science.

The Bible says "For since the creation of the world [God's] invisible attributes are clearly seen, being understood by the things that are made, even His eternal power and Godhead" *(Romans 1:20)*. It is in *"the things that are made"* that we see the natural laws we have discussed in operation. Those who attempt to clarify the universe deprived of God are confronted with the insurmountable problem that these laws, simply observable by theist and atheist alike, themselves repudiate any 'natural' clarification for the "origin of the universe" and life. They tell us unambiguously: our origins are supernatural, naturally.

**Corresponding information:**

1. Bernoulli's theorem is a different case of the law of conservation of energy, the First Law of Thermodynamics.
2. Notably the famous physicist and mathematician Stephen Hawking, who admits that the equations he works with collapse as his theoretical musings approach the actual point of 'beginning'.
3. For example, the 'brane' theory undertakes the existence of two-dimensional 'branes' (an extension of 'string' theory), though the 'landscape' theory assumes a pre-existing universe so huge that our observable universe is only a tiny corner of it, and the 'quasi steady state' theory assumes an oscillating universe of infinite age.
4. This is why you soak your dishes in water before washing up—it breaks down the proteins.
5. See Gitt, W., In the Beginning was Information, CLV, 1997 [or sequel Without Excuse, Creation Book Publishers 2011]. Return to text.
6. Sarfati, J., Why does science work at all? Creation 31(3): 12–14, 2009; Biblical roots of science, Creation 32(4):32–36, 2010.

## Einstein, and the Universe

Selected by Time magazine to be their *'Person of the Century'*, Albert Einstein is well-known for numerous things --apart from his disheveled appearance, Figure 5.7. His theories of distinct and general relativity and his formulation for the correspondence of mass and energy,

**Fig.5.7:** *Professor Albert Einstein*

$$E = mc^2,$$

altered endlessly our interpretations on time/space, light/gravity, and matter/energy. He is to some extent less well-known for his statement '*God does not play dice with the universe.*' But what did Einstein actually mean by 'God'?

Was his 'God' anything like the God of the Bible?

## Einstein's Childhood

While born in 1879 of German-Jewish parents, Albert was not brought up in the Jewish belief. He attended a nearby Catholic elementary school in Munich and then the local high school. A somewhat unhurried and dreamy student, Albert was bored uninterested with non-scientific subjects, and acquired little learning under the severe military-style 19th century German schooling system. He was raised up with a dislike to discipline, and a life-long distrust of all authority.

At age 11 he went through an concentrated religious phase during which he ate no pork and composed songs to God, which he sang to himself on the way to school.

From age 12 Albert read popular books on science, educated himself algebra, geometry and calculus, and studied Immanuel Kant's anti-theistic "Critique of Pure Reason." Regarding this time in his life, Albert later wrote, '*Through the reading of popular scientific books I soon reached the conviction that much in the stories of the Bible could not be true. The consequence was a positively fanatic (orgy of) [sic] freethinking coupled with the impression that youth is intentionally being deceived by the state through lies; it was a crushing impression. … It is quite clear to me that the religious paradise of youth, which was thus lost, was a first attempt to free myself from the chains of … an existence which is dominated by wishes, hopes, and primitive feelings.*'4

Albert's anti-authoritarianism, and perhaps also his wish to discharge compulsory military service at age 17, led him to relinquish his German citizenship. On January 28, 1896 he turn into a stateless person at the age of 16. His submission for Swiss citizenship was approved February 21, 1900.

## UNDERGRADUATE STUDIES

### Fatherhood and Marriage

From 1895 to 1900 Albert was admitted to the Zurich Polytechnic in Switzerland, then the finest technical school in Europe. He rarely attended lectures, however, consumed much of his time doing his own experiments in the outstanding physics laboratory, and reading about the latest advances in physics by Hertz, Helmholtz, and other pioneers in science.

Also, Einstein learned about revolutionary socialism from his friend, "Friedrich Adler," who in 1918 accomplished celebrity by assassinating the Prime Minister of Austria.

### Einstein Married Meliva Meric

Albert fell in love with Mileva Maric, a Hungarian and the only woman student in his class who, nevertheless rather plain, stricken with a limp, and not in the slightest flirtatious, learnt adequate physics to be able to have gifted discussions with Albert. In 1901 he fathered an illegitimate child with her. He married Mileva in 1903, after he had secured a job as "patent examiner" at the Swiss Patent Office in Berne.

### The Greatest Genius on Earth

In 1905, the impressive Berlin journal "Annalen der Physik" published four papers written by Albert between March 17 and June 30 of that year in his spare time!

He received the Nobel Prize 16 years later for the first paper, for which, he described how light could behave as both a wave and a stream of particles.

The second paper, on the size of atoms, earned him a doctorate from Zurich University.

The third paper, on Brownian motion, is the groundwork of modern statistical mechanics.

The fourth paper turn out to be the foundation for his "Special Theory of Relativity." This was founded on Albert's 'thought experiments', such as what he might or might not see if he were in a space ship travelling at the speed of light.

In 1916 Albert published 'The Foundation of the General Theory of Relativity.' This was founded on more 'thought experiments' that gravity and acceleration yield same influence, and that this is a result of gravity warping (distorting) both space and time. Scientists were both bewildered and confused. Then the theory seemed to be deep-rooted during an eclipse of the sun in the West Indies, on May 29, 1919. The world's media started mentioning to Albert as 'the greatest genius on earth,' Figure 5.8.

**Fig.5.8**: *Einstein became a wanted man as Hitler and the Nazis began a campaign against 'Jewish science', offering a 20,000 Mark reward for his assassination.*

## EINSTEIN'S BELIEF IN THE DIVINITY OF NATURE

Pantheists believe that the whole thing is God. It means that 'God' just becomes another word for '*everything*' and loses any real meaning. Albert Einstein openly shared the pantheism of Spinoza, of whose views "The Hutchinson Softback Encyclopedia," 1996, writes: "*Mind and matter are two modes of an infinite substance that [Spinoza] called God or Nature, good and evil being relative.*" Like New Age and Eastern thought, this is a '*monistic*' belief, which explicitly repudiates a Creator in the traditional meaning of the word, i.e., one who pre-existed -and is thus transcendent, Independent of or 'outside' - that which was created.

### Love on the Rock

The marriage of Albert and Mileva's had progressively tumbled apart and in 1914 they had separated. In 1918, divorce proceedings were set in motion, grounded on the adultery of Albert with his divorced cousin Elsa Löwenthal, who had taken care of him during a period of sickness.

The Zurich court granted the divorce on February 14, 1919, and ordered "inter alia" that Albert should give the monetary reward from a Nobel Prize, if and when he should receive it, to Mileva.

Albert married Elsa on June 2, 1919, but then again he was unfaithful. He wrote that he venerated a departed friend for having lived for numbers of years in peace and *'lasting harmony with a woman—an undertaking in which I twice failed rather disgracefully.'*

## The Nobel Prize

In 1922 Albert received authorized news that he had been awarded the "1921 Nobel Prize for Physics" for his work in theoretical physics and his photoelectric law.  Relativity, still highly controversial, was specifically excluded.

Numerous people from all over the world corresponded with Albert.   Some of his answers exposed his wry sense of humor. In Berlin, he received a letter from New York asking,

*'Would it be reasonable to assume that it is while a person is standing on his head—or rather upside down—that he falls in love or does other foolish things?'*

Albert wrote,

'To fall in love is by no means the most stupid thing man does—gravitation cannot be held responsible, however.'

On another occasion, he was asked his formula for success. He replied,

'If A is success, I should say the formula is:

$$A = X + Y + Z,$$

X: being work and

Y: being play.'

'And what is Z?'

'Keeping your mouth shut!'

In 1933, after Adolf Hitler had come to power, the Nazis launched a campaign against 'Jewish science' and offered a 20,000-mark reward for Albert's assassination. He migrated to the USA and established himself in Princeton, New Jersey, a scientific super-celebrity, becoming a US citizen on October 1, 1940.

## Relativity and Morality

Many people have erroneously held responsible Einstein's theory of relativity for the deterioration in morality seen today. In actual fact, Einstein wished-for a view of nature in which "absolute space" and time were substituted by "absolute velocity of light." He favored to call his theory the *'invariance'* theory, but the term *'relativity'* stuck.

The foundation for morality is the absolute truth of the Word of God, which encompasses God's directions for holy living. These have been damaged, not by Einstein's theory of relativity, nonetheless, by the teaching of evolution, in which man discards the absolute truth about God and our necessity to live in a right relationship with Him, and man himself chooses how he desires to live.

Science can only express to us what is, not what ought to be. For example, science expresses to us that shooting a man in the heart will kill him.  Also, science tell us that  particular sexual practices indorse the spread of AIDS, but it cannot tell us whether these actions are right or wrong. For this we need a divine Lawgiver.

## Nuclear Bomb

Albert was a gentle pacifist man for all of his life. However, on August 2, 1939, after his knowledge that German scientists were occupied on splitting the uranium atom, he signed a letter to President F. D. Roosevelt which stated, *'This new phenomenon would also lead to the construction of bombs,'* and urged *'quick action'* on the part of the United States in atomic bomb research.

The world's first atomic bombs was developed through The Manhattan Project, which was produced under two years later. Albert, viewed it as a security risk. He was omitted from participation in this. After the bombs had detonated on Hiroshima, Nagasaki and Yokoyama he painstaking considered this letter one of his utmost mistakes.

In November 1952 David Ben-Gurion, the Prime Minister of Israel offered Albert Einstein the Presidency of Israel. Albert declined.

Albert tried for most of the last 30 years of his life, to establish a mathematical correlation between electromagnetic forces, such as light, and gravity. He was unsuccessful. His goal was to derive a single formula to explain the behavior of everything in the universe, varied from electrons to stars, called a "Unified Field Theory." He died in his sleep on April 18, 1955, from a ruptured defect in the main abdominal artery.

## Einstein and 'God'

Albert Einstein was not a Christian. Sadly, he had no perception of the God of the Bible or trust in Jesus Christ as his Lord and Savior. His interpretations on religion and 'God' were evolutionary and pantheistic. Einstein wrote:

'I cannot conceive of a God who rewards and punishes his creatures, or has a will of the kind that we experience in ourselves. Neither can I nor would I want to conceive of an individual that survives his physical death; let feeble souls, from fear or absurd egoism, cherish such thoughts.'

*'The desire for guidance, love, and support prompts men to form the social or moral conception of God. ... The man who is thoroughly convinced of the universal operation of the law of causation cannot for a moment entertain the idea of a being who interferes in the course of events. ... A God who rewards and punishes is inconceivable to him ... .'*

'During the youthful period of mankind's spiritual evolution human fantasy created gods in man's own image. ... The idea of God in the religions taught at present is a sublimation of that old concept of the gods. ... In their struggle for the ethical good, teachers of religion must have the stature to give up the doctrine of a personal God ... .'

Answering a Japanese scholar who asked Einstein about *'scientific truth'*, Albert wrote:

'Certain it is that a conviction, akin to religious feeling, of the rationality or intelligibility of the world lies behind all scientific work of a higher order. This firm belief, a belief bound up with deep feeling, in a superior mind that reveals itself in the world of experience, represents my conception of God. In common parlance this may be described as "pantheistic" (Spinoza).'[25]

It is thus clear that when Albert mentioned 'God', e.g. 'God does not play dice with the universe', and 'The Lord God is subtle, but malicious he is not',[26] he was referring to something like rationality in the universe. He is recorded as saying that a 'deeply emotional conviction of the presence of a superior reasoning power, which is revealed in the incomprehensible universe, forms my idea of God'.[27]

However, he surely was not referring to anything like the God of the Bible, who is Creator, Lawgiver, Judge and Savior.

Addressing Princeton Theological Seminary on May 19, 1939, Albert said: *'[A] conflict arises when a religious community insists on the absolute truthfulness of all statements recorded in the Bible.'*

Christian apologist Dr Hugh Ross claims that, despite not believing in the biblical God, 'Einstein held unswervingly, against enormous peer pressure, to belief in a Creator.' However, in the normal meaning of these terms, Einstein believed no such thing (see aside above on starlight). Thus, Christians who incorrectly invoke Einstein in their evangelization, writing or witnessing do so to the loss of their cause.

Note: As Einstein wrote his scientific papers and most of his correspondence in German, translations used above vary slightly among his biographers.

## Relativity and Distant Starlight

Einstein's General Theory of Relativity (GR) is the greatest experimentally justified theory of gravity ever in existence. It has strengthened Newton's laws, however, has captivated them within a greater framework, being a more precise account under certain conditions.

GR includes many counter-intuitive notions, such as black holes—regions of space with so much mass that even light rays cannot escape. Additional of its inferences is that gravity distorts time itself, so there is no such thing as 'absolute time'.

An important creationist model, developed by physicist "Dr Russell Humphreys," of how distant starlight could reach the Earth in a young universe, depend on heavily on GR. This is clarified in detail for the layperson (with technical appendices) in Dr Humphreys' book, "Starlight and Time."

Numerous 'progressive creationists' have strongly criticized "Dr Humphreys' work," however, to date he has been able to soundly answer them.

## The light That Rules the Night

Our Moon. God created it. Man reached it. Poets have written about it. Lovers gazed at it. Find out about some fascinating truths behind our great 'lesser ligh.'

The moon—an object of wonder since the dawn of mankind. It lights up the night sky like nothing else in the heavens, and appears as if it regularly changes shape, Figure 5.9. As we shall see, it is well designed for life on Earth, while its origin baffles evolutionists.

## The Origin of the Moon

Although there are many different ideas on how and when the moon formed, no scientist was there at the time. So we should rely on the witness of One who was there (cf. *Job 38:4*), and who has revealed the truth *in Genesis 1:14–19*:

*Fig.5.9: The Moon Affects Earth Property*

And God said, "Let there be lights in the expanse of the heavens to separate the day from the night. And let them be for signs and for seasons, and for days and years, and let them be lights in the expanse of the heavens to give light upon the earth." And it was so. And God made the two great lights—the greater light to rule the day and the lesser light to rule the night—and the stars. And God set them in the expanse of the heavens to give light on the earth, to rule over the day and over the night, and to separate the light from the darkness. And God saw that it was good. And there was evening and there was morning, the fourth day.

This passage clearly states that God made the moon on the same day as the sun and stars—the fourth day of Creation Week. It was also created one day after the plants. This order of events is impossible to reconcile with evolutionary/billions of years ideas.

## Light – Gravitational Purpose

The answer resides in Genesis! A key purpose is to light up the night. The moon mirrors the sun's light on to us even when the sun is on the other side of the earth. The amount of reflected light depends on the moon's surface area, consequently, we are privileged to have a moon that is very large. It is over a quarter of Earth's diameter—far larger relative to its planet than any other in the solar system. Similarly, if it were much smaller, it would not have sufficient gravity to maintain its spherical shape.

Another purpose for the moon is to show the seasons. The moon orbits the earth about once a month triggering regular phases in a 29½ day cycle, Figure 5.10. Consequently, calendars could be made, so people could plant their crops at the best time of the year.

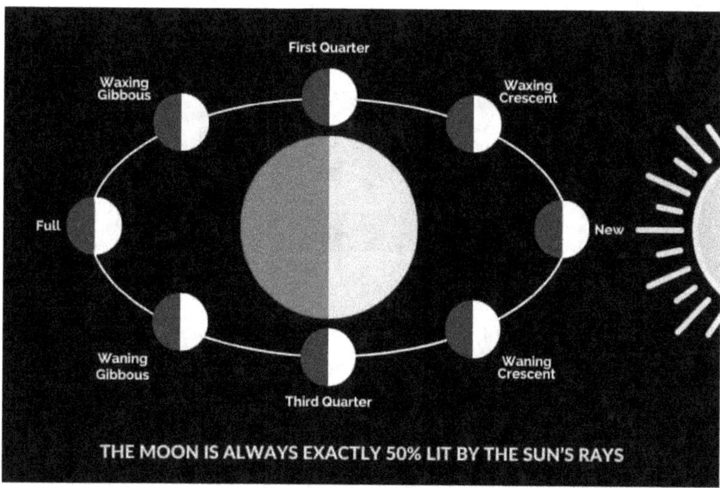


**Fig.5.10**: *The Moon Cycles around the Earth*

A vital feature is that the moon continuously keeps the same face towards the earth, Figure 5.10. If diverse parts were visible at diverse times, the moon's brightness would be contingent on which part was pointing towards the earth. Then the 29½ day cycle would be far less obvious.

The moon's size and closeness to Earth provides the greatest tidal effect on Earth. Even the sun has less than half this effect, and the effect of the other planets is negligible. When the sun and moon are aligned, their combined gravity results in strong spring tides. When they are at right angles, their gravity partly cancels, resulting in weak neap tides.

Gravitational force between two objects is given by

$$F = Gm_1m_2/R^2,$$

where G is the gravitational constant, $m_1$ and $m_2$ are the masses of the objects, and R is the distance between their centers of mass—an inverse square law.

Nonetheless, the tidal consequence drops off far more rapidly, with $R^3$—an inverse cube law. If more people had recognized this, they wouldn't have been frightened by learning all the planets would be almost aligned in 1982, when many people predicted, this would lead to catastrophe.

## TIDES

The earth's gravity retains the moon in orbit, and is very robust that it would require a steel cable 850 km (531 miles) in diameter to offer an equal binding force without breaking. The moon applies the equivalent force on the earth. Nonetheless, the force is somewhat larger on the part of the earth closest the moon, subsequently, any water there will bulge in the direction of it—a high tide. The part outermost from the moon is attracted the least by the moon, so flows away from the moon (and Earth's center)—additional high tide on the opposite side of the earth, Figure 5.9. In between, the water level must drop—the low tides. As the moon orbits the spinning earth, there is a cycle of two high tides and two low tides about every 25 hours.

*Fig.5.11: The Moon Cycle Affects Earth's Tide*

Tides are essential to life on Earth, Figure 5.11. Tides wash the ocean's shorelines, and aid preserve the ocean currents circulating, averting the ocean from decaying. They benefit man by scrubbing out shipping docks and diluting sewage expulsions. In some places, people adventure the enormous energy of the tides to produce electricity.

*Fig.5.12: The Apollo moon landing. Such achievements may be a logical extension of the dominion mandate given to mankind in Genesis 1:28. The moon's utter barrenness should remind us of our planet's unique design for life.*

## LIFELESS MOON

One of the most histrionic events of our time was the landing of men on the moon, Figure 5.12. Though, they long-established that it is an inert, airless world, with enormous temperature immoderate and no liquid water. From the moon, Earth seems as a bright blue-and-white object in the black sky. Earth is the planet God has constructed for life. Man may be able to live on other worlds one day, but it will be hard to make them livable.

Many people don't realize that the man behind the Apollo moon mission was the creationist rocket scientist "Wernher von Braun." And another creationist, Jules Poirier," designed some vital navigational equipment employed in the space program.

### The Moon Receding

Friction by the tides is decreasing the earth's rotation, so the span of a day is growing by 0.002 seconds per century. This is evidence that the earth is losing "angular momentum." The "Law of Conservation of Angular Momentum" states that the angular momentum the earth loses must be added by the moon. Accordingly, the moon is gradually receding from Earth at about 4 cm (1½ inches) per year, and the rate would have been larger in the past. The moon could never have been nearer than 18,400 km (11,500 miles), known as the "Roche Limit," as Earth's tidal forces, i.e., the result of diverse gravitational forces on diverse parts of the moon, would have devastated it, Table 5.3.

However, even if the moon had begun receding from being in contact with the earth, it would have taken only 1.37 billion years to reach its present distance. NB: this is the "maximum" possible age—far too young for evolution, and much newer than the radiometric *'dates'* assigned to moon rocks—not the actual age.

**Table 5.3:** Moon Facts

| | |
|---|---|
| Mean istance from earth | 384,404 km or 239,000 miles |
| Diameter | 3,476 km or 2172.5 miles (0.273 Earth, 1/400 Sun) |
| Mass | $7.35 \times 10^{22}$ kg (0.0123 Earth) |
| Density | 3.34 g/cm$^3$ (0.6 Earth) |
| Surface Temperature | 204°C (400°F) day, -205°C (-338°F) |
| True (sidereal) orbital period | 27.322 Earth days (29.531day phase cycle)14 |
| Orbital angular momentum | $2.68 \times 10^{34}$ kg m$^2$/s (82.9% of earth-moon system) |
| Inclination of equator to orbital plane | 6° 41' (*cf.* Earth 23° 27') |
| Earth-moon gravitational attraction | $1.98 \times 10^{20}$ N ($2.23 \times 10^{16}$ tons) |

## DOES THE MOON CREATE ITSELF?

Evolutionists and progressive creationists refute the moon is direct creation by God. They have come up with numerous hypotheses, however, they all have serious flaws, as many evolutionists themselves acknowledge. For example, lunar researcher "S. Ross Taylor" said:

*'The best models of lunar origin are the testable ones, but the testable models for lunar origin are wrong.'* Another astronomer said, *"half-jokingly, that there were no good (naturalistic) explanations, so the best explanation is that the moon is an illusion!"*

1. ***Fission theory,*** invented by the astronomer George Darwin, son of *Charles Darwin*. He suggested that the earth twirled so fast that a chunk broke off. However, this theory is universally rejected today. The

earth could never have twirled sufficiently fast to toss a moon into orbit, and the escaping moon would have been devastated while within the *Roche* Limit.

**Day Becomes Night**

One of the most captivating views in the sky is a total eclipse of the sun. This is conceivable as the moon is virtually precisely the exact angular size, ½ a degree, in the sky as the sun—it is both 400 times smaller and 400 times closer than the sun. This appears like design. If the moon had actually been receding for billions of years, and man had been existing for a tiny segment of that time, the probabilities of mankind alive at a time so he could see this precise size matchup would be far-off.

2.  *Capture theory*—the moon was wandering through the solar system, and was apprehended by Earth's gravity. However, the probability of two bodies passing near enough is minute; the moon would be more likely to have been 'sling-shotted' like artificial satellites than apprehended. Lastly, even a successful apprehension would have resulted in an elongated comet-like orbit.

3.  *Condensation theory*—the moon grew out of a dust cloud attracted by Earth's gravity. Nevertheless, no such cloud could be sufficiently dense, and it doesn't represent the moon's low iron content.

4.  *Impact theory*— the presently trendy idea that material was blasted off from Earth by the impact of another object. Computations illustrate that to have enough material to formulate the moon, the impacting object would necessitate to have been twice as enormous as Mars. Then, there is the unresolved problem of losing the excess angular momentum.

The moon is a virtuous example of the heavens declaring God's glory *(Psalm 19:1)*. It does what it's designed to do, and is vivacious for life on Earth. It is also a headache for evolutionists/uniformitarians.

The most stable configuration for a massive body is for all parts of the surface to be the same distance from the center of mass, i.e., a sphere. The pressure inside the moon is ten times the crushing strength of granite, so any large unevenness would be crushed into shape. Such a sphere may bulge at the equator if the body is revolving fast enough.

That is, its rotational period is identical to its (synodic) orbital period. This is true of many moons in the solar system, because the planet's gravity is always stronger on the nearest side (a tidal interaction), and this will eventually lock one side so it will always face the planet. The effect is enhanced if one side is denser than the other.

## The Earth's Magnetic Field

*Fig.5.13: The Aurora Borealis (Northern Lights). This is caused by charged particles from space striking the earth's atmosphere. These particles have been deflected towards the poles by the presence of the earth's magnetic field (which also diverts many such particles harmlessly into space).*

The earth possesses a magnetic field pointing almost north-south—Deviated only 11.5°. This is one of the most outstanding design characteristics of our planet: compasses are utilized to provide accurate navigations. This magnetic field shields people from harmful charged particles from the scorching sun. Additionally, it is also unique indication that the earth must be young Earth as it is stated in the Scriptures.

Dr Thomas Barnes the creationist physics professor studied the Earth magnetic field by accumulating data from 1835 till 1970. In the 1970s, he reported that measurements have illustrated that the magnetic field is decaying at 5% per century. Also, the archaeological measurements illustrated that the magnetic field was 40% stronger in AD 1000 than today. Dr. Barnes, the author of a well-respected electromagnetism textbook, projected that the earth's magnetic field was triggered by a "decaying electric current" in the earth's metallic core. Dr. Barnes computed that the current data could not have been decaying for more than 10,000 years, or otherwise, its original strength would have been significant enough to melt the earth. Consequently, the earth must be younger than that, Figure 5.13.

## Magnetic Decay – Young Earth

The decaying of the present data is clearly unharmonious with the billions of years required by evolutionists. Therefore, the evolutionists favored a model of a "self-sustaining dynamo," which is referred to electric generator.

It is known that the earth's rotation and convection is hypothetically circulates the molten nickel/iron of the outer core. Accordingly, positive and negative charges in this liquid metal are theoretically circulate unevenly, which are producing an electric current. Thus, creating the magnetic field. Nevertheless, scientists have not produced a workable model in spite of 60 years of research, during which the rise of many problems have risen.

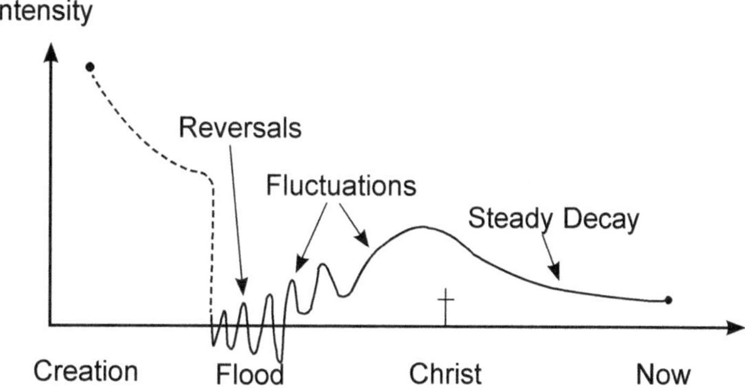


*Fig.5.14: How the earth's magnetic field has changed. The intensity could not have been much higher than the starting point shown, indicating a young age.*

However, the main criticism of Barnes' young-earth disagreement concerns evidence that the magnetic field has "inverted" many times—i.e. compasses would have pointed south instead of north. When substances of the known magnetic mineral "magnetite" in volcanic lava or ash flows cool below its "Curie point" of 570°C (1060°F), the magnetic spheres partly align themselves in the direction of the earth's magnetic field "at that time." As soon as the rock has fully cooled, the magnetite's alignment is fixed. Accordingly, we have a permanent record of the earth's field through time, Figure 5.14.

Although evolutionists have no good clarifications for the reversals, they uphold that, because of them, the straightforward decay assumed by Dr Barnes is unenforceable. Additionally, evolutionists' model necessitates at the minimum thousands of years for a reversal. According to the evolutionists dating assumptions, they believe that the reversals happen at recesses of millions of years, and point to an old earth.

## Response

Dr Russell Humphreys, renounced physicist held his views that Dr Barnes had the correct concept, and he also acknowledged that the reversals were real. Dr. Hu,phreys adapted Barnes' model to represent special influences of a liquid conductor, as the molten metal of the earth's outer core.

If the molten liquid ran upwards, due to convection—hot fluids rise, cold fluids sink-- this could occasionally make the field reverse quickly.

Now, 1997, Dr John Baumgardner propositions that the plunging of tectonic plates was a cause of the Genesis Flood. Dr Humphreys says these plates would have sharply cooled the outer parts of the core, due to the flood, driving the convection. This clarifies that most of the magnetic reversals happened in the Flood year, every week or two. And after the Flood, there would be huge fluctuations due to residual motion. Nevertheless, the reversals and fluctuations could not stop the overall decay pattern—rather, the total field energy would decay even quicker, Figure 5.14.

This model also clarifies the reason the sun reverses its magnetic field every 11 years. The sun is a enormous ball of hot, actively moving, electrically conducting gas. Conflicting to the dynamo model, the overall field energy of the sun is declining.

Dr Humphreys also planned a test for his model: magnetic reversals must be discovered in rocks recognized to have cooled in days or weeks. For example, in a thin lava stream, the outside would cool initially, and record earth's magnetic field in one direction; the inside would cool later, and record the field in another direction.

Three years after Dr. Humphreys' prediction, the principal researchers "Robert Coe" and "Michel Prévot" discovered a thin lava layer that must have cooled within 15 days, and had 90° of reversal recorded continuously in it. And it was no accident—eight years later, they conveyed an even faster reversal, Figure 5.15. This was stunning news to the creationists and the rest of the evolutionary community; however, robust support was hailed for Humphreys' model.

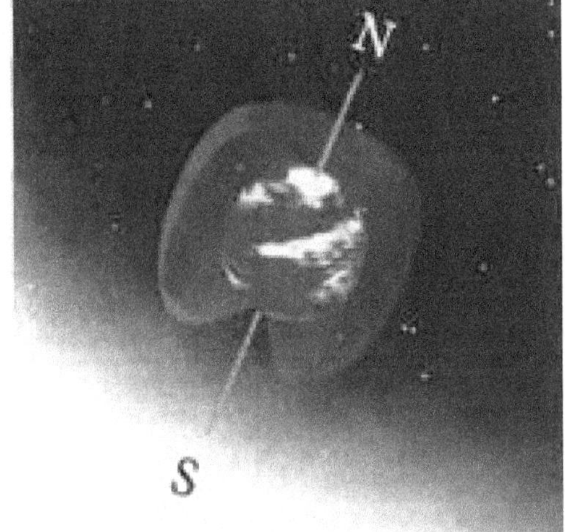


*Fig.5.15: Reversal of Magnetic Field in Rocks was Discovered*

A 'force-field' around the earth.

All human beings on earth are safe no matter how the earth's magnetism is running down. Undoubtedly, this world-wide phenomenon could not have been going on for more than a few thousand years, despite exchanging direction many times. Evolutionary hypotheses are not able to explain adequately how the magnetism could withstand itself for billions of years, Figure 5.15!

The earth's magnetic field is not only a good navigational aid and a shield from space particles, it is powerful evidence against evolution and billions of years. The clear decay pattern shows the earth could not be older than about 10,000 years.

Recently, geophysicist "David Stevenson" at the California Institute of Technology acknowledged the difficulties that the earth's magnetic field poses for long-age dogma, he says:

"Right at this moment, there is a problem with our understanding of Earth's core and it's something that's emerged only over the last year or two. The problem is a serious one. We do not now understand how the Earth's magnetic field has lasted for billions of years. We know that the Earth has had a magnetic field for most of its history. We don't know how the Earth did that. We have less of an understanding now than we previously thought we had a decade ago of how the Earth's core has operated throughout history.

## Origin of Earth's Magnetic Field

*Fig.5.16*: Earth Magnetic Field

## Magnet in Hydrogen Atoms

Dr Humphreys postulated that the infinite Creator God first created the earth out of water. He founded this on several Scriptures, e.g., *2 Peter 3:5* which achieves that the earth was formed out of water and by water. Thereafter, God would have converted much of the water into other materials like rock minerals. Now water encompasses hydrogen atoms, and the nucleus of a hydrogen atom is a miniscule magnet. Typically, these magnets cancel out so water as a whole is almost non-magnetic, Figure 5.16.

However, Humphreys suggested that God created the water with the nuclear magnets aligned. Directly after creation, the nuclear magnets would shape a more random arrangement, which would cause the earth's magnetic field to decay. This would create current in the core, which would then decay according to Barnes' model, apart from many reversals in the Flood year as Humphreys' model states.

## Neptune Magnetic Field

Dr Humphreys also computed the fields of other planets, including the sun based on this model. The vital characteristic is the mass of the planet, the size of the core and how well it conducts electricity, in addition to the assumption that their original material was water, Figure 5.17. Dr. Humphreys' model clarifies properties which are profound puzzles to dynamo theorists. For example, evolutionists denote 'the enigma of lunar magnetism'—the moon once had a powerful magnetic field, though it rotates only once a month. Additionally, based to evolutionary models of its origin, it never had a "molten" core, essential for a dynamo to work. Also, Mercury has a far powerful magnetic field than dynamo theory expects from a planet rotating 59 times slower than Earth.

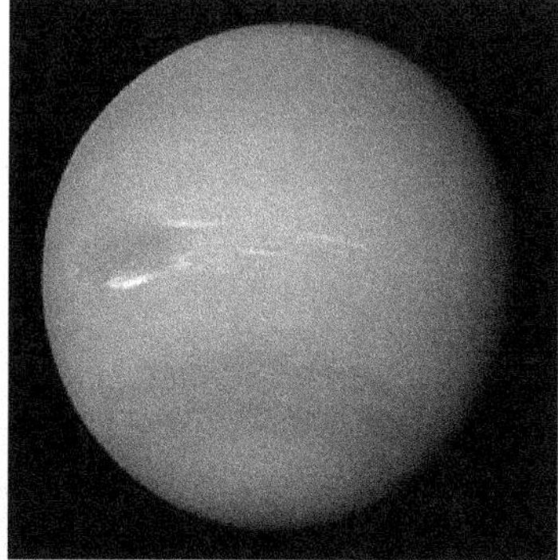

*Fig.5.17*: The planet Neptune as photographed by the Voyager Probe

Even more prominently, in 1984, Dr Humphreys foretold that the field strength of Uranus was about 100,000 times the evolutionary predictions from their 'dynamo' theory. The two competing models were tested when the Voyager 2 spacecraft flew past these planets in 1986 and 1989.

The magnetic field for Uranus was exactly as Humphreys had foretold. Dr Humphreys' forecast for the field strength of Neptune aligned with the evolutionary prediction, because this planet has a high heat outflow. But Neptune's field was still dissimilar in other vital ways from the evolutionary dynamo prediction: the tilt and offset. However, many anti-creationists call creation 'unscientific' because it allegedly makes no predictions!

Humphreys' model also clarifies the reason the moons of Jupiter that have cores have magnetic fields, while Callisto, which lacks a core, also lacks a field

## EARTH'S MAGNETIC FIELD SUBSTANCE

Materials as iron are comprised of minuscule magnetic domains, which each behave like tiny magnets. The domains themselves are composed of even tinier atoms, which are themselves microscopic magnets, lined up within the domain. Normally the domains cancel each other out. But in magnets, like a compass needle, more of the domains are lined up in the same direction, and so the material has an overall magnetic field.

Earth's core is primarily iron and nickel, therefore, could its magnetic field be caused the same way as a compass needle's? The answer is no. Above a temperature called the "Curie point," the magnetic spheres are disturbed. The earth's core at its coolest region is about 3400–4700°C (6100–8500°F), much hotter than the Curie points of all recognized materials.

Nonetheless, in 1820, the Danish physicist "H.C. Ørsted" revealed that an electric current yields a magnetic field. Without this, there could be no electric motors. Therefore, could an electric current be accountable for the earth's magnetic field? Electric motors have a power source, but electric currents normally decay almost instantly once the power source is turned off, except in superconductors substances. Thus, how could there be an electric current inside the earth, without a source?

Michael Faraday, the great creationist physicist responded to this question in 1831 with his detection that a changing magnetic field "induces" an electric voltage, the basis of electrical generators.

Visualize the earth immediately after creation with a large electrical current in its core. This would yield a strong magnetic field. Deprived of a power source, this current would decay. Therefore, the magnetic field would decay too. As decay is change, it would induce a current, lower but in the same direction as the original one.

Accordingly, we have a decaying current generating a decaying field which creates a decaying current … If the circuit dimensions are sufficiently large, the current would take sometime to die out. The decay rate can be precisely computed, and is constantly exponential. The electrical energy doesn't vanish—it is changed into heat, a process detected by the creationist physicist "James Joule" in 1840. This is the basis of Dr Barnes' model.

Thomas Barnes, who first pointed out magnetic field decay as a problem for evolutionists, was a regarded authority in electromagnetism and composed some well-regarded textbooks on the subject. But most of his critics are tactlessly ignorant of the subject.

## GRAVITY THEORY INVALIDATES BIG BANG

Often, we read in the reputable press, particularly online, that somebody has come up with a new theory of gravity. The obvious reason is that the current theory defining the evolution of the universe is questionable.

The standard "ΛCDM" big bang cosmology is resulting from an application of certain non-biblical border tailored to the physics of "Einstein's general relativity theory." However, as that was inserted to the universe as a whole, two problems advanced for the materialistic model. The first one is the necessity to incorporate in dark energy, or the cosmological constant, "Λ" (Lambda), to "Einstein's field equations." The second is the necessity for a substantial amount of "invisible cold dark matter (CDM)."

On the scale of galaxies and even clusters of galaxies "Newtonian physics" is employed as it is the small gravity limit of general relativity. However, deprived of the addition of dark matter the outcome theory, utilizing the known density of visible matter in galaxies, Figure 5.16 and clusters, does not tie observations. However, for more than 40 years now dark matter has been pursued in various lab experiments with unfailing negative results. This has created the senses called "the dark matter crisis."

Infrequently, an entitlement is made that a theorist has additional suspicion of what dark matter particles could be; nonetheless, the predicament remains. Dark matter particles have been sought after deprived of any triumph in the galaxy utilizing very sensitive detectors almost bottomless in underground mines, or with the "Large Hadron Collider (LHC)" over 15 years of experiments searching for the lowermost mass stable particle in a theorized class of as-yet-undiscovered supersymmetric particles.

## Big Bang Wrong!

The observational data from countless of galaxies combined with the negative results of all the experiments probing for dark matter particles illustrate that either something is erroneous with the physics we employ or that the predictable dark matter is much more indefinable than even hypothetical, or, astonishing, does not, in fact, exist in the material sense—which obviously allow us back to something being erroneous with the physics.

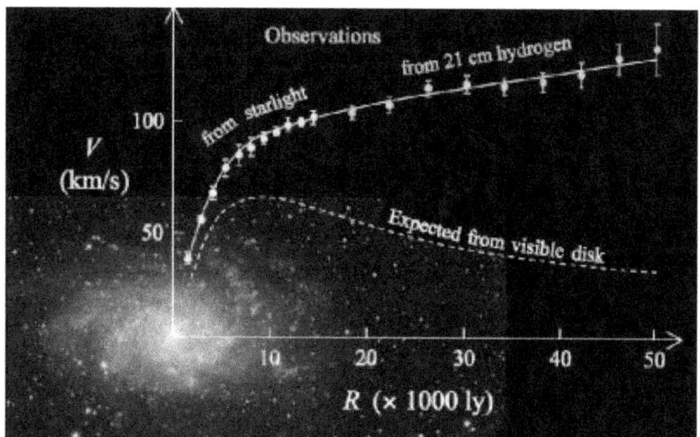


**Fig.5.18:** *Typical rotation curve of a spiral galaxy. Speeds (V) in km/s units as a function of distance from the center of the galaxy (R) in 1000 light-year (LY) units. The upper curve shows the speeds of the stars in disk region determined from their visible light and the gasses beyond that determined from radio frequency emissions. The lower curve shows what standard Newtonian physics predicts should be observed. The discrepancy is made up by positing the existence of invisible dark matter.*

This exact kind of difficulty also spread over to the concept of dark energy, which was presented into the big bang to clarify observations that were understood to illustrate that the expansion rate of the universe was really cumulative over cosmic time, that is, the expansion is accelerating.

The fundamental issue of the identity of this 'anti-gravity' dark energy has also gone unrequited. It has been proposed that the answer may lie with a particle called a chameleon. Only in empty space does the conjectured chameleon particle put on the features which deliver a novel anti-gravity force in the universe, henceforth dark energy. However, as you search for it nearby Earth it puts on a new identity and hides itself so you may not locate it.

To clarify dark matter, novel physics has been wished-for. These comprise "Milgrom's Modified - Newtonian Dynamics (or MOND)" and "Bekenstein's Tensor-Vector-Scalar" gravity (or TeVeS) a relativistic theory of modified gravity.

Although not originally envisioned to deal with the dark matter crisis, I found that "Carmeli's Cosmological General Relativity (CGR)" clarifies the existing observations with no necessity for either dark matter or dark energy, Figure 5.18.

## Cosmology with no Acceleration

Presently, some scientists have proposed that, as the large-scale observational data for the entire universe are all considered into account, dark energy does not require to be involved, as the observations applied to the standard cosmology are consistent with no acceleration. The author of one such study said:

"... it is quite possible that we are being misled and that the apparent manifestation of dark energy is a consequence of analyzing the data in an oversimplified theoretical model..."

*Moreover, it was noted even earlier that*:

*"an alternative possibility ... can explain the observations as a fluke of cosmological geometry. It avoids invoking dark energy as an* ad hoc *cause but at the price of throwing out the Copernican principle: roughly speaking, it puts the Earth, or at least our galaxy, back at the center of the observable universe."*

The Copernican principle is, obviously, an random condition forced on the interpretation of cosmological observations so as to evade the likelihood that the Earth has a exclusive location in the universe, something one might conclude from reading Genesis.

## NOVEL IDEA

"In 2010, "Erik Verlinde" astonished the world with a totally novel theory of gravity. According to Verlinde, gravity is not a essential force of nature, however, an emergent marvel. Equally, that temperature rises from the movement of microscopic particles, gravity emerges from the changes of essential bits of information, stowed in the very assembly of spacetime."

This novel theory describes gravity as "emergent", meaning that it is not essential but 'emerges' from the universe as it cooled from the original hot big bang fireball.

*"At large scales, it seems, gravity just doesn't behave the way Einstein's theory predicts,"* Verlinde said. His method is novel physics. That may be a much more sensible method than searching for all the dark sector entities that are now invoked, mainly to plug in where the standard physics has failed.

## Big Bang in Doubt

My intention here is not to try to choose which theory of gravity is correct. I would like to ask a different question. If what we are told is factual—i.e., that big bang cosmology is fundamentally comprehensive and the "big bang" origin of the universe factual—why do these physicists even look for new physics?

Is it not since as time goes by the investigation problem of the non-detection of dark matter, dark energy, dark radiation, dark photons, chameleons and many more dark entities starts to illustrate that the entire paradigm itself is in doubt?

Confidently, the very notion of a universe that created itself from nothing, without a Creator God, must be tossed into distrust. And does this exploration for a novel physics not authenticate other suggestions of novel physics that try to clarify the observed phenomenon in a manner in harmony with the biblical creation account in the Bible, such as those projected by Dr. "Russell Humphreys?" The universe requires a First Cause, a Creator, and that merely will not do for the incredulous world.

The secular chase of novel physics to resolve big bang cosmology problems, as their 'god of the gaps' – "dark matter crisis," is actually an endeavor to stave off reception of what the Infinite Creator expressed to us thousands of years ago, "In the beginning God …". *(Genesis 1:1)*

## APPENDIX

1. For the technical reader: since tidal forces are inversely proportional to the cube of the distance, the recession rate (dR/dt) is inversely proportional to the sixth power of the distance. So $dR/dt = k/R^6$, where k is a constant = (present speed: 0.04 m/year) × (present distance: 384,400,000 m)$^6$ = $1.29 \times 10^{50}$ m$^7$/year. Integrating this differential equation gives the time to move from $R_i$ to $R_f$ as $t = \frac{1}{7k}(R_f^7 - R_i^7)$. For $R_f$ = the present distance and $R_i$ = the Roche Limit, $t = 1.37 \times 10^9$ years. There is no significant difference if $R_i = 0$, i.e. the earth and moon touching, because of the high recession rate (caused by enormous tides) if the moon is close. See also Don DeYoung, 'The Earth-Moon System', Proc. Second International Conference on Creationism **2**:79–84, 1990.

   The sidereal period is the time for a complete orbit of the moon around the earth, relative to an observer outside the solar system. The phase cycle (synodic period) is the time taken for the moon to return to the same orientation towards the sun. It is longer because the earth moves about 1/13$^{th}$ of the way in its orbit around the sun, so the moon must travel further than one true lunar orbit for a given orientation to recur.

2. According to Einstein's theory, light from distant stars should be deflected when it passed through the strong gravitational field of the sun. During the eclipse, the light from stars previously not visible because of the sun's brilliance appeared to be bent, i.e., the stars appeared to be in a different position from when their light did not pass close to the sun. The amount that the starlight was bent by the sun's gravity agreed with Einstein's predictions.

3. Isaac Newton overlooked Psalm 14:1.

## REFERENCES


1. Storr, A., Isaac Newton, British Medical J. (Clinical Research Edition) 291(6511):1779–1784, 1985; p. 1779.

2. Brewster, D., The life of Sir Isaac Newton, Harper & Brothers, New York, p. 18, 1840.

3. Keynes, J.M., Newton, the man; in: Johnson, E. and D. Moggridge (Eds.), The collected writings of John Maynard Keynes, Royal Economic Society, Cambridge, pp. 363–364, 1978.

4. Bedford, A., Animadversions upon Sir Isaac Newton's book, intitled the chronology of ancient kingdoms amended, printed by Charles Ackers, London, p. 1, 1728.

5. Russell, B., The history of western philosophy, Simon & Schuster, Inc., New York, p. 535, 2008. Russell claims that Newton's "triumph" in the sciences was "so complete", that he "was in danger of becoming another Aristotle."

6. Rickey, V.F., Isaac Newton: man, myth, and mathematics, The College Mathematics J. 18(5):362–389, 1987; p. 362.

7.  Manuel, F.E., Isaac Newton: Historian, Belknap Press, Cambridge, MA, p. 2, 1963.

8.  In the original: "Collocavit igitur Deus planetas in diversis distaniis a sole, ut quilibet pro gradu densitatis calore solis majore vel minore fruatur" (Newton, I., Philosophiæ naturalis principia mathematica, 1st edn, Streater, Joseph, London, p. 415, 1687)

9.  Newton, I., The Principia: Mathematical principles of natural philosophy, University of California Press, Berkeley, CA, p. 814, 1999.

10. Cohen, I.B., Introduction to Newton's 'Principia', The University Press, Cambridge, pp. 155–156, 1971.

11. Newton, I., Newton's Principia: The mathematical principles of natural philosophy, ed. Motte, A., Daniel Adee, New York, p. 504, 1846. Return to text.

12. Newton, ref. 16, p. 506.

13. Newton, I., The third book of Opticks (1718), University of Sussex, East Sussex, UK, p. 345, 2006; newtonproject.sussex.ac.uk/view/texts/normalized/NATP00051.

14. Newton, I., Opticks: Or a treatise of the reflections, refractions, inflections and colours of light, 4th edn, Printed for William Innys, London, p. 344, 1730.

15. Newton, I., A short schem of the true religion, Publisher, Cambridge, p. 1r, 2002; newtonproject.ox.ac.uk/view/texts/normalized/THEM00007.

16. Manuel, F.E., The Religion of Isaac Newton, Oxford University Press, Oxford, p. 35, 1974.

17. Newton, I., Original letter from Isaac Newton to Richard Bentley, dated 10 December 1692, Trinity College Library, vol. 189.R.4.47, folio no. 4A-5, Cambridge, p. 4r, 2007; newtonproject.ox.ac.uk/view/texts/normalized/THEM00254.

18. Newton, I., Seven statements on religion, The Newton Project, Cambridge, p. 1, 2002; newtonproject.sussex.ac.uk/view/texts/normalized/THEM00006.

19. Snobelen, S.D., The theology of Isaac Newton's Principia Mathematica: a preliminary survey, p. 377, 2010. Similarly, Westfall: "Having studied the entire corpus of his theological papers, I remain unconvinced that it is valid to speak of a theological influence on Newton's science." (Westfall, R.S., Newton's theological manuscripts; in: Bechler, Z. (Ed.), Contemporary Newtonian Research, Studies in the history of modern science, D. Reidel Publishing Company, London, p. 139, 2012).

20. Snobelen, S.D., Isaac Newton and Apocalypse Now: A response to Tom Harpur's "Newton's strange bedfellows", University of King's College, Canada, p. 2, 2004.

21. Hearne, T., Remarks and Collections of Thomas Hearne, Salter, H.E. (Ed.), vol. 11, Clarendon Press, Oxford, pp. 100–101, 1921; See also: Hearne, T., Remarks and Collections of Thomas Hearne, Salter, H.E. (Ed.), vol. 10, Clarendon Press, Oxford, p. 412, 1915; Pfizenmair, T.C., Was Isaac Newton an Arian? J. History of Ideas 58(1):57–80, 1997; p. 80.

22. Snobelen, S.D., Isaac Newton, heretic: the strategies of a Nicodemite, The British J. History of Science 32:381–419, 1999; pp. 383, 388.

23. Austin, W.H., Isaac Newton on Science and Religion, J. History of Ideas 31(4):521–542, 1970; p. 53. Keynes, ref. 3, pp. 368–369. Mandelbrote, S., A duty of the greatest moment: Isaac Newton and the writing of biblical criticism, The British J. History of Science 26(3):281–302, 1993; pp. 283–284. Pfizenmair, ref. 27, p. 79. Rickey, ref. 9, p. 384. Storr, ref. 1, p. 1780. Westfall, ref. 25, p. 130. Hughes, M., Newton, Hermes and Berkeley, The British J. Philosophy of Science 43(1):1–19, 1992; pp. 1–4.

24. Manuel, ref. 21, p. 58. In a fairly recent dissertation on the theology of Isaac Newton, Van Alan Herd has argued that Newton was not an Arian, but instead a "mainstream Puritan" (Herd, V.A., The theology of Isaac Newton: a dissertation submitted to the graduate faculty in partial fulfillment of the requirements

for the degree of Doctor of Philosophy, The department of the history of science, University of Oklahoma, OK, p. 2, 2008; https://pqdtopen.proquest.com/pubnum/3304232.html?FMT=AI). But as will be evident from the ensuing discussion, this thesis is hardly compelling.

25. Newton, I., Miscellaneous notes and extracts on the Temple, the Fathers, prophecy, Church history, doctrinal issues, etc., National Library of Israel, Publisher, Jerusalem, pp. 25r, 173r–173v, 2012; newtonproject.sussex.ac.uk/view/texts/normalized/THEM00057. Newton has been quoted as saying to a Mr Hopion Haynes that, "the time will come, when the doctrine of the incarnation as commonly received, shall be exploded as an absurdity equal to transubstantiation!" (Baron, R., The preface, A cordial for low spirits, vol. 1, London, pp. xviii–xix, 1763.).

26. Newton, I., Irenicum, or Ecclesiastical Polyty tending to Peace, Publisher, Cambridge, p. 33, 2002; newtonproject.ox.ac.uk/view/texts/normalized/THEM00003.

27. Newton, I., Treatise on Revelation (Section 1), National Library of Israel, Jerusalem, pp. 19v–21v, 2002; newtonproject.sussex.ac.uk/view/texts/normalized/THEM00216. Manuel, ref. 10, p. 149. Manuel, ref. 21, p. 84. Snobelen, ref. 28, p. 388. Snobelen, ref. 26, p. 160. It is possible that Newton's demonology was derived from Thomas Hobbes (1588–1679). (Hobbes, T., Hobbe's Leviathan, Clarendon Press, Oxford, p. 502, 1929.).

28. King, L., The life and letters of John Locke, Henry G. Bohn, London, p. 230, 1858. Popkin, R.H., Newton, Spinoza and the biblical scholarship of the day; in: Osler, M.J. (Ed.), Rethinking the scientific revolution, Cambridge University Press, Cambridge, pp. 299–300, 2000.

29. Newton, I., Observations upon the prophecies of Daniel, James Nisbet and T. Stevenson, London, pp. 4–5, 1831. Simon, R., A critical history of the Old Testament, Walter Davis, London, p. 37, 1682. Newton also "collected, arranged, and strengthened Simon's arguments" in a dissertation on 1 John 5:7 (King, ref. 36, p. 230).

30. Hoffecker, W.A., Enlightenments and awakenings: the beginnings of modern culture wars; in: Revolutions in Worldview: Understanding the flow of western thought, Hoffecker W.A. (Ed.), P&R Publishing Company, New Jersey, p. 244, 2007. Korab-Karpowicz, W.J., A history of political philosophy: from Thucydides to Locke, Global Scholarly Publications, New York, p. 291, 2010.

31. Newton, I., Letters from Sir Isaac Newton; in: King, L. (Ed.), The life and letters of John Locke, Henry G. Bohn, London, p. 217, 1858. King, ref. 35, p. 231. Westfall, R.S., Never at rest: a biography of Isaac Newton, Cambridge University Press, Cambridge, pp. 490–491, 2010.

32. Like Newton, Le Clerc disliked the trinitarian formulation in Nicene Creed and probably shared Newton's preference for Arius over Athanasius (Klauber, M.I., Between protestant orthodoxy and rationalism: fundamental articles in the early career of Jean LeClerc, J. History of Ideas 54(4):611–636, 1993; pp. 630–631. Manuel, ref. 21, p. 58).

33. Israel, J.I., Radical enlightenment: philosophy and the making of modernity 1650–1750, Oxford University Press, New York, p. 464, 2001. Return to text.

34. Burnet, T., The sacred theory of the earth: containing an account of the original of the earth, and of all the general changes which it hath already undergone, or is to undergo, till the consummation of all things, 6th edn, vol. 1, Printed for J. Hooke, London, pp. xxiii, pp. 206–207, 1726.

35. Burnet, T., Correspondence with Thomas Burnet, The Newton Project, Cambridge, p. 2, 2008; newtonproject.sussex.ac.uk/view/texts/normalized/THEM00014.

36. Newton, I., Correspondence with Thomas Burnet, The Newton Project, Cambridge, pp. 10–12, 2008; newtonproject.sussex.ac.uk/view/texts/normalized/THEM00014.

37. Revelation to be the "fundamental book" of the Bible by which the other 65 books should be interpreted (Westfall, ref. 25, p. 131). Although I have not been able to corroborate this claim from Newton's writings, Newton did consider the study of Revelation to be "a duty of the greatest moment" for the church living in these last days (Newton, I., Untitled treatise on Revelation (section 1.1), National Library of Israel, Publisher, Jerusalem, p. 1r, 2004; newtonproject.ox.ac.uk/view/texts/normalized/THEM00135).

38. Contra Drake, E.T. and P.D. Komar, Speculations about the earth: the role of Robert Hooke and others in the 17th century, Earth Sciences History 2(1):11–16, 1983, p. 12; Jones, F.N., The Chronology of the Old Testament: A return to the basics, 21st edn, Master Books, Arizona, p. 22, 2019; Morris, H.M., Men of Science, Men of God: Great scientists of the past who believed the Bible, Master Books, Arizona, p. 32, 2012.

39. Contra Janiak, A., The book of nature, the book of Scripture, The New Atlantis 44:95–103, 2015; pp. 95–96, 99.

40. Burnet, T., Archeologiae Philosophicae: Or, the ancient doctrine concerning the originals of things, written in Latin: to which is added Burnet's theory of the visible world by way of commentary on his own theory of the earth, being the second part of his archiologiae philosophicae, J. Fisher, London, pp. 11–12, 17–18, 21–23, 35, 50, 1736.

41. Although Newton affirms the general historicity of Babel, he postulates that the division and dissemination of nations was primarily caused by "the rebellion of Nimrod". Newton, I., The Chronology of Ancient Kingdoms Amended, Printed for J. Tonson in the Strand, and J. Osborn and T. Longman in Pater-noster Row, London, pp. 186–187, 1728.

42. Newton, I., Of the Chronology of the First Ages of the Greeks & Latines, Jerusalem, p. 38r, 2013; newtonproject.ox.ac.uk/view/texts/normalized/THEM00406. Return to text.

43. Gaukroger, S., The collapse of mechanism and the rise of sensibility: science and the shaping of modernity, 1680–1760, Oxford University Press, Oxford, p. 376, 2010; Jones, ref. 53, p. 22; Lincoln, B., Isaac Newton and Oriental Jones on myth, ancient history, and the relative prestige of peoples, History of Religions 42(1):1–18, 2002; pp. 4–5.

44. Westfall, R.S., Newton's scientific personality, J. History of Ideas, 48(4):551–570, 1987; pp. 568–569.

45. Although he concedes that the "Greeks and Latines have made their first kings a little older than the truth."

46. Israel, ref. 41, p. 605. Redwood, J.A., Charles Blount (1654-93), Deism, and English free thought, J. History of Ideas 35(3):490–498, 1974; p. 490.

47. Whiston, W., A discourse concerning the nature, style, and extent of the Mosaic history of the creation; in: A New Theory of the earth, Printed for John Whiston at Mr Boyle's Head, London, pp. 52, 55, 87–109, 325, 1737.

48. Whiston, ref. 71, p. 4. Return to text.

49. Snobelen, S.D. and L. Stewart, Making Newton easy: William Whiston in Cambridge and London; in: Knox, K.C. and R. Noakes (Eds.), From Newton to Hawking: A history of Cambridge University's Lucasian professors of mathematics, Cambridge University Press, Cambridge, p. 140, 2003.

50. Halley, E., A short account of the cause of the saltness of the ocean, and of the several lakes that emit no rivers; with a proposal, by help thereof, to discover the age of the world. Produced before the Royal-Society by Edmund Halley, R. S. Secr., Philosophical Transactions (1683–1775) 29:296–300; p. 296.

51. Cohen, I.B., Newton in light of recent scholarship, Isis 51(4):489–514, 1960; p. 491.

52. Sibley, A., Deep time in 18th-century France—part 1: a developing belief, J. Creation 32(3):85–92, 2018; pp. 86–87. Rappaport, R., Fontenelle interprets the Earth's history, Revue d'histoire des sciences 44(3/4):281–300, 1991; p. 282.

53. Voltaire, The important examination of the holy Scriptures: attributed to Lord Bolingbroke, but written by M. Voltaire, R. Carlile, London, pp. 8, 12, 1819.

54. Buffon, G.-L.L.C.d., Buffon's Natural History: Containing a theory of the Earth, a general history of man, of the brute creation, and of vegetables, minerals, etc., vol. 1, Printed for the proprietor, London, p. 112, 1797.

55. Mayr, E., The Growth of Biological Thought: Diversity, evolution, and inheritance, Belknap Press of Harvard University Press, Cambridge, p. 336, 1982.

56. Frame, J.M., A History of Western Philosophy and Theology, 1st edn, P&R Publishing, New Jersey, p. 252, 2015.

57. Kant, I., Universal natural history and theory of the heavens or essay on the constitution and the mechanical origin of the whole universe according to Newtonian principles; in: Watkins, E. (Ed.), Natural Science, Cambridge University Press, Cambridge, pp. 200–201, 2012.

58. Israel, J., The early Dutch and German reaction to the Tractatus Theologico-Politicus: foreshadowing the Enlightenment's more general Spinoza reception?; in: Melamed, Y.Y. and M.A. Rosenthal (Eds.), Spinoza's 'Theological-Political Treatise': A critical guide, Cambridge University Press, Cambridge, p. 20, 2010.

59. Ford's assessment corroborates this: "Young Earth literalism was a minority viewpoint from 1660 onwards. A historical view of science and religion which does not embrace this view is unrealistic, and the founding of geology in the 1790s cannot be seen in its proper context." (Ford, B.J., Shining through the centuries: John Ray's life and legacy. A report of the meeting 'John Ray and his successors', Notes and Records of the Royal Society of London 54(1):5–22, 2000; p. 8.)

60. Halley, E., Ode on this spendid ornament of our time and our nation, the mathematico-physical treatise by the eminent Isaac Newton; in: Cohen, I.B. and A. Whitman (Eds.), The Principia: Mathematical principles of natural philosophy, University of California Press, Berkeley, p. 380, 1999.

61. Manuel, F.E., The Enlightenment, Prentice-Hall, Inc., New Jersey, p. 4, 1965.

62. Popkin, ref. 35, pp. 309–310. This point is made contra Manuel who maintains that it was Newton's science rather than his religion which had a lasting impact on western society (Manuel, ref. 21, p. 4).

63. Time, pp. 42–67, December 1999.

64. Albert was born in Ulm, Germany, on March 14, 1879. The family moved to Munich in 1880.

65. His interest in science stems from age five. His father gave him a compass. 'Why does the needle always point one way?', Albert wanted to know. 'Magnetism.' 'How does an invisible force pass through space?' Albert lay awake that night pondering the mystery. His life-long interest in asking and solving scientific questions had been awakened.

66. Pais, A., Einstein Lived Here, Oxford Uni. Press, New York, NY, USA, pp. 114–15, 1994.

67. Eidgenossische Technische Hochschule or ETH (the Federal Institute of Technology). This lasted from June 1902 to July 1909.

68. Translated titles: i. On a Heuristic Viewpoint Concerning the Production and Transformation of Light. ii. A New Determination of the Size of Molecules. iii. On the Motion of Small Particles Suspended in a Stationary Liquid According to the Molecular Theory of Heat. iv. On the Electrodynamics of Moving Bodies.

67. He received the first of his many honorary doctorates in 1909.

68. Albert was nominated for the Nobel Prize for Physics every year from 1910 to 1921, except 1911 and 1915.

69. This he did in 1923. It was then worth about US$32,000. Mileva bought a house in Zurich and lived there for most of the rest of her life. She died in 1948.

70. Ironically, this anti-Semitism may have kept the Nazis from developing an atomic bomb.

71. Albert Einstein signed another letter about the atomic bomb, intended for Roosevelt, on March 7, 1940.

72. Fölsing, A., Albert Einstein, Viking, New York, NY, USA, p. 734, 1997.

73. Ideas and Opinions by Albert Einstein, Crown Publishers, New York, NY, USA, pp. 36–39, 1954.

74. He was objecting to the random unpredictable element in quantum mechanics, where one cannot calculate what will happen, only what will probably happen.

75. Barnett, L., The Universe and Dr. Einstein, Victor Gollancz Ltd, London, UK, p. 95, 1953.

76. The Creator and the Cosmos, Navpress, CO, USA, p. 49, 1993.

77. Fred Pearce, 'Catching the tide', New Scientist 158(2139):38–41, June 20, 1998.

78. See Ann Lamont, 21 Great Scientists who Believed the Bible, Creation Science Foundation, Australia, 1995, pp. 242–251.

79. Angular momentum = mvr, the product of mass, velocity and distance, and is always conserved (constant) in an isolated system.

80.  S. Ross Taylor, paraphrased by geophysicist Sean Solomon, at Kona, Hawaii, Conference on Lunar Origin, 1984; cited in: Hartmann, Wm. K., The History of Earth, p. 44, Workman Publishing Co., Inc., Broadway, NY, 1991.

81. Irwin Shapiro in a university astronomy class about 20 years ago, cited by Lissauer, J.J., It's not easy to make the moon, *Nature* 389(6649):327–352, 25 Sep 1997 | doi:10.1038/38596 (comment on Ida *et al.*,, Ref. 11). Lissauer affirms that the first three theories have insoluble problems.

82. Shigeru Ida *et al.*, 'Lunar accretion from an impact generated disk', Nature 389(6649):353–357, 25 Sep 1997 | doi:10.1038/3866.

83. See also D.R. Faulkner, 'The angular size of the moon and other planetary satellites: An argument for Design', Creation Research Society Quarterly 35(1):23–26, June 1998.

84. From John C. Whitcomb and Donald B. DeYoung, The Moon: Its Creation, Form and Significance, Baker Book House, Grand Rapids, Michigan, 1978.

85. K.L. McDonald and R.H. Gunst, 'An analysis of the earth's magnetic field from 1835 to 1965', ESSA Technical Report, IER 46-IES 1, U.S. Govt. Printing Office, Washington, 1967.

86. R.T. Merrill and M.W. McElhinney, The Earth's Magnetic Field, Academic Press, London, pp. 101–106, 1983.

87. T.G. Barnes, Foundations of Electricity and Magnetism, 3rd ed., El Paso, Texas, 1977.

88. Measurements of electrical currents in the sea floor pose difficulties for the most popular class of dynamo models—L.J. Lanzerotti et al., Measurements of the large-scale direct-current earth potential and possible implications for the geomagnetic dynamo, Science 229:47–49, 5 July 1986. Also, the measured rate of field decay is sufficient to generate the current needed to produce today's field strength, meaning that there is no dynamo operating today, if it ever did.

89. D.R. Humphreys, Reversals of the earth's magnetic field during the Genesis Flood, Proceedings of the First International Conference on Creationism, Creation Science Fellowship, Pittsburgh, 2:113–126, 1986. The moving conductive liquid would carry magnetic flux lines with it, and this would generate new currents, producing new flux in the opposite direction. See also the interview of Humphreys in Creation 15(3):20–23, 1993.

90. Humphreys, D.R., Physical mechanism for reversals of the earth's magnetic field during the flood, Proceedings of the Second International Conference on Creationism, Creation Science Fellowship, Pittsburgh, 2:129–142, 1990. Dr Barnes, who had opposed field reversals because no mechanism could be demonstrated, responded (p. 141): 'Dr Humphreys has come up with a novel and physically sound approach to reversals of the magnetic field.'

91. D.R. Humphreys, Discussion of J. Baumgardner, Numerical simulation of the large-scale tectonic changes accompanying the Flood, Proceedings of the First International Conference on Creationism, Creation Science Fellowship, Pittsburgh, 2:29, 1986.

92. The field intensity (B) fluctuated up and down during and after the Flood, but the total field energy always decreased. For the technically minded, the energy is the volume integral of B.

93. R.S. Coe and M. Prévot, Evidence suggesting extremely rapid field variation during a geomagnetic reversal, Earth and Planetary Science 92(3/4):292–298, April 1989. See also the reports by Dr Andrew Snelling, Fossil magnetism reveals rapid reversals of the earth's magnetic field, Creation 13(3):46–50, 1991 The Earth's magnetic field and the age of the Earth, Creation 13(4):44–48, 1991.

94. R.S. Coe, M. Prévot and P. Camps, New evidence for extraordinarily rapid change of the geomagnetic field during a reversal, Nature 374(6564):687–692, 1995; see also A. Snelling, The principle of 'least astonishment', Journal of Creation 9(2):138–139, 1995.

95. Cited in: Folger, T., Journeys to the Center of the Earth: Our planet's core powers a magnetic field that shields us from a hostile cosmos. But how does it really work? *Discover*, July/August 2014.

96. D. Russell Humphreys, The creation of planetary magnetic fields, Creation Research Society Quarterly 21(3):140–149, 1984.

97. L.L. Hood, The enigma of lunar magnetism, Eos 62(16):161–163. Return to text.

98. The Voyager measurements were 3.0 and 1.5 x $10^{24}$ J/T for Uranus and Neptune respectively. N.F. Ness et al., Magnetic fields at Uranus, Science 233:85–89, 1986; A.J. Dessler, Does Uranus have a magnetic field? Nature 319:174–175, 1986; R.A. Kerr, The Neptune system in Voyager's afterglow, Science 245:1450–51.

99. Dr Humphreys had predicted field strengths of the order of $10^{24}$ J/T—Creation Research Society Quarterly 27(1):15–17, 1990. The fields of Uranus and Neptune are hugely off-centered (0.3 and 0.4 of the planets' radii) and at a large angle from the planets' spin axis (60° and 50°). A big puzzle for dynamo theorists, but explainable by a catastrophe which seems to have affected the whole solar system (see Revelations in the solar system).

100. Magnetic moon findings support creationist's theory Creation 19(4):8, 1997.

101. Humphreys, D.R., Physical mechanism for reversals of the earth's magnetic field during the flood, Proceedings of the Second International Conference on Creationism, Creation Science Fellowship, Pittsburgh, 2:129–142, 1990.

102. Barraclough, D.R., Geophy. J. Roy. Astr. Soc., 43:645–659, 1975.

103. Kroupa, P., The dark matter crisis: problems with the current standard model of cosmology and steps towards an improved model. adsabs.harvard.edu (accessed November 2016).

104. Hartnett, J.G., Claimed dark matter 'find' won't help end 'big bang' crisis, creation.com, January 25, 2014.

105. Hartnett, J.G., Dark matter search comes up empty, biblescienceforum.com, July 2016.

106. Hartnett, J.G., SUSY is not the solution to the dark matter crisis, biblescienceforum.com, August 2016.

107. Identity means what type of agent causes it. Some suggest a particle, the chameleon (see Ref. 6), others a slowly evolving scalar field. The latter then begs question of what that is.

108. Hartnett, J.G., Dark energy and the elusive chameleon—more darkness from the dark side, creation.com, October 2015.

109. Milgrom modified gravity to make the strength of its force inversely proportional to distance in weak gravitational fields. This contrary to standard Newtonian gravity which is inversely proportional to distance squared. Milgrom's theory can account for the observed orbital speeds of stars and gasses as shown in Fig. 1 without the need for dark matter. But Milgrom has no underlying fundamental reason why such a modification should be made except by tuning it to observations.

110. Bekenstein introduced a relativistic generalisation (hence more fundamental) of Milgrom's MOND to produce a mechanism why gravity is modified in the limit of weak accelerations. In the weak-field approximation of the spherically symmetric, static solution, TeVeS reproduces the MOND acceleration formula.

111. From type Ia supernovae, which led to the Nobel prize in physics in 2011, for detection of an accelerating universe.

112. Hartnett, J.G., Now the expansion of the universe is not accelerating, biblescienceforum.com, November 2016.

113. Rennie, J., Nothing special, Scientific American 300(4):8, 2009 Return to text.

114. Verlinde, E., Emergent Gravity and the Dark universe, Preprint at arxiv.org.

115. New theory of gravity might explain dark matter, phys.org, November 2016.

116. Hartnett, J.G., Where materialism logically leads, creation.com, May 2016.

117. Hartnett, J.G., 'Dark photons': another cosmic fudge factor, creation.com, August 2015.

118. Hartnett, J.G., Dark radiation in big bang cosmology, creation.com, November 2014.

119. Hartnett, J.G., Why is Dark Matter everywhere in the cosmos?, creation.com, March 2015.

120. Humphreys, D.R., The 'Pioneer anomaly', Creation 31(1):37, December 2008.

121. Humphreys, D.R., New creation cosmology, creation.com, June 2009.

122. Hartnett, J.G., Starlight Time and the New Physics, 2nd Ed., Creation Book Publishers, 2010.

123. Hartnett, J.G., A biblical creationist cosmogony, Answers Research Journal 8:13–20, 2015.

124. Hartnett, J.G., Is 'dark matter' the unknown god?, Creation 37(2):22-24, April 2015.

# 6

# THE MYSTERY OF OUR UNIQUE EARTH

## CREATED EARTH

### Astonishing Design

In July 1969, a man safely landed on the moon surface.  I heard the history making news through a transistor radio, while I was standing outside the office of the Dean of Science and Technology awaiting to be called for my final-year oral interview prior to my graduation.  Although my knees were shaking, the exciting news of the moon-landing overwhelmed me, that temporarily subsided my fear.  As my name was called to assemble before the prominent committee of professors.  My big smile and my visible excitement spilled over the examiners.  Since the Dean of Science knew me quite well, he landed the first question to test my knowledge. Surprisingly, the first question was: "Soloman why do you appear so happy and excited."  My answer came out like a flood of sentences, yet well composed describing the great news: "A man has just landed safely on the surface of the moon, how exciting." The group of examiners started to whisper among themselves, until another examiner asked me, "What does it mean to you?" I said: "It is a triumph to human beings; I feel as if the moon was waiting the arrival of the astronaut to take the first step on its surface to welcome him!" I continued to express my though that the milky way galaxy was designed and placed in this particular location in the universe for the benefit of mankind, to simultaneously explore it - and enjoy it.  The Dean leaned over to whisper with the committee members.  He raised his head and said: "Soloman – no further questions." I was very pleased to hear that -- also my knees stopped shaking.  However, before I made any comment, I heard Dean saying "Congratulations!" The committee of examiners said the same.  Till today, I wondered why none of the committee members ask me any further question regarding any of the subjects I had studied for 5 years! Nonetheless, I have graduated with distinctive honors.

*Fig.6.1:* Astonishing Earth Design

When observing the Earth from the moon, the Apollo astronaut James Irwin said, "*When you lean far back and look up, you can see the earth like a beautiful, fragile Christmas tree ornament hanging in the blackness of space.*"  The gentle blue shell of the atmosphere, the deep blue of the sea, the brown continents, the spectacular white polar caps and scattered cloud, all in stark contrast with the pitch darkness of space with its multitudes of stars, undoubtedly make the earth the most beautiful place in the universe.

There is a concealed beauty about our planet that makes it obvious that Earth is amazingly well matched to be the home for mankind, just as it has been designed to be. Let us discover just a few of the astounding characteristics of our planet that make it so well appropriate for life, Figure 6.1.

## THE ORBIT OF THE EARTH

In a nearly perfectly circular orbit, our planet Earth moves in the "circumstellar habitable zone," or 'Goldilocks' region, around the sun. This region is where liquid water is present allowing it be not too hot, to avoid all the water from boiling away, and not too cold, to avoid all the water from freezing solid. In order to allow liquid water to exist on a planet, that planet must have a solid surface and an atmosphere providing sufficient pressure at the surface to avoid all the water evaporating. In actual fact, on earth, water does and will exist in all three states, liquid, solid, and water vapor.  Water on planet Earth moves effortlessly on planet Earth. Nevertheless, if the earth's orbit were exceedingly elliptical, there would be huge differences in temperature, making the atmosphere unsuitable for life.

# ROTATION OF THE EARTH

The earth rotates on its axis once per solar day, giving difference of night and day and providing colorful exhibitions in the clouds at sunrise and sunset. The spinning of the earth aids to normalize the temperature around the globe allowing no one part becomes too hot or too cold. If the Earth were tidally sealed to the sun, as the moon is sealed to the Earth, one side would be always facing the sun, and would be searingly hot, with the other in permanent frozen darkness.

## Axis of the Earth

The axis of the Earth is tilted about 23.5 degrees with respect to the plane in which the Earth orbits the sun, consequently, we experience a dissimilarity of seasons each year. In the northern hemisphere summer, the North Pole is tilted towards the sun.  Therefore, the sun is higher in the sky and the days are longer than the nights. At the same time the southern hemisphere is experiencing its winter. The reverse is true six months later. As the sun passes through Earth's equatorial plane, the days and nights are of equal length. This is called equinox and occurs in late March and late September.

The difference of seasons is significant for numerous forms of life to thrive. The annual cycle of cold to warm seasons revives plants and animals and assists to measure the passing of time with diversity in the weather circumstances around us. The warmth of summer provides method to the glorious colors of autumn, then to the restfulness of winter shadowed by the explosion of new life in the spring.

## The Moon

*Fig.6.2*: *Full Moon – March 2023*

The moon orbits around the earth every 29.5 days, the moon aids in a significant part in making the earth fit for inhabitation. At around one-eightieth of the mass of the earth, our moon is far larger with respect to its planet than any other of the more than 60 moons in the solar system. One of the greatest manifested influences of the moon, apart from being a source of light at night, reflected sunlight, is to be the primary cause of tides in the oceans of the world, Figure 6.2. Each day manifests two high tides and two low tides, which repeat on an approximately 25-hour cycle. These tides are indispensable for circulating and oxygenating the coastal waters in bays and river estuaries around the world to encourage marine ecosystems and circumvent stagnation. If the moon was much smaller, like other moons in the solar system, the tides would be useless in supporting coastal life. If it were much larger, the coasts would be subject to enormous destructive tides twice a day.

## Solar Eclipses

Conceivably the most magnificent natural marvel on earth is a total solar eclipse. Many people experience a total solar eclipse and, although sometimes, the sun perhaps obscured by cloud, the eerie darkness and chill wind that immediately arose could be spine-tingling. Although the sun is 400 times larger than the moon, the sun is 400 times further away, and so both appear to be virtually exactly the same size in the sky, Figure 6.3. This indicates that on rare occasions, when the alignments are precisely correct, the moon will block out the light from the penetratingly bright photosphere of the sun for almost two minutes, which allows the observer to view the sun's thin faint chromosphere and the remarkable corona with its huge eminences. The area of totality is no more than about 200 km (120 miles) across, and it quickly races across the surface of the earth from west to east.

*Fig.6.3: Most Famous Solar Eclipse in History in 2017*

Astronomers have recognized much about the nature of the sun, and consequently, the stars, because of total solar eclipses. If the moon were much bigger, the chromosphere would only be fleetingly visible at the onset and end of an eclipse. If it were just a little smaller, totality would not occur and eclipses would hardly even be observed. the probability that such an amazing match of apparent size would just happen by chance is nonetheless, miniscule.

## THE EARTH'S ATMOSPHERE

The Earth's atmosphere is consisting of 78% nitrogen and 21% oxygen, Earth's atmosphere forms a thin cover around the globe, detained there by gravity, shielding and nourishing life on the planet Earth. The atmosphere is confined within about 100 km (60 miles) of the Earth's surface, which is only about 1.5% of its radius, about the same proportions as the skin of an apple, Figure 6.4.

*Fig.6.4: Thin Line of Earth's Atmosphere and the Setting Sun*

Scattering the sun's light, the oxygen and nitrogen molecules yield a blue canopy which allows us to typically view only the sun and moon by day, however, at night the atmosphere becomes transparent to divulge the planets and the stars. Oxygen is vital for life for all air-breathing creatures, nonetheless, too much oxygen would make the air hazardously flammable, and too little would not deliver sufficient for life to thrive. Furthermore, there is an adjustable amount of water vapor, around 1% at sea level, and fewer than 0.04% carbon dioxide, along with hints of other gases. Carbon dioxide is indispensable for plant life which, through the astonishing process of photosynthesis, receipts in carbon dioxide and delivers out oxygen.

The atmosphere aids normalize the temperature of the earth and conveys water vapor to enable the hydrological cycle of evaporation and precipitation, rainfall and snow, etc.. to dispense water around the earth. The collaboration between the energy radiated from the sun and the atmosphere influences the weather patterns around the world, which in turn impact living things, Figure 6.5.

*Fig.6.5: Earth Atmosphere Supports Plant Life*

## WATER

*Fig.6.6: All the Water on Earth is Older Than Our Sun, Say Scientists*

The most plentiful substance on the planet is water, the chemical formula for which is $H_2O$. The exceptional geometry of the $H_2O$ molecule provides water a number of assets vivacious for life, Figure 6.6. Water, dissimilar to most liquids, enlarges on freezing, so ice floats on water. This averts lakes and rivers from freezing from the bottom up, so conserving numerous forms of marine life during winter. The amount of energy essential to freeze, melt, boil or condense one gram of $H_2O$ is higher than for virtually all other substances, which tells us that water is very effective at moderating the earth's climate and behaves as a coolant for larger animals. Likewise, the high surface tension of liquid water makes it effective in capillary action in soils, plants and biological systems. Around 72% of the Earth's surface is covered in water. If the mountains were lowered and the ocean basins raised so the Earth was a perfect sphere, the oceans would cover the Earth to a depth of around 3 km (2 miles).

## How Amazing is that?

The more we acquire knowledge about our planet Earth the more astounded we are at how extremely well it suited us for life. The evidence is not consistent with "natural processes" (the absence of creator) occurring randomly over vast periods of time.

Those who have faith in the infinite Creator God, as revealed in the book of Genesis, are not astonished to find evidence of remarkably intricate design reflecting the power, intelligence and care of the God who made us. Rather, when we view such things, we realize there is a "Designer" who made planet Earth to be our home. *Isaiah 45:18* says, *"For this is what the Lord says—he who created the heavens, he is God; he who fashioned and made the earth, he founded it; he did not create it to be empty, but formed it to be inhabited … ."*

# EARTH MAGNETOSPHERE

## Makes life on Earth Possible

The perseverance of life strongly depends largely on the precise, non-coincidental arrangement of numbers of natural phenomena, Figure 6.7. Some of these life-sustaining features of Earth are invisible to the human eye. For example, the conformation of the gases in our atmosphere, which permits beings to breathe, and plants to manufacture foodstuffs precisely 'out of thin air'. Or the ozone layer that guards' life from lethal ultraviolet light.

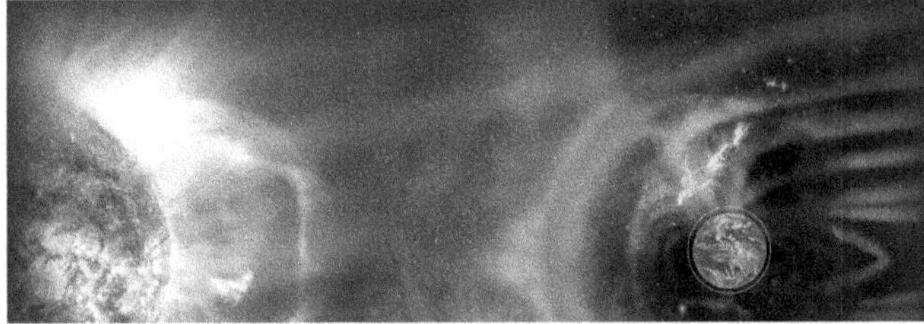

*Fig.6.7*: Earth Magnetosphere - Makes life on Earth Possible – Curtsy - © Elen33 | Dreamstime.com

One such invisible marvel is our planet's radiation shield, called its magnetosphere. Its life-sustaining characteristics strongly illustrate thoughtful design.

A magnetosphere is the domain that encompasses an astronomical object, like a planet, or a star, in which charged particles are influenced by that object's magnetic field. Many planets in our solar system have magnetospheres, however, Earth has the most remarkable and vigorous one of all the rocky planets. Earth's magnetosphere plays a vital part in our planet's capability to withstand life.

## Designed Barrier

Parallel to the 'force field' of science fiction, the magnetosphere creates a commanding deterrent to cosmic radiation blasting our planet continuously, Figure 6.8.

*Fig.6.8*: The Magnetosphere: under the Influence of the Earth and the Sun – Curtsy Space Environment

Cosmic radiation comprises of numerous active charged particles—electrons, and the nuclei of numerous elements stripped bare of electrons—and has two components:

## Solar

It is often called the solar wind, this unceasing powerful stream of charged particles emitted from the sun constitutes the bulk of cosmic radiation. It comprises mostly of electrons, protons (hydrogen nuclei) and alpha particles (helium nuclei), Figure 6.9. To offer some idea of merely how powerful this is, the entire mass of the particles expelled by the sun each second is around 1.5 million tons!

*Fig.6.9*: *Solar Wind Hits the Earth – Curtsy – Earth.com*

## Galactic

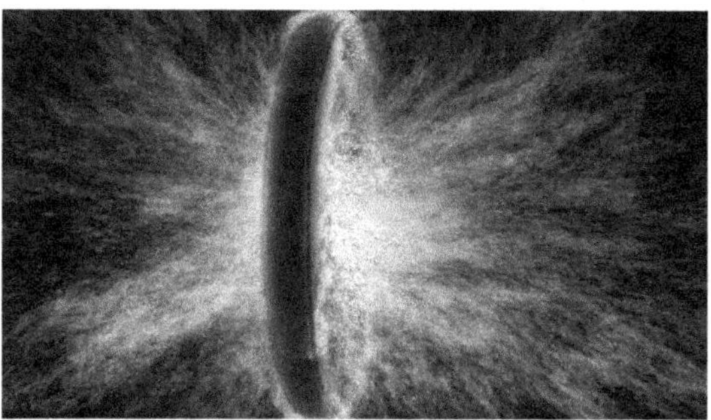

*Fig.6.10*: *Galactic Wind – Curtsy – Evan E Schneider*

Majority of this radiation is considered to come from within our own galaxy from supernovas, enormous explosions as some stars 'die'. In prior study especially, the term 'cosmic radiation' frequently mentions to this component, Figure 6.10. It encompasses numbers of the bare nuclei of elements heavier than hydrogen and helium, frequently travelling at much faster speeds than the solar wind. Since such high-energy cosmic rays influence our atmosphere, they release showers of other, secondary, particles as well.

If this radiation were permitted to bombard the surface of our planet, it would have tremendously damaging effects on all life. Though, the existence of the magnetosphere rebounds most of this from ever reaching us. This is contributed to of the manner in which a magnetic field applies a force on a charged particle moving through it, Figure 6.10.

Continuous bombardment by the solar wind 'compresses' the sun-facing side of our magnetic field. The sun-facing side, or dayside, spreads about six to ten times the radius of the earth. The side of the magnetosphere facing away from the sun—the nightside—stretches out into an enormous magnetotail, which fluctuates in length and can measure hundreds of Earth radii, far past the moon's orbit at sixty Earth radii.

## SUSTAINING THE FORCE FIELD

*Fig.6.11: Earth's Magnetic Field – Curtsy © Vjanez | Dreamstime.com*

Earth's magnetic field, which gives such astonishing defense. It is caused by a circulating electric current in the planet's outer core. As it flows through the metallic core, this current making the field characteristically loses energy, or 'decays,' according to the natural resistance to its flow. The energy of its movement is dissipated as heat. This decay has been computed in real time over many years, Figure 6.11.

"Long-agers (evolutionists)" necessitate uncertain 'dynamo theories' to attempt to clarify how this magnetic field could still be in existence after their alleged '4.5 billion years' age for the Earth. Within about 10,000 years at the maximum, even the strongest field should have decayed to nothing, permitting cosmic rays to wipe out all life on Earth.

## Protecting Life

Deprived of this protective sheath, our planet Earth would be unceasingly showered by these damaging and harmful rays. Astronauts in space are far less protected. Therefore, they are potentially uncovered from high levels of such radiation. This is a substantial work-related health hazard that space agencies pursue to alleviate. The most common effects of cosmic rays and solar wind on the unshielded human body are impairment to the cardiovascular system, i.e., the heart and blood vessels, causing hardening and tightening of the arteries, for example, Figure 6.12.

By destroying DNA within cells, radiation exposure can cause cancer and significantly hamper neurogenesis, the process of generating new cells in the brain.

Space agencies are anxious, too, about the effects of this radiation on the reproductive health of both male and female astronauts. The damaging effects could comprise an increased danger of passing on genetic damage (mutations) to the next generation.

Possibly the most substantial shielding aspect provided by the magnetosphere is the defense of our atmosphere—the envelope of gases surrounding the earth. The solar wind is exceptionally powerful, it could effortlessly rip through Earth's atmosphere consuming and stripping off the gases necessary to support life. The powerful shielding effect generated by the magnetosphere blankets these atmospheric gases and shields them from harmful decay.

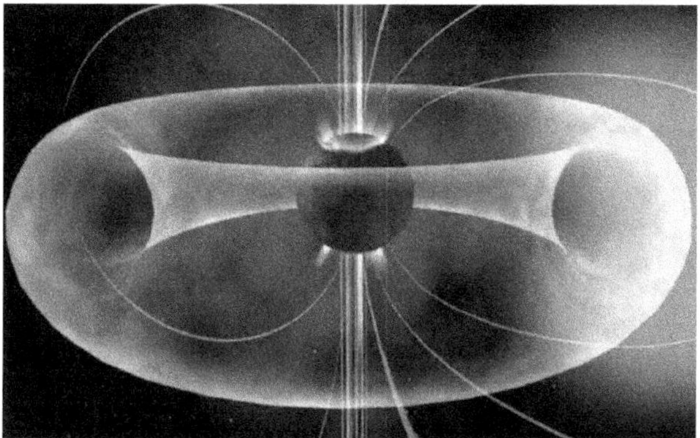

*Fig.6.12: Protective Magnetic Fields Increase the Chance for Life on Earth-Like Planets – Curtsy Science Tach Daily March 2023*

Interestingly, the solar wind itself delivers another life-protecting force field. In 2019, the Voyager 2 probe illustrated that the solar wind delivers a strong barrier to the most powerful interplanetary radiation. Solar wind generates a huge bubble called the heliosphere, which extends 27 billion km (17 billion miles) from the sun—quite a lot of times further than our planets. Past the boundary, called the heliopause, Voyager 2 discovered that the radiation was far more intense. A report clarifies, "The protective heliosphere shields everything inside it, including our fragile human DNA, from most of the galaxy's highest-energy radiation."

## The Auroras

Sporadically, there are further bursts of energy from the sun recognized as solar storms. If one kind of these, a "*coronal mass ejection*," is directed in our direction, some of the charged particles will travel down the magnetic field lines towards the north and south polar regions. As they reach the atmosphere, they interact with gases to give the beautiful colored light displays called the "*aurora borealis*" and "*aurora australis*" aka the northern and southern lights, respectively, Figure 6.13. Their distinctive color characteristics are:

Oxygen = red and green,

Nitrogen = blue and purple.

*Fig.6.13: The Aurora Displays its Splendor – Curtsy © Surangaw | Dreamstime.com*

## Directive Design

Albeit, this world is falling down *(Genesis 3)*, indication of design repeatedly overwhelms our minds. It does not take much coherent thought to conclude that our material existence is the consequence of design in difference to chance, evolutionary causes. From the macro to the micro, the elements of our material existence are finely composed and operate with elaborate, shared precision, Figure 6,14.

*Fig.6.14: The Universe and Its Laws – Curtsy - Photo: © Pixels (Main Image)*

Some of the most extraordinary discoveries of modern physics and cosmology, particularly the evidences that our universe and its laws seem amazingly custom-made to withstand life, backing this point of view. The Apostle Paul writes about the uniqueness of the Designer, the Lord Jesus Christ:

"For by him all things were created, in heaven and on earth, visible and invisible, whether thrones or dominions or rulers or authorities—all things were created through him and for him. And he is before all things, and in him all things hold together" *(Colossians 1:16–17)*.

## Earth's Water

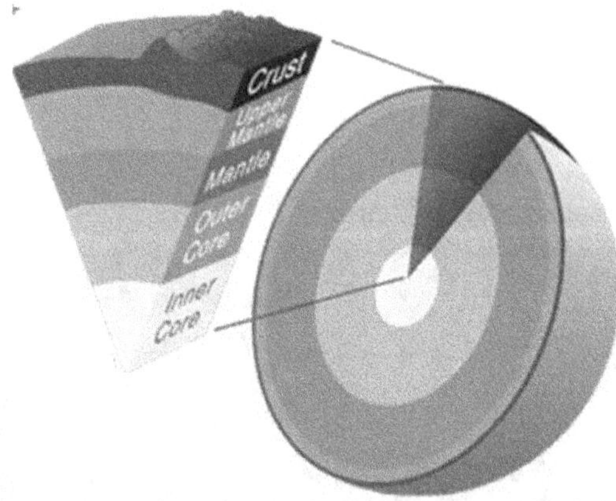


*Fig.6.15: Earth's Cores, Mantles and Crust*

The earth's mantle exceeds 2,800 km thick, extending from the core to the crust, Figure 6.15. The upper mantle is recognized as the 'depleted mantle' as geologists believe material has come out of it onto the earth's crust. There is still sufficient water in the mantle to fill the oceans at least ten times over. Precisely as the scriptures describe that water for Noah's Flood came from two sources.

It is stated in *Genesis 7:11* that '*all the springs of the great deep burst forth*'. It's possible that this water came from the earth's mantle. Geologists compute that the rocks in the mantle still encompass within their mineral structure sufficient water to fill the oceans at least ten times over. Geologists also identify that material has come out of the outer part of the mantle, which they call the 'depleted mantle'. Also, some geologists suggest that uplift of the ocean floor spilled water onto the continents.

*Genesis 7:11* chronicles that water also came from the sky: '*the floodgates of the heavens were opened*'. This was not usual rain because it sustained for 40 days until the Ark began to float. The water kept increasing on the earth for five months and eventually covered all the high mountains *(Genesis 7:19)*.

How could it rain for so long? It is not known for certain, however, creation scientists have proposed some possibilities, stated below:

- Breakdown of a thick water-vapor canopy which encompassed the pre-Flood earth high in the atmosphere. Computations show this could not have detained much water, but it may clarify some.
- Jets of water shooting high into the atmosphere from under the earth and falling back as rain.
- Intense cyclones called hyper-canes that established over warm ocean water, heated by underwater volcanic eruptions.
- Water dumped on the earth by a swarm of comets. The craters on the moon point to an powerful solar system bombardment, and some creationists propose this occured during the Flood.
- A combination of all the above.

Noah's Flood was a one-off occurrence and we did not observe it occur. All the same, the sources of water chronicled in the Scripture are in harmony with our understanding of the structure of the earth.

## Where did it Go?

According to the Scripture, the water that water covered the whole world during Noah's Flood; I have often wondered where did it has all go?

Even if the glaciers and ice caps melted, the oceans would merely rise some 70 m (230 ft, however, Mt Everest rises 8,848m (29,029 ft) above sea level.

It may surprise you to know that we don't need any more water to cover the earth. There is already plenty.

The main reason water doesn't enclose the globe now is that the earth's surface is uneven. The ocean basins sit low and the continents sit high. Some mountains are particularly high and some ocean ditches are very deep, but these extravagances do not account for a large percentage of the earth's surface, Figure 6.16.

If the earth's surface were even, then there is sufficient water in the oceans to cover the globe to a depth of about 3 km.

**Fig.6.16**: *The Water of the Globe – Curtsy Photo Mountain High Maps® www.digiwis.com*

This advocates that, during the Flood, the ocean floor moved straight up relative to the continents, as stated in the Scripture *(Psalm 104:8 NASB)*. In the first half, the pre-Flood ocean basins rose and the pre-Flood continents eroded down until water covered everything.

That does not illustrate the earth's surface would have to be completely even. The ocean basins would have only had to rise *enough* for the water to cover everything.

Then, in the second half of the Flood, other parts of the earth's crust sank. The water flowed off our continents into new ocean basins. Also, the movement of the earth's crust at this time also pushed up new mountain ranges, including the one that is home to Mt Everest.  Accordingly, where did all the water go? It is in the ocean.

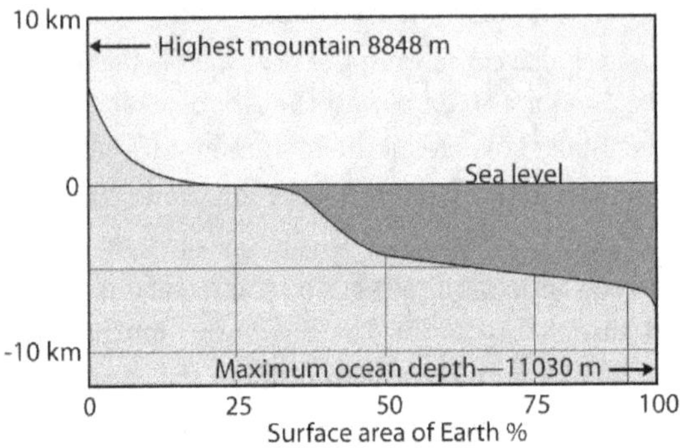


*Fig.6.17: The Amount of the Earth's Surface at Various Elevations*

The hypsographic (Gk. hypsos=height) chart shows the amount of the earth's surface at (or higher than) various elevations.  It exemplifies that if the ocean basins were pushed up 5 km and the mountains shaved off, water would cover the whole earth, Figure 6.17. Such tectonic movements appear huge to us, but compared with the radius of the earth, (6,378 km), the movement is tiny, less than 0.1%.

## Drain the Water of Noah's Flood

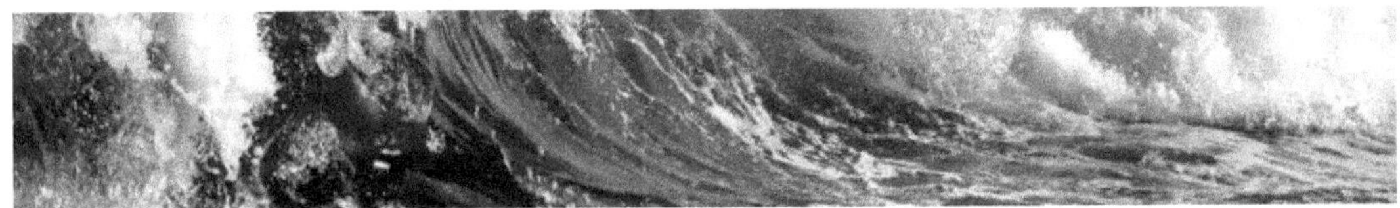

*Fig.6.18: Noah's Flood Covered the Whole Earth – Curtsy iStockphoto*

Many questions: "If Noah's Flood actually covered the entire earth, then where did the water go?"

This question has a simple answer.  As soon as we comprehend what and how the Flood occurred, we may see the authenticity of the Biblical Flood all around the world, Figure 6.18.

## The Floodwater in the Oceans

In Fact, the Scripture tells us where the water dissipated. By Day 150 of the Flood catastrophe the floodwaters had risen until they covered "all the high mountains under the whole heaven" (Genesis 7:19). After that "the waters receded from the earth continually" *(Genesis 8:3),* a process that took about seven months.

As the water receded from the continents it must have poured into the oceans. It only takes a swift look at a globe of the earth to appreciate that the water indeed sits in the oceans. The Pacific Ocean alone takes up almost half the earth's surface, Figure 6.18.

## The Earth's Crust Rises or Recedes

Figure 6.19, this view of the globe illustrates how the waters of Noah's Flood receded into the oceans. The Pacific Ocean alone covers almost the whole hemisphere.

Rationally, the only method for the water to drain from the continents into the oceans is for the continents to rise and the ocean floors to sink. As our understanding of the structure of the earth has advanced, we can appreciate how that could have occurred, Figure 6.19.

The top part of the earth, called the crust, sits on top of the mantle, which is about 3,000 km - 1,900 miles thick, which in turn sits on the earth's iron core. The continental crust is about 40 km, 25 miles thick, though the thickness of the oceanic crust is only around 7 km, 5 miles. Up-and-down movement of the crust during Noah's Flood, called *"differential vertical tectonics,"* clarifies how the waters drained from the continents. On a smaller scale, mountain ranges would have risen and valleys sunk.

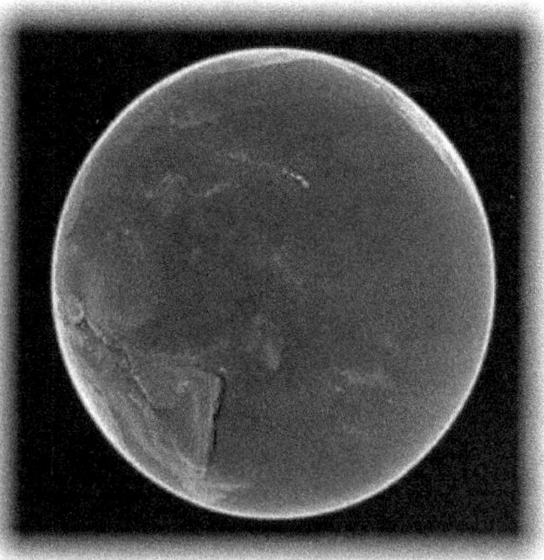

**Fig.6.19**: *The Pacific Ocean Covers Virtually the Entire Hemisphere – Curtsy commons.wikimedia.org*

As the continental crust rose and the ocean floors sank, the floodwater covering the globe drained off, allowing massive erosion of the continents. As the time the floodwaters had fully receded, the surface had been transformed into its present shape. Then, Noah and all those with him left the Ark, 371 days after the Flood had begun *(Genesis 8:18–19)*.

When the ocean basins began to sink, the water flowed across the continents in wide sheets, shaving the surface flat. Geologists call such features 'planation surfaces.' The runoff eroded the uplifting mountains, moving the rock debris across the continent, and rounding any hard, resistant rocks into boulders and gravel. Large deposits of well-rounded quartzite rocks are discovered at numerous places in the northwest United States and adjacent Canada.

Figure 6.20. Schematic of a guyot, a volcano possibly shortened at sea level creating a flat top. Geologists discovered thousands of guyots on the ocean bottom, particularly in the western Pacific, representing that the ocean basins have sunk.

Toward the end of the Flood, mountain ranges started to emerge above the water and the runoff became more channelized. These flowed across mountain ranges, ridges, and plateaus, eroding gorges from one side of the barrier to the other, a feature called a water gap, through which a river or stream now passes, Figure 2.21.

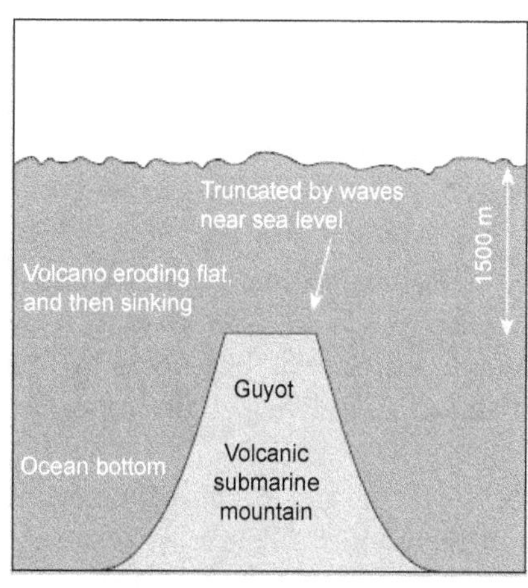


**Fig.6.20**: *Schematic of a guyot, a volcano likely truncated at sea level producing a flat top*

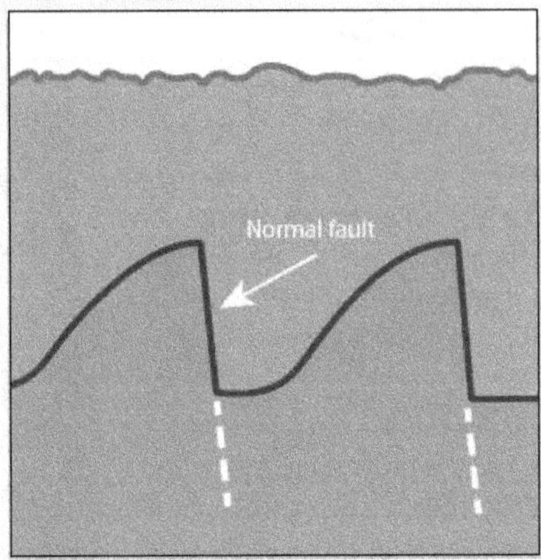


*Fig.6.21: Most likely Explanation for the Formation of Abyssal Hills by normal Faulting*

## Evidence for up-and-down Movements

Indeed, there is plentiful evidence for differential vertical tectonics of mountains and valleys, and continents and oceans. This is exposed through the study of geomorphology, i.e. the shape of the earth's surface. Mountains demonstrate indication of upward movement along faults, while the adjacent valleys illustrate evidence they have sunk down, and then collected sediments. The sediments prove that the movement started while the land was still under the floodwater.

When the ocean basins sank, thousands of meters of sediment washed off the continents, creating the continental margin. These margins are indication the ocean basins near continents sank. Additional evidence of ocean basins sinking are flat-topped undersea volcanoes called *"guyots"* Figure 6.20, discovered far from land. Water currents shaved these smooth, and they now sit at an average of about 1,500 meters (5,000 ft) below sea level. Secular geomorphologist and world traveler, Lester King, exclaimed:

"Marine volcanic islands which have been truncated by the waves and since subsided below sea level are called guyots. Most of them seem to have sunk by 600 to 2,000 m [2,000 to 6,500 ft] and it is evident that they afford a measure of the amount by which "the ocean floor has sunk in late geologic time". … All the ocean basins afford evidence of subsidence (amounting to hundreds and even thousands of meters) in areas far from land."

Even the fine details of the ocean bottom, deep underneath the surface, display signs of differential vertical tectonics in the crust—that areas rose and sank. *"Abyssal hills"* are discovered over most of the deep ocean crust, though they are typically covered by sediments. They are narrow ridges characteristically 10 to 20 km (6 to 12 miles) long, 2 to 5 km (1.2 to 3 miles) wide, and 50 to 300 meters (160 to 1,000 ft) or higher Figure 6.21.

## Worldwide Phenomenon

In harmony with the Genesis Flood, the upward and downward movement in the earth's crust is a *"worldwide phenomenon."* Lester King summarizes:

Consequently, the "fundamental" tectonic mechanisms of global geology are vertical, up or down: and the "normal and most general" tectonic structures in the crust are also vertically disposed …

However, one must bear in mind that "every part of the globe"—on the continents or in the ocean basins—provides "direct" geological evidence that formerly it stood at different levels, up or down…" Indeed, there is abundant evidence of differential vertical tectonics during Flood runoff.

## The covering of Mount Everest

Many think the height of Mount Everest at 8,848 meters (29,029 ft—Figure 4) a fatal flaw for the Genesis Flood. How could the floodwater have topped the mountains, they ask? Even if the ocean floor was raised to sea level, the present water on the earth would be only 2,700 meters (8,800 ft) deep, one-third the depth needed to cover Mount Everest, Figure 6.22.

*Fig.6.22*: Mount Everest in the Himalaya Mountains - Curtsy - wikimedia.org

The answer lies in the "differential vertical tectonics" as the floodwater was draining provides the answer, because the mountains were pushed up as a result of the Flood, by way of "upward vertical tectonics."

It is distinctively clear that the mountains were once under the ocean because the sedimentary rocks that form the tops of most mountains contain marine fossils. For example, Mount Everest is topped with marine crinoid, sea lily, fossils embedded in limestone.

This illustrates that Mount Everest and the other high mountains in our present world, with their sedimentary rocks and fossils, "rose up out of the floodwater" during the later stages of the Flood. Figure 6.23, illustrates how, as the floodwater drained, the Uinta Mountains of the western United States rose about 12,000 meters (40,000 ft) compared with the same type of rock in the basins to the north and south, which sank with other sedimentary rocks filling the basins (formations 5–7, Figure 6.24d).

It was mountains rising and valleys sinking that were the prime causes the floodwater to drain at the end of Noah's Flood. The water flowed toward the low spots on the planet and the rising land was exposed. As a result of vertical movements in the earth's crust, the continents and mountains rose at the same time as the valleys and ocean floors sank. The mountains were first to rise above the water, which explains why the Ark grounded on the "mountains of Ararat" *(Genesis 8:4)*. If the Genesis Flood was a local event, the Ark would have been carried downstream. Assuming it didn't sink, it would have landed on a flood plain or been washed into the ocean. Scripture is strikingly consistent in asserting the Genesis Flood as a global event, and the topographies on the surface of the earth are in full agreement.

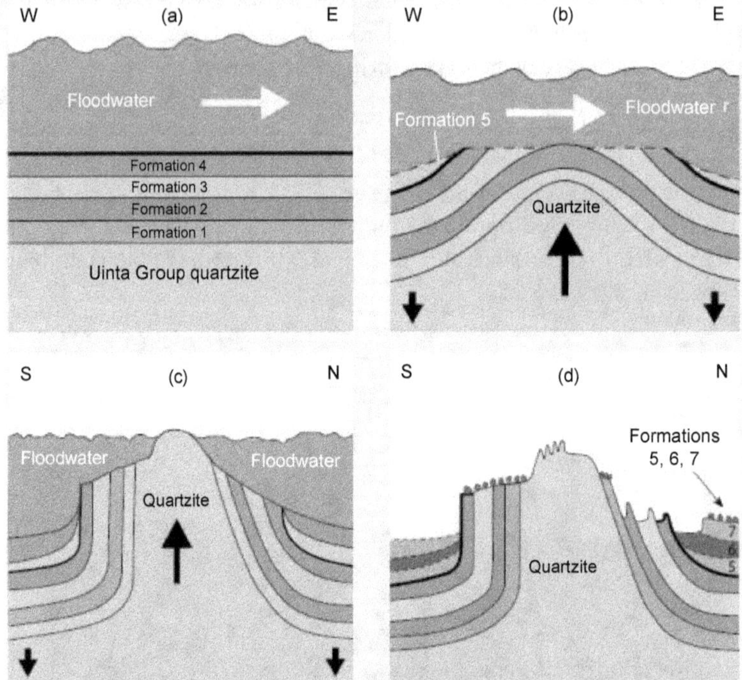


*Fig.6.23*: *A Schematic of the Uinta Mountains Rising about 12,000 m (40,000 ft) out of the Floodwater.*

## Oceans Deep Inside Earth

Figure 6.24, schematic section of the Earth's interior, illustrating upper mantle, transition zone, lower mantle and core. The crust of the earth, continental and oceanic, is not visible at this scale. Contrary to some reports, the transition zone is not near the core, as illustrated, Figure 6.25.

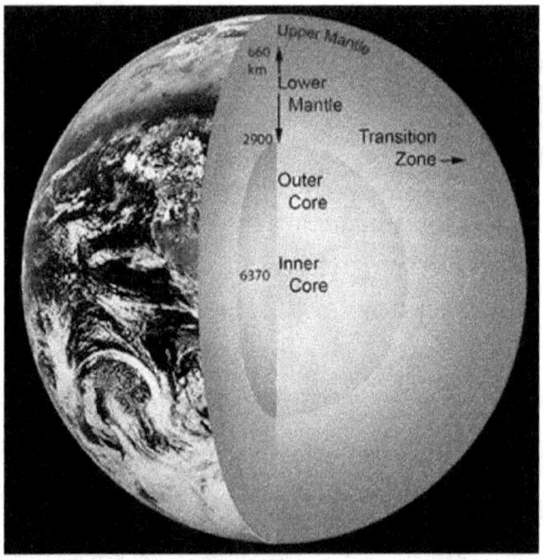

Recent news reports of "oceans of water locked 400 miles inside Earth" have trapped people's imagination, touching mental pictures of water sloshing around in massive underground reservoirs. One report said "Massive 'ocean' discovered towards Earth's core." Not so bright was the title of the pertinent academic paper in the journal Science—"Dehydration melting at the top of the lower mantle."

This is not a novel finding. Scientists have been venturing for decades that the earth's transition zone grips plentiful water within the mantle, in a mineral called Ringwoodite.



*Fig.6.24*: *Schematic Section of the Earth's Interior*

The transition zone extends from 410 to 660 km below the surface—the upper mantle sits above it, and the lower below, Figure 6.24.

The academic paper co-author "Steve Jacobsen" has been researching for years in his laboratory with ringwoodite, Figure 6.26, the mineral well-thought-out to be the most plentiful in the lower transition zone.

He has been able to synthesize the blue, sapphire-like mineral by reacting, at high-pressure, the mineral olivine with water. Olivine is green, and abundant in the earth's upper mantle.

Concerning ringwoodite, Jacobsen said, "*It's rock with water along the boundaries between the grains, almost as if they're sweating.*" At the depth of the transition zone, the pressure and temperature are suitable to release the water from the ringwoodite.

The other portion of the story was provided by "Brandon Schmandt." Through analyzing seismic waves from hundreds of earthquakes and thousands of seismic recorders, he determined there are massive pockets of magma (molten rock) beneath the North American continent at the base of the transition zone.

Based on these discoveries, the researchers proposed that the water in the ringwoodite in the transition zone had been forced out, while the rock partly melted. Jacobsen said, "*Once the water is released, much of it may become trapped there in the transition zone.*"

With respect to oceans of water under the earth raised questions in some people's minds: is this water connected to Noah's Flood? It is conceivable that there is a connection, but there are many questions that would need to be answered. We must remember that the features of magma oceans and their mineral composition are dependent on circumstantial evidence.

*Fig.6.25: Crystals of Blue Ringwoodite Synthesized from Olivine*

Connecting water in the mantle to the biblical Flood has been suggested earlier. Other scientists suggested that water from minerals in the earth's mantle may have been the source for some of the water of Noah's Flood. Geologists are still attempting to find a scientific explanation describing the mechanism where Flood rocks begin within the geological record

Some geologists advocated that the pre-Flood boundary is towards the base of the earth's transition zone. They suggested that decompression of the mantle originated mantle melting, magma development, and the release of volatiles, counting copious volumes of water, which arrived to the surface during the Flood. Also, it would indicate that the Flood was an enormous planetary cataclysm.

Another method the water may have been released from the mantle is for the rocks to experience a change in mineral structure during Noah's Flood. Such changes in ringwoodite may release water, and this would rise through the mantle with the magma and be expelled onto the surface through volcanic eruptions. However, there is much evidence for abundant volcanic eruptions during the Flood. Also, other geologists suggested that water from ringwoodite ended up in the oceans. They stated also that there is good evidence the Earth's water came from within, Figure 6.25.

The claims of vast quantities of water within the mantle in the transition zone are entirely plausible, however, we must keep in mind that they are based on "interpretations of indirect evidence." It is possible that such quantities of water signify the remains of a major planetary differentiation that occurred during the global Flood cataclysm, but there are questions and issues that would require further investigation.

## MANTLE MINERAL RINGWOODITE FOUND IN DIAMOND

In March 2014 journal "Nature" described a diamond from Brazil that contained a small speck of the mineral ringwoodite, Figure 6.26. The significance of the discovery stems from the concept that diamonds are blasted explosively from deep inside the earth to the surface in "vertical volcanic tubes" called "kimberlite pipes," this would have been early during the catastrophe of Noah's Flood. The explosive eruptions 'sample' the rocks in the mantle and enclosed the ringwoodite inclusion in the diamond.

Based on theoretical computation it has long been assumed that ringwoodite exists in the mantle. Scientists have been able to simulate ringwoodite in the laboratory by combining the mineral olivine with water under high temperatures and pressures. The mineral has also been discovered in meteorites. Though, the ringwoodite in this diamond is the first time that the mineral from the mantle has been found in on the surface.

Ringwoodite encompasses 1.5% water, not in the form of a liquid but as hydroxide ions, which are particles with a negative charge encompassing of one oxygen and one hydrogen atom. Ringwoodite is actually the confirmation that there is an enormous amount of water that's trapped in a really distinct layer in the deep Earth.

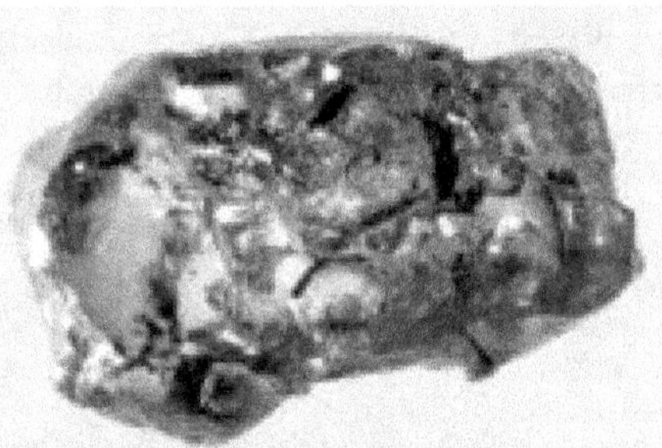

*Fig.6.26: A diamond from Juína, Brazil, found to contain a tiny inclusion of ringwoodite. Curtsy – Richard Siemens- University of Alberta*

## EARTH CATACLYSM OF NOAH'S FLOOD

The scientific comprehension of the Biblical Flood must incorporate the hydrology and sedimentation that happened during the Flood and in the following years as the Earth settled down. A number of scientific models previously proposed for the Flood are discussed and evaluated. Further progress may require an integrated methodology from numbers of scientific facets. As well as the traditional contribution from the geological sciences, coordinated inputs from a number of other disciplines may be necessary such as fluid flow, heat transfer, plate tectonics, vulcanology, planetary astronomy, and mathematics, in order to construct a possible Flood hypothesis. Any model for the Flood can only be hypothetical. A coordinated approach may influence present Flood models that have accepted the sequential nature of the geological column and that have put the Flood/post Flood boundary far down in such conjectural reconstructions.

Characteristically, for scientists operating from an evolutionary perception, the geological sciences have offered the chronological framework to permit other scientific disciplines to place their data in an historical context. The primary principle of uniformitarianism has motivated research into current geological processes so that rocks these scientists regard as ancient can be interpreted in terms of such processes.

In the past thirty years there has been a fundamental shift in thinking amongst evolutionary geologists with the development of plate tectonics—all modern geological processes are presently viewed as part of a global interaction of plate tectonics, which itself has been adopted as the interpretative geological paradigm.

By contrast, scientists investigating from a creation viewpoint all vital geological events within a Biblical chronological framework. Nonetheless, there is still a necessary for scientific models of these events as the

Biblical record is not comprehensive, nor is it tailored as a scientific dissertation. Especially, creation scientists require to comprehend the Biblical Flood by tackling the hydrology and sedimentation that happened during the cataclysm and in the subsequent years as the Earth settled down. Modern geological processes, though informative, do not have the same standing as for long-age uniformitarian scientists. This is since geological processes during Creation and the Flood were dissimilar from what we observe today. Consequently, creationists have a greater necessity to advance an integrated method from many scientific disciplines. As well as the geological sciences, inputs from many other disciplines are required, such as fluid flow, heat transfer, plate tectonics, vulcanology, planetary astronomy, and mathematics. In this regard we summarize the current state of a number of scientific models that have been suggested to explain the world-wide Flood and to integrate our comprehension of science from the Biblical perspective.

## THE VAPOR CANOPY MODEL

The vapor canopy model of the Flood is the one that has maintained greatest influence in scientific creationism since meaningful research started in the 1960s, Figure 6.27. The book "The Genesis Flood" by Whitcomb and Morris. The book was first published in 1961, and Whitcomb's later "The World that Perished" (1996) explain this view. The vapor canopy theory is that the Earth's atmosphere was encompassed by a water vapor blanket that collapsed at the beginning of the Flood. "Dillow" has expansively studied this concept theoretically. This model has guided the field for a number of years, but has problems in considering the large amount of catastrophic disturbance in the Earth at the beginning and through the Flood year.

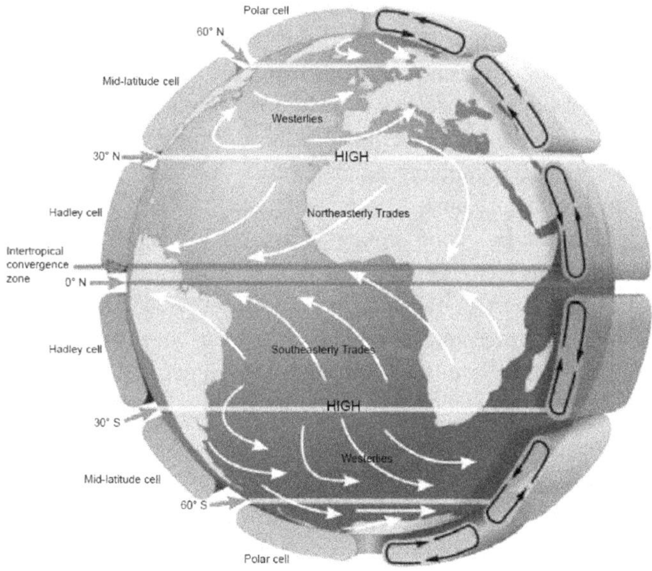


*Fig.6.27: Noah's Flood Problem - Heat Problem – Vapor Canopy Model*

Catastrophic upheaval is a manifest, for instance, at the Old Red Sandstone rock formation from Loch Ness to the Orkneys in Scotland where an area 2500 m deep and 160 km across, encompasses countless fish, buried in twisted and constricted positions, as though in spasm. There is all the proof of catastrophic burial by processes, it would appear, of superior power than that provided by the vapor canopy theory, Figure 6.27. Though there may be some element in these objections to the vapor canopy suggestion, it should be observed that this model of the Flood, though it envisages late drowning of creatures by rising floodwaters, should not be viewed as serene. Indeed, in this model, the rising waters would be enormously turbulent, and perhaps include vast surging tidal waves. Nevertheless, it is still problematic to clarify the main fossil strata by this way. Figure 6.28.

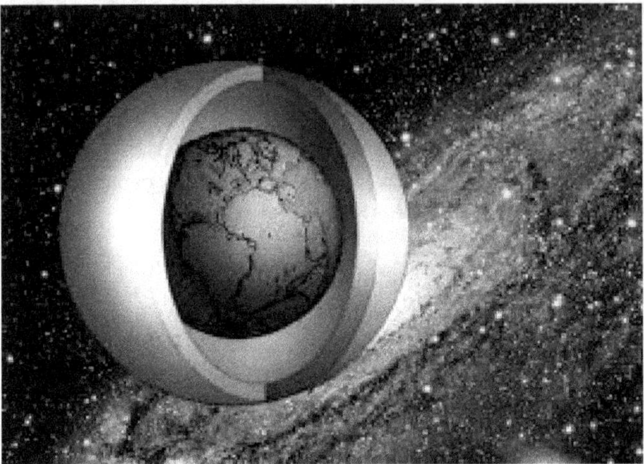

*Fig.6.28: Theory of Water Canopy Surrounding the Earth*

Subsequently, some, such as Garner, Garton, Tyler, and Robinson, disapprove, not only to the vapor canopy model of the Flood, but also, more fundamentally to the rudimentary evidence that the Flood produced most of the fossils. Their opposition arises from their belief that the geological column signifies a real time sequence, though on a fast time-scale of the one-year Flood permitted by many post-Flood tragedies. Since there is proof deep in this geological column that many animals were alive on land, and yet are buried above waterborne sediments, they suggest that most of the geological column was deposited after the Flood. Thus, they suggest that the Flood erased all trace of land air-breathing creatures and that most of the fossils discovered on the Earth were buried by post-Flood catastrophes. Known as the 'European Flood model', we have wanted to illustrate in a companion writing that Biblically, this is importantly draining the straightforward meaning of *Genesis 6–9*. Here we pursue to illustrate that to esteem the geological column as a true chronological record is at best a questionable supposition. We agree with "Froede" that there needs to be a complete rethink of how to interpret the geological layering so evident in the rocks. It was correctly showed that the way the suggested ten periods are allocated can be quite subjective. In this writing we question whether we actually yet have any secure understanding of the method all the strata have been laid down. Even the rudimentary notion that 'bottom is oldest' is not proven.

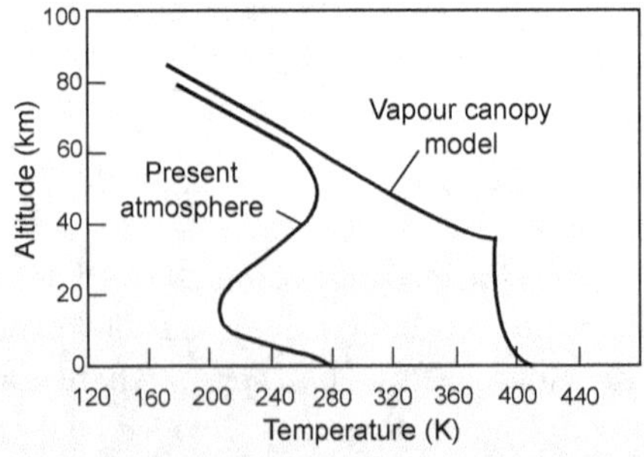


*Fig.6.29: The Canopy Model*

One of the main problems elevated by Flood models of fossilization, which is including the canopy theory. It is the problem of dinosaur nest locations within the fossil record. These undoubtedly posture quite a difficult problem to resolve in the setting of Flood sedimentation.

Computed vertical temperature profile for a vapor canopy model of the Earth's atmosphere equated with the temperature profile today (after Rush and Vardiman). Theoretical models of assumed pre-Flood vapor canopies are utilized to discover whether it is possible to assume massive quantities of water in the atmosphere above the Earth, Figure 6.29. In this example, only 50 cm of precipitable water is kept but this raises the surface temperature of the Earth to above 100°C.

"Garner" illustrated that the eggs are naturally in neat patterns, proposing that they have to be considered as "in situ," and cannot be accounted for by sediments deposited elsewhere and transported in before final fossilization. "Garton" illustrates that there are dinosaur tracks all the way from the Cretaceous to the Tertiary and Quaternary rocks. He came to the conclusion that this must be proof of post-Flood activity. Tyler understands the vast chalk deposits, normally taken to be the crushed remains of marine shells, require decades to form, and also considers that the Cretaceous is post-Flood.

According to such evidence, opponent of Flood fossilization in general, and the "Whitcomb and Morris" model of the Flood in particular, have upheld that the Flood/post-Flood boundary is low down in the geological record, in the Paleozoic, as clarified by "Tyler." This geological column term is utilized merely for communication uses. The order of the strata may well be incorrect for reasons discussed later. Such opponents have upheld that all Flood models which attribute most fossils to the Flood, are incorrect, and suggest that the Flood left no trace whatsoever of all air-breathing land creatures—the so called 'blot out' theory.

In companion research, we provide vital Biblical vital causes why fossils are the most natural evidence produced from the Flood. However, these authors are right to oppose the vapor canopy model if it does not give sufficient sedimentation to accomplish such a significant thickness of fossil-containing strata. This is the reason we discuss other models here, which, we trust, result a more accurate picture of the Flood year.

## THE HYDRO-PLATE MODEL

The hydro-plate theory has the benefit of clarifying great catastrophe in the first 40 days. This theory for the disastrous formation of the sedimentary rock layers during the Flood has been suggested by "Dr Walter Brown," former chief of Science and Technological Studies at the Air War College, and Associate Professor at the U.S. Air Academy.

The major assumption for the origin of the Flood waters is enormous catastrophism in the first 40 days of the Flood. We agree with the European Flood proponents that the initial devastation was excessively huge, but we dispute that there still no evidence of the "mabbul" and its influence on creatures in the geological record. The "Brown hypothesis is that the Earth's crust was fractured, perhaps by an impact, discharging massive subterranean waters, the *'fountains of the great deep'* under great pressure into the atmosphere, may be as high as 30 km. "Brown's" model basically deals with water, however, in the following continental drift phase incorporates volcanic activity as a consequence of the rapid tectonic reavel caused by the widening break in the Earth's crust. Thus, he states:

*'In some regions, the high temperatures and pressures formed metamorphic rock. Where this heat was intense, rock melted. This high-pressure magma squirted up through cracks between broken blocks, producing other metamorphic rocks. Sometimes it escaped to the earth's surface producing volcanic activity and "floods" of lava outpourings such as we see on the Columbia and Deccan Plateaus. This was the beginning of the earth's volcano activity.'*

## Brown states further:

*'Shifts of mass upon the earth created stresses and ruptures in and just beneath the earth's crust. This was especially severe under the Pacific Ocean, since the major continental plates all moved toward the Pacific. The portions of the plates that buckled downward were pressed into the earth's mantle. This produced the ocean trenches and the region called the "ring of fire" in and around the Pacific Ocean. The sharp increase in pressure under the floor of the Pacific caused ruptures and an outpouring of lava which formed submarine volcanoes called seamounts.'*

Thus, the original rupture of the Earth's crust under this perception would throw rocks and sediments in gargantuan muddy fountains of water which then cause forceful precipitation, in harmony with *Genesis 7*, for the 40-day period. These fountains would ultimately be trailed by many large volcanic outbursts in the 'Ring of Fire' around the Pacific, all with the force of Krakatoa. This volcano detonated in 1883 shooting rocks and dust into the atmosphere to a height of 55 km. The explosion was so forceful that it could be heard 4,600 km away. Dust fell at a distance of 5,327 km ten days after the explosion and a tsunami, tidal wave, 30 meters high moved right across the Indian Ocean at 720 km/h. Likewise, during the Flood, on top of the water borne sediments, and occasionally mixed with them, huge layers of magma would be discharged out or disastrously blasted into the atmosphere, Figure 6.30.

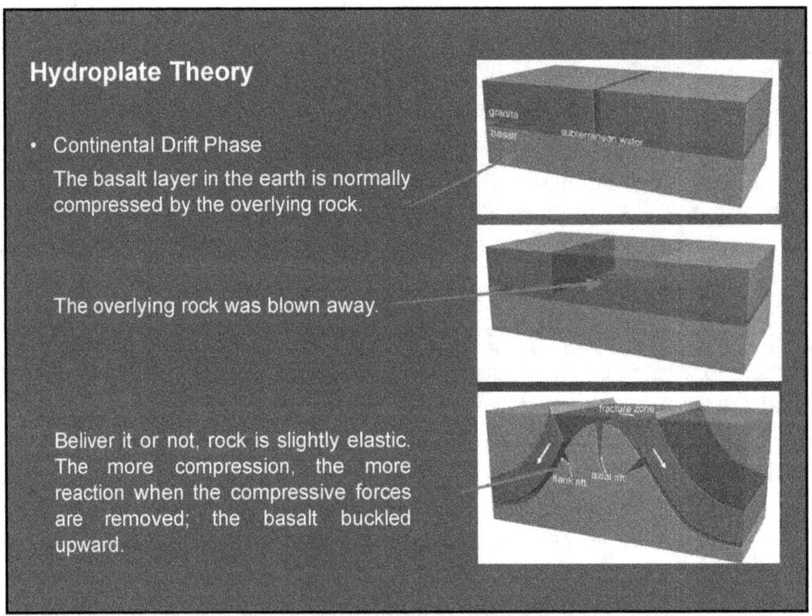


*Fig.6.30: Hydro-plate Model of Noah's Flood*

The rain in the first 40 days of the Flood complicated not only the return to the Earth of the jets of superheated steam expelled into the atmosphere, which would partially fall as hail and snow, but also excessive quantities of rock debris. Numerous fossils could have shaped within the first few weeks of the Flood in this model. In the subsequent 110 days, further massive layering, scrubbing and re-layering of the continents would happen under the ravages of the Flood waters. The last catastrophic drainage of the waters happened at the end of the continental drift phase when, after massive tectonic disturbance, the land ultimately re-appeared as the Earth's crust create a new steadiness. It is significant that *Genesis 8:3* expresses of the waters *'returning from off the earth,'* exactly *'going and returning'* in the Hebrew.

Recoil phase of the hydro-plate model for the geological events of the flood, "from Brown." Rupture of the crust allows steam and sediment to be ejected as a fountain into the atmosphere, returning to the Earth as rain, while the continents start to move apart.

Several Geologists have criticized the rupture phase of the hydro-plate model with its massive quantities of hot steam expelled at enormous speeds into the atmosphere, resulting huge rainfall. Though, the 'explosive mixing of water and lava' aimed by these oppositions, is very possibly how the 'windows of heaven' were opened as described in the Flood account.

Within the framework of the hydro-plate model, it is completely possible that many creatures would flee without success to survive. We would imagine to discover fossil evidence of this, such as tracks in mud then covered rapidly by sediment. Also, since it was a full year before Noah came out of the Ark, there is surely room within the Genesis account for some late-Flood and post-Flood tragedies as the waters receded. Thus, the Grand Canyon may well have been shaped when a massive ordinary inland lake, left behind after the Flood receded, rupture its banks and scoured out the canyon. In this process, massive quantities of silt and debris would be lifted to the Pacific coast-line. "Brown," relating the aftermath of the hydro-plate catastrophe, in harmony with "Austin" that the "Grand Canyon" formed in this way. The "Toutle Canyon" was detected to form catastrophically in a like manner, but on a much smaller scale, after the "Mount St Helen's" outbreak in 1980. Such disastrous processes may explain for the hideaways of small marine creatures in rocks at one horizon, but which are now covered by further sediments.

## CATASTROPHIC PLATE TECTONICS THEORY

The theory of catastrophic plate tectonics (CPT) was originated by "Baumgardner," and later advanced in coordination with other creation scientists. "Reed et al." offered a good appraisal of plate tectonics as understood within a catastrophic outline, however, offered in their findings that the original driving mechanism behind continental plate movement and subduction is not known, Figure 3.31. CPT theory begins with the supposition that the Flood was originated when chunks of oceanic crust raptured loose and subducted along thousands of kilometers of pre-Flood continental boundaries. It is supposed that subducting slabs of material locally distorted and heated the mantle, locally decreasing its viscosity. With decreasing viscosity, the subduction rate upsurged—and this in turn triggered the mantle to heat up even more. This, it is argued, produced a thermal runaway

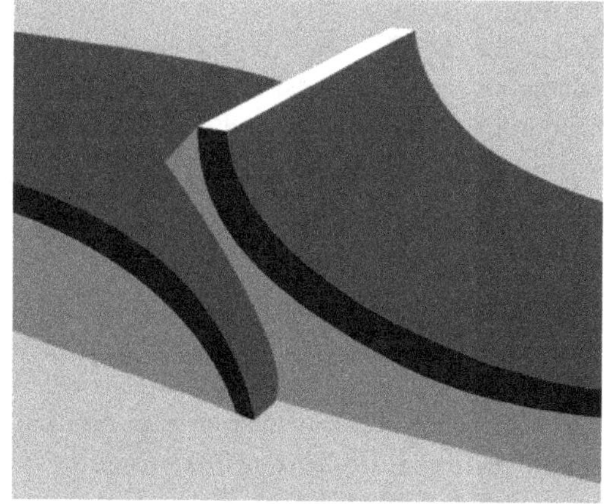

*Fig.6.31:* Catastrophic Plate Tectonics Theory

unsteadiness, and permitted subduction rates of meters per second. "Baumgardner" illustrates that fast, large-scale subduction would additionally pledge global-scale flow of the mantle beneath the Earth's crust. This in turn would result in strong convection currents in the Earth's outer core and clarifies how geomagnetic reversals took place. Magnetic reversals of course had been considered to have taken place gradually over millions of years on the evolutionary geological timescale. Nevertheless, the extension by "Humphreys" of the CPT theory of "Baumgardner" to account for the Earth's magnetism provides a fundamental reason for the rapid reversals. In that indication for rapid reversals has been revealed in thin lava flows, the magnetic field deductions from CPT theory provides substantial assurance in the theory of continental plate collision and subduction as being a principal mechanism for major global disturbance during the Flood.

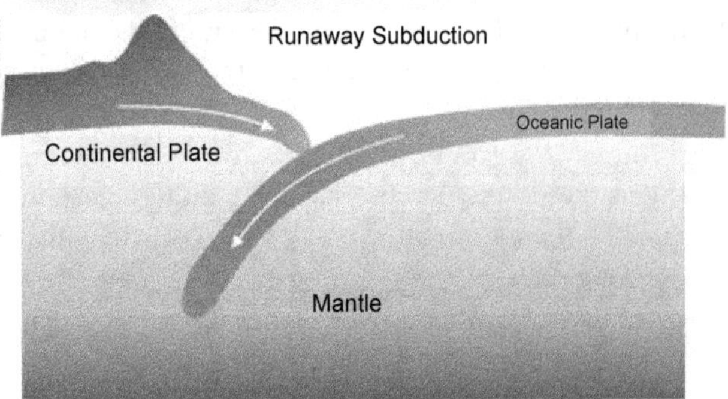



*Fig.6.32: Runaway Subduction of the Oceanic Plate into the Earth's Mantle*

The runaway subduction of the oceanic plate into the Earth's mantle drives meters-per-second motion of the rigid lithospheric plates in the catastrophic plate tectonic model of the Flood, Figure 6.32.

It is only lately that the inferences of the mathematical modelling of CPT have been fruitfully comprehended. It was essential to resolve numerically the rigid "partial differential equations" governing the conduct of silicate rock material, fully considering of the great necessity of effective rock viscosity on temperature and strain rate. The highly nonlinear relationship between viscosity and stress suggests that the effective viscosity decreases substantially once the material is exposed to a strong shear stress. This liquefying influence upsurges intensely as the temperature increases, even though it may only be at 60% of its melting temperature. A vital response mechanism then comes into play. As the cold upper boundary layer of the Earth's mantle descends into the hot mantle underneath, contributed to the liquefying stress, it heats the mantle locally. This decreases the viscosity even further, thus permitting the plate to sink quicker. The two effects, strong shear stress and the peeling away of the upper cool boundary of the mantle, successfully reinforce each other, and therefore thermal runaway begins. As "Baumgardner" states in his paper:

*'A compelling logical argument in favor of this mechanism [subduction] is the fact that there is presently no ocean floor on the earth that predates the fossiliferous strata. In other words, all the basalt that comprises the upper five kilometers or so of today's igneous rocks has cooled from the molten state since sometime after the Flood cataclysm began.'*

"Baumgardner" then enquires where the pre-Flood seafloor went. The model persuasively proposes the answer that the original sea-floor was disastrously subducted, so that we now have a comparatively novel sea-floor—shaped as igneous flows from the Earth's mantle dumped in very thick, five kilometers or so, layers at the bottom of the present-day oceans.

The hydro-plate theory beforehand deliberated and CPT are typically viewed as mutually exclusive. Nonetheless, this necessity not be so. There is substantial room for volcanic action during the continental drift phase of the hydro-plate theory. The breaking of the Earth's crust, perhaps by an impact, may well have freed large volumes of subterranean waters into the atmosphere, and yielded the fast movement of the broken continental plates from the impact center. Then, a subduction mechanism may then have taken over from the initial catastrophe, causing unceasing disturbances in the Earth's mantle under the seas, and supporting the disaster for the rest of the Flood year.

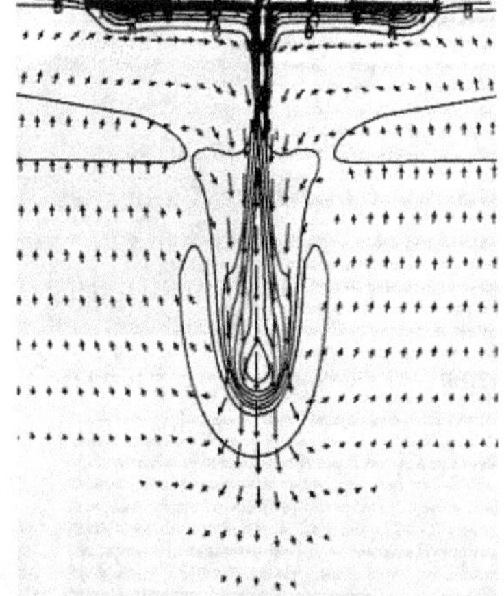

*Fig.6.33: Temperature Profiles Associated with Subduction of Oceanic Slab into the Mantle of the Earth*

Temperature profiles related to subduction of oceanic slab into the mantle of the Earth "Baumgardner. "Computer models of the Earth's mantle show the conditions that would be essential to initiate runaway subduction, Figure 6.33.

## SEDIMENTOLOGY RESEARCH

***Fig.6.34:*** Sedimentary Rocks are Formed from Particles Eroded from Older Rocks –
*Curtsy – Leeds University - UK*

"Guy Berthault" has yielded some milestone investigation into sedimentology. First on a small scale, but more recently on a larger scale, he has investigated the deposition of heterogeneous mixtures from flowing water. His outcomes illustrate that diverse sediment layers do not deposit one after the other in a vertical direction, but all at the same time horizontally, Figure 6.34. Implementing these discoveries to the "Grand Canyon," the dissimilar layers would have been dumped under robust water currents and laid down horizontally, not vertically. Accordingly, many of the layers of the canyon would have been dropped concurrently, and do not unavoidably signify dissimilar periods of time. If showed, this has enormous suggestions for the entire theory of sediment development world-wide. Clearly, we must evade saying that all sediments were arranged down this way. The massive coal seams would be one example of deposition in a non-flowing environment. Though, "Berthault's" sedimentology experiments have reversed previous credence that layers form one after the other in stages. Such astonishing outcomes may help us comprehend the reason the seeming order of the so-called geological column is inverted in some regions of the world.

Moreover, it is motivating to observe that the initial outcomes of the work by "Baumgardner" and "Barnette" sustain "Berthault's" rudimentary evidence. Baumgardner and Berthault well-thought-out the simplified problem of a shallow, homogeneous and inviscid fluid (water) flowing over a rotating sphere. Their completely temporary solution to this problem yielded some unpredictably rapid flowing regions of strong cyclonic coils with velocities of 40–80 m/sec. The outcome of such rapid flowing currents on deposits of material lifted with the water is not yet comprehended, however, this illustrates that there is an excessive arrangement to be done with heterogeneous flows where the shallow water supposition is lifted. Basic studies, both experimental and numerical, are required.

A manner for categorizing rock developments without direct petition to the geological column, with collapsed time-scales, has been suggested by "Walker." This manner promotes diverse kinds of flood developments as the waters rose and subsided. The hydro-plate method or "Baumgardner's" approach can both be utilized as likely driving mechanisms for the Flood within such a classification.

# FLOOD FOSSILIZATION

*Fig.6.35: Flood Fossilization – Curtsy (Image credit: Getty)*

It is fundamental to adopt that no one theory is perhaps wholly passable to consider with all the data, but however, one can venture about likely answers to an apparent problem. For example, "Garner" has correctly stated the trouble with certain basalt flows seeming 'late on' in the hypothetical geological column. Meanwhile, these appear to necessitate a sub-aerial setting, one can comprehend his deduction that the post-Flood boundary must be prior than the basalts. Therefore, with the water drained from the land, the succeeding volcanic action in the Mesozoic and Cainozoic would be sub-aerial. Nonetheless, if we agree to take the hydro-plate model of the start of the Flood, then the first 40 days would include enormous destruction in agreement with the Paleozoic, (some even include most of the Precambrian) record, Figure 6.35. The waters of the oceans were still rising, parts of the land were still not sheltered entirely by water—there may even have been a brief lull. Undoubtedly, this is not unreliable with the account in *Genesis 7:17–24*. In the next 110 days, huge volcanic disturbance happened on the land masses, nonetheless, still not all the land was lastly covered. Simultaneously, cataclysms of the land masses were also happening, therefore, that some of the land that had been covered was exposed, albeit temporarily—of the order of weeks.

It is imaginable that dinosaur tracks could have been accomplished in this time. "Garton" correctly illustrated that these dinosaur tracks go right through the Mesozoic and into the Cainozoic. Under our scenario, tracks in the Mesozoic are according to ground still being accessible at the late stage of the 150 days. Some tracks may by now have been achieved earlier, just after the 40 days' initial onslaught, and then pushed upwards when the mountains rose. Likewise, tracks illustrating no sign of chaotic motion in the Pyrenees in Spain may also be at the late stage of the 150 days, again pushed upwards as the mountains rose. Lastly, the waters with massive amounts of debris and sediment overcame these huge creatures which, not astonishingly are buried in the same part of the strata as the later tracks and usually 'higher' up the column. We do not state that such a scenario clarifies everything. There is a massive amount of work still to be done to comprehend the mechanisms involved. however, we propose that a preparedness to expect and look for the uncommon is continuously vital for development in scientific research.

# DINOSAUR TRACKS AND NESTS

Egg-laying by dinosaurs in Mesozoic strata, well above what seems to be the early fossils of marine creatures in the lower strata, tests the opinion that most of the fossils were shaped by the Flood. Robinson provides additional indication of other seemingly "in situ" fossils counting plant roots in the Jurassic in addition to marine fossils seemingly "in situ" right up over Cretaceous into Tertiary rock, Figure 6.36. "Robinson" contends against the vapor canopy model, stating:

*Fig.6.36*: *Fossilized Dinosaur Eggs – Discovered During Roadwork in Heyuan City, China Curtsy – Imaginechina/AP Photo*

*'The sudden death of the dinosaurs and other animals at the end of the Cretaceous is a phenomenon for which the received Flood model [i.e., the vapor canopy model of Whitcomb and Morris] has no explanation.'*

Though, the model recommended by "Robinson," "Garner," "Garton" and others connecting several post-Flood disasters provides no real response either to the unexpected death of dinosaurs in the Cretaceous. Their post-Flood fossilization hypothesis, in our opinion, turn out to be a serious scientific difficulty.

Marine fossils are discovered in height up in mountains in the Alps, frequently dropped with excessive violence, (as recommended by the Jurassic marine fossils at lower altitude on the North East Coast of Yorkshire near Whitby). The burial of huge dinosaurs, by their thousands in Alberta and Montana, South Dakota, Kansas and Colorado with massive continental sedimentation, (in some places thousands of feet thick), would not be likely without triggering gigantic disturbance in other regions of the Earth. It appears unthinkable that post-Flood catastrophes could dump such thick strata without producing ferocious influences all round the world. The scale, depth and the sheer quantity of fossils contends powerfully that these must be part of the Flood. Rather than compelling the clarification of "mabbul" to mean the elimination of all possible indication of any creatures, (the 'blot out' theory—to allow suggested post-Flood activity), it appears more proper to query whether we have correctly comprehended the scientific indication.

"Robinson" declares that Oard's post-Cretaceous model for the Flood/post-Flood boundary is *'not a straightforward clarification of Scripture'*. He contends that the situation on the geologic column whereby the Flood killed the dinosaurs is *'a paradigm constraining the interpretation of Scripture'*. Though, the other position he trusts, of completely blotting out all animal remnants without trace is, in our opinion, compelling a weak denotation on the word "mabbul." This and the prerequisite of post-Flood catastrophes on a continental scale are becoming a much bigger difficulty in the natural clarification of Scripture.

It is proposed that the burial of the dinosaurs by Flood waters is in harmony with the indication. We have proposed, as one option, that the dinosaurs and their nests were buried late in the Flood. Other circumstances could be likely. For example, sea creatures may have been buried by enormous submarine landslides, which were then pushed above sea-level in the first few days of the Flood. Simultaneously, sediments encompassing dinosaurs buried in the early stages of the Flood may have then been moved only a short distance across the newly uncovered submarine deposits. It appears distinct that some dinosaurs must have been buried by disastrous waterborne sediment, at the minimum in the case of the Mongolian examples of burial in the Cretaceous layers.

A third option, and perhaps the most reasonable alternative, for the incidence of dinosaur tracks late in the strata is that supported by "Garton. He proposes that big creatures (counting dinosaurs) were stuck in the floating Carboniferous forests. The indication for these massive islands of vegetation carried by the heaving seas appears to be predominantly strong. "Garton" upholds that these creatures hovered the harsh land in the last stages of the Flood, Figure 3.37. (In that he permits a few creatures to have endured the first 40 days, we assume he does not consider the 'blotting out' to be all-inclusive.) This option clarifies the seeming irregularities and proposes that there may have been some defense from the initial barrage from above in the early stages of the Flood. "Scheven's" outstanding accomplishment on floating mats of vegetation appears to clarify the Carboniferous coal measures very skillfully. It is imaginable that as these mats collide with land, the sustained beating of the seas as the waters rose to their maximum height could cause forceful deposition of sediments with massive ocean waves crisscrossing the continents. The waters at this stage were not essentially calm. In actual fact, this is most doubtful. Great geological action appears to have been

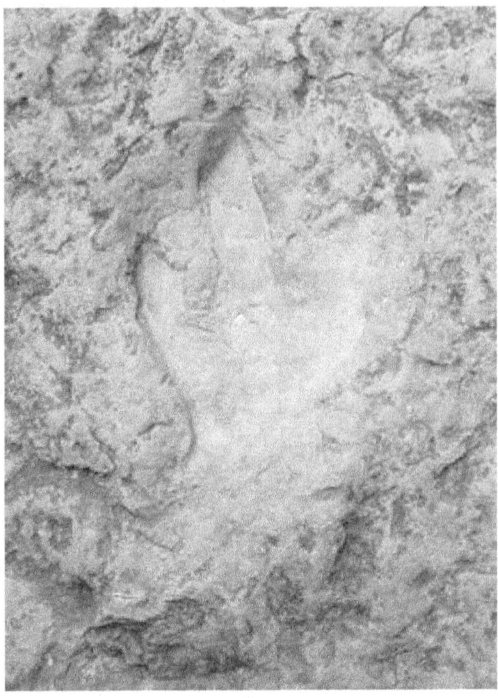

*Fig.3.37:* Dinosaur Footprints Reservation, Holyoke, MA

going on still, even though the rains had stopped. The bringing up of the mountains and the sinking of the valleys (*Psalm 104:8* '*The mountains ascend, the valleys descend*') happened directly after the earth was lastly covered (*Psalm 104:6*).

It is consequently, totally possible that additional huge mudslides entombed the dinosaurs as the rafts struck land in the final stages of the 150 days, or that some run-away onto land, only to be entombed as the rising waters lastly covered the land. The burial of birds in the later strata is all consistent with the final stages of the 150 days, everywhere no land was accessible.

## INTEGRATED APPROACH

It is vital that all scientific facets be employed to comprehend the likely processes of the Flood. It is not only geology that should be well-thought-out. Hydrodynamics also must play an important part in comprehending sedimentation processes. Geologist "Berthault" has correctly specified '*Determination of initial hydraulic conditions from sedimentary structures, resulting from sedimentological data is, therefore, a research priority.*' Nowadays, in the expertise of the principal author, (in fluid dynamics and thermodynamics research), a multi-disciplined method is typically necessary before scientific developments can be achieved in the comprehension of complex and rare phenomena. Development is not usually conceivable when it is asserted that only specialists of one discipline can resolve the fundamental physics of a specific problem.

The modelling of the flow of heterogeneous mixtures with the full laws of conservation of energy, mass, momentum, is one of the utmost tests that computational fluid dynamics has confronted. A very cautious and detailed approach is necessitated when the particle size of the material carried with the water differs extensively. The problem includes materials of diverse densities, diverse viscosities, with very huge differences in local "Reynolds" number (convection divided by viscous diffusion) and hydraulic circumstances. Additionally, boundary layers have to be modelled with specific courtesy to the likely modification from tempestuous to laminar flow. The battery of tests of "Berthault" have by now obviously illustrated that astonishing lamination can happen in the sediment deposits from such flows. These situations now requ to be modelled by fluid dynamitists and mathematicians, so a comprehension of the greater picture can arise by prudently created mathematical models.

On the greater scale modelling of solid earth geophysics, we recognize the inspiring accomplishment already in progress with the research of geologists "Baumgardner" and "Barnette." Collaboration among geologists and other scientists (mainly those investigating in fluid dynamics), is indispensable if there is to be advancement in Flood geology, beyond (the not redundant) rudimentary explanation of what rocks and fossils are found at specific settings. Only as there is superior interaction between the pertinent scientific disciplines will some of the unanswered problems of the Biblical Flood models be resolved.

## THE BEIJING ANOMALY

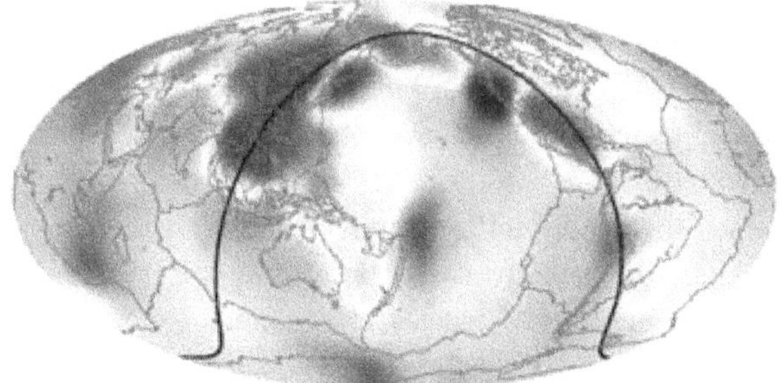

*Fig.6.38: Huge Underground Ocean Spotted Beneath Asia*

Unfathomable depth beneath Asia, the existence of an enormous body of water has lately been discovered at a depth between about 700,000 and 1,400,000 meters below the Earth surface. It is roughly in the middle of the mantle, Figure 6.38. This enormous 'seismic anomaly,' a segment of the mantle that weakens seismic waves from earthquakes, was discovered by examining some 600,000 seismograms, which is graphic recordings of shock waves moving through the interior of the planet Earth. As said by the discoverers, "M.E. Wysession" and "J. Lawrence" *'the volume of water in this anomaly is at least that of the Arctic Ocean.'*

## VIEW INSIDE THE EARTH

Insignificant amount of research can be directly conducted at the inner parts of the Earth. The deepest mine in the world is a gold mine in the Witwatersrand area in South Africa. It descends 3,500 meters into the lithosphere (crust), Figure 6.39.

The deepest humans have ever drilled into the earth is on the Kola Peninsula in Russia, where drill-core was recovered from 12.26 km below the surface. From that point on to the center of the planet Earth, some 6,365 km, is all 'unknown territory'. All we can accomplish is to deduce what may be down there from the incomplete data we may collect.

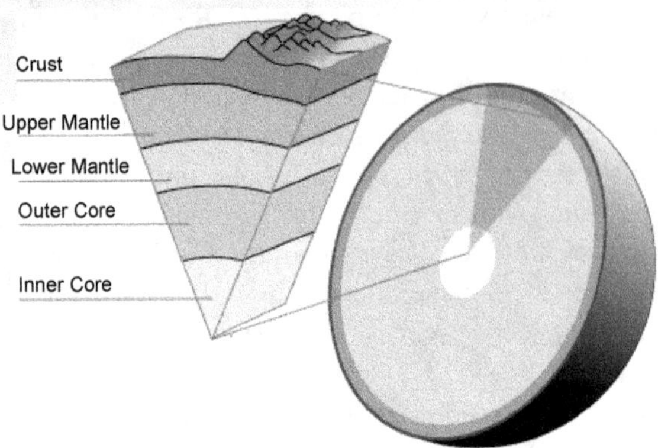


*Fig.6.39: The internal structure of the earth.*

In making such extrapolations, the first issue we well recognize is the mass of our planet, which was calculated by "Henry Cavendish" in 1789 utilizing Newton's laws. We may also guess the average density of the highest 'layer' known as 'crust' or 'lithosphere,' from boreholes and from the rocks that outcrop on the surface. Nonetheless, how thick is the crust and what lies underneath it? This is where seismic investigation derives the activities as a result of shock waves. The shock waves may occur by earthquakes or explosions. These shock waves have the ability to travel over the entire Earth. Their speed varies according to the density of the environment they travel through. Frequently, these shock waves experience reflections and refractions whenever the speed changes. Founded on these and many other consequent characteristics, an image of the inner region of the earth has appeared, signifying a series of concentrical spheres: the inner core, the outer core, the mantle and the crust.

The core characteristic distinguishing these regions is density, which, as a rule, upsurges with depth. Starting at slightly above 2, where the density of water is 1 at the surface. The density is estimated to reach 11 in the core, which is believed to be made of nickel and iron. Each concentrical sphere is detached from the previous one via a thin region at which the speed changes significantly; such a region is called a "discontinuity." The Kola Peninsula superdeep borehole was designed to reach the "Mohorovičić ('Moho') discontinuity." It named after the Croatian seismologist Andrjia Mohorovičić who discovered it. The Moho is situated at about 15 km down under the oceans, which lack the uppermost layer of the lithosphere. The Moho shock-wave rises in places to only 6 km below the seafloor. Other major discontinuities are "Wiechert–Gutenberg" - at 2,900 km- and "Lehmann" - at 5,100 to 5,200 km.

## HIDDEN KNOWLEDGE

The concentric-sphere structure developed from seismic data and the properties of resident minerals, i.e., from the rocks, meteorites and lab experiments, Figure 6.40. It is understood that the material encompassing each of the inner spheres also has different chemical compositions. The prime settings of pressure, temperature and viscosity at different depths have thus been projected which has allowed certain calculations concerning behavior and dynamics. However, none of those predictions made any correlation to anything like the "Beijing Anomaly" (BA). This illustrates that water at such a depth was unbelievable! There is one exclusion though, explicitly called a prediction derived from a creationist model of the inner dynamics of the Earth and the manner it clarifies how plate tectonics began. The model was created by "Dr John Baumgardner," then of the "Los Alamos National Laboratory."

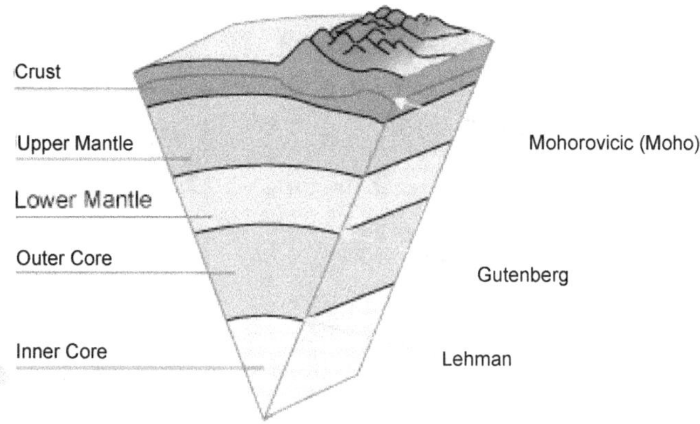


*Fig.6.40: Some major seismic discontinuities within the earth.*

## Noah's Flood – Creationist Model

Also, the "Catastrophic Plate Tectonics (CPT)" model offers a good creationist model for Noah's Flood. According to it, as the first segments of the crust, i.e., seafloor, began sinking into the earth's interior, they were moving "meters per second" rather than "millimeters per year," i.e., the stride at which plates move today, Figure 6.41. At that speed, the sinking plates could have reached the bottom of the mantle, i.e., 2,900,000 meters below surface in 15 days. Certainly, seismic tomography, i.e., a sort of "CAT" scan of the innards of the planet, offering clear evidence that there are colder-than-the-surroundings slabs of rock the size of **continents** at the bottom of the mantle which could not have been there for too long, i.e., else they would have been completely melted and mixed beyond recognition with the mantle material.

## SIGNIFICANCES

Mantle density structure of the eastern (top) and western (bottom) hemispheres derived from seismic tomography. Blue represents low temperature rock - and red high temperature rock. Bright green contours represent present-day subduction areas, Figure 6.42.

*Fig.6.41: In the 17th Century, Scientists Used Physics to Explain the Noah's Flood Miracles Described in the Bible.*

If tectonic plates were subducted at such a speed, i.e., '*runaway subduction*,' sediments on the seafloor and enormous amounts of water would have been pulled down with them. Once they touched the areas of high pressure and temperature inside the mantle, that water and the waterlogged sediments would have transformed immediately into very active chemical solutions and gases, i.e., occasionally mentioned as volatiles, which, being significantly lighter, would incline to rise towards the surface. It is fascinating that recent experiments have discovered that, when give-in to pressures and temperatures parallel to those in the mantle, calcite ($CaCO_3$), in the presence of iron, turns into methane gas.

In light of this, a huge and spectacular array of other similar chemical variations would be likely when massive volumes of seafloor speedily sank into the mantle. At the slow pace subduction reveals today, the seafloor melts as it descends and the volatiles separate early on, much closer to the surface. Consequently, they will have diverse chemical properties and most of them reach the surface in its place of remaining inside the mantle.

Runaway subduction may well be accountable for the water in the "Beijing Anomaly." However, even greater volumes of mineral-laden fluids, i.e., hydrothermal fluids could have reached the upper mantle and the crust. Anywhere they would have filtered through liberated sediments, i.e., Noah's Flood undoubtedly, formed incredible amounts of them. The dissolved minerals would precipitate, strengthening the liberated sediments into hard rock. About 90% of all sedimentary rocks are well-thought-out to be 'terrigenous,' i.e., made of fragments of earlier rocks worn out from the continents and bound together by chemical cements. The basis of these chemical cements, i.e., their total volume is immense, has long been a mystery. Not anymore, today.

## NOAH'S FLOOD – SOLVED

The sediments dumped during Noah's Flood contained enormous vegetal debris and immeasurable carcasses. The same hydrothermal fluids could have speedily fossilized them, i.e., substituting the organic matter with mineral matter to variable degrees. Consequently, the entire of the recognized fossil record could have been shaped in a short time. Such profound hydrothermal solutions are not identified to be present on the surface now, i.e., all existing ones created by infiltration of water from the surface to a depth no greater than a few kilometers, which would clarify the reason widespread fossilization is not being observed today. The vast majority of the recognized fossils are the consequence of an exceptional geological process—Noah's Flood—which the newly discovered "Beijing anomaly" appears to confirm.

*Fig.6.42: East and West Mantle density structure – Curtsy – Photo John Baumgardner.*

## MASSIVE LAKE UNDERGROUND

The globe-covering Flood of Noah complicated, among other things, the breaking up of '*all the fountains of the great deep*' *(Genesis 7:11)*. This proposes a world-wide cracking of the earth's crust and the ferocious eviction of water, vapors and other underground material such as volcanic lava. New evidence from inner the earth pops-up captivating thoughts on this global-scale scenario, Figure 6.43.

Scientists have deeply deliberated that the hot inner regions of the Earth would be substantially dry, since the heat would have vaporized and driven off any water. However, consistent with a report in "New Scientist," particular minerals, even under the concentrated intense heat and pressure deep underground, can accumulation lots of water. Models of the mantle, that part of the earth between the molten core and the solid crust, and particularly the 'transition zone' between the upper and lower mantle, now define it as 'sopping wet'.

*Fig.6.43: There is still a massive lake of underground water inside out planet Earth.*

Moreover, it appears that hot wet rocks are more unstable than hot dry rocks. This novel evidence may now clarify 'why enormous volcanic eruptions abruptly flooded hundreds of thousands of square kilometers of 'land with lava,' as detected in a number of diverse parts of the geological record.

Who could enquire for a more graphic explanation of the behavior of the *'fountains of the great deep'*? It is thought-provoking that even today, up to 70% or more of what comes out of volcanoes is water, typically in the form of vapor.

So how much water is warehoused in the mantle? Approximations differ from 10 to 30 times the volumes in all the earth's present-day oceans! Is it likely for the mantle to abruptly liberate this water and for the earth to be *'drowned from below'*? The geologists determined that a 'sudden outpouring of water, Noah-style', was highly improbable. This deduction is in harmony with God's solemn promise to Noah, sealed by the sign of the rainbow, not ever to destroy the earth with water again (Genesis 9:11–17).

We distinguish from Scripture that water once dispensed from the *'fountains of the great deep'* for five months, so that the early world was once literally *'drowned from below'*.

Even after all that, there is still a immense oceans of underground water deep inside our planet Earth. Its very survival, and the histrionic effect it has on the behavior of the Earth's hot interior, shows once more that the earliest Bible stories are not mythological but are historic facts and mere descriptions of a sobering physical genuineness.

## WATER AND BIBLICAL SCIENCE

Water! We drink it, wash in it, cook with it, swim in it and generally take it for granted. This clear, tasteless and odorless liquid is so much share of our lives that we barely ever think about its astounding characteristics. We would die in a few days deprived of water—and our bodies are 65% water. Water is essential to dissolve vital minerals and oxygen, flush our bodies of waste products, and convey nutrients around the body where required. Water is the lone constituent that has these properties. It has numerous more fascinating structures that propose that it has been designed *'just right'* for life.

### Liquid Water

There are three statuses of matter: solid, liquid, and gas. All three are indispensable for living things.

- The solid state upholds its form.
- Liquid is capable to flow and take up the shape of its container, while reserving the same total volume.
- A gas expands to fill both the shape and size of its container.

For molecules to interact together, it is best to have them adjacent to each other, but free to move around. This is just what the liquid state delivers, so it is ideal for all the thousands of chemical reactions occurring in every cell of every organism.

However, of all the temperatures in the universe from the –270°C (–454°F) of outer space to the tens of millions of degrees inside the hottest stars, water is liquid in a very narrow range. At typical atmospheric pressure, water is only liquid from 0–100°C (32–212°F). It should not then be astonishing that Earth is the solitary place in the universe recognized to have liquid water. And this is contingent upon having the right kind of adjacent star—neither too bright nor too dim, and thus neither too big nor too small. And the planet Earth must be at the right distance from the adjacent star, i.e., the sun is our special adjacent star.

## SLIPPERY ICE

Countless people relish winter sports such as ice skating and skiing. Why ice is very slippery, permitting these fun activities? Numerous people trust that it originates from pressure melting the ice and forming a lubricating liquid layer. Correct, it is well-recognized to physical chemists that applied pressure inclines to aid in forming the water which takes up the least volume. Therefore, pressure will favor manufacture of water from ice (melting), so its melting point will decrease, Figure 6.44.

*Fig.6.44: The Surprising Science of Why Ice is So Slippery*

However, the influence is much lesser than many people think—it involves over 100 times normal air pressure to lessen the melting point by only one Celsius degree. So, there is no way that this effect could be responsible for ice skating, and certainly not for skiing where the pressure is far less. Nor could it have caused airplanes to melt ice and sink 75 meters (250 feet).

The factual cause is yet another unfamiliar property—the molecules on the surface of ice vibrate much more than usual in a solid, although they don't move around. This gives the surface a '*quasi-liquid*' character, i.e., liquid-like but not liquid.

## BUFFER TEMPERATURE

Another very important characteristics of water is its high specific heat. This means it consumes a lot of energy to heat it, i.e., about ten times as much as the same mass of iron, and it must dissipate a lot of energy to cool down. Consequently, the massive bodies of water on earth assist keep the earth's temperature fairly steady. Alternatively, land masses heat up and cool down more quickly. When combined with the fairly steady temperature of water bodies, this is a favorable thing. It means diverse parts of the atmosphere are heated differently, which creates wind. This is indispensable for keeping the air fresh.

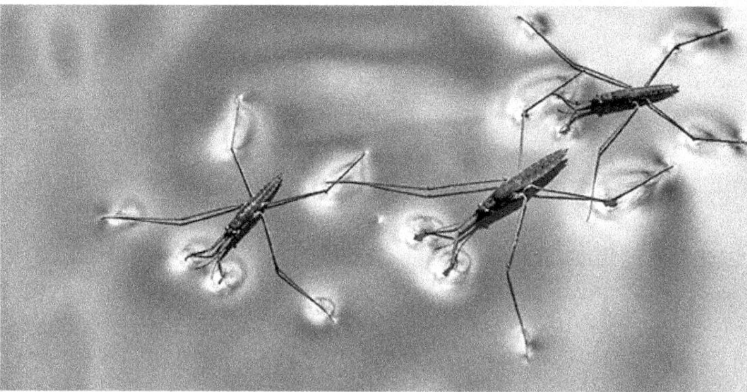

*Fig.6.45*: *Surface Tension on Water is Easily Seen When Insects Walk Across the Surface Without Sinking.*

## SURFACE TENSION

Water has a distinctive high surface tension, the force attempting to retain the surface area as small as possible. It is higher than that of a syrupy liquid like glycerol. Surface tension inclines to generate bubbles and drops spherical, and is robust enough to support light objects, counting some insects. More importantly, this means that biological composites can be concerted near the surface, speeding up many of life's significant reactions, Figure 6.45.

### Power in Water

Granting, water typically give the impression to be placid, if a lot of it is moving fast enough, it can move car-sized boulders and carve deep canyons, even cutting into solid rock. Once flowing very fast, a particularly destructive process called "cavitation occurs," Figure 6.46.

*Fig.6.46*: *Water Powers Life*

Similarly, on a chemical level, it rapidly breaks down several significant big molecules in living cells. Even though living cells have numerous inventive restoration mechanisms, DNA cannot last very long in water separate from its cell, Figure 6.46. Also, a recent writing in "New Scientist" defined this as a '*headache*' for scientists and researchers occupied with evolutionary concepts on the origin of life. Additionally, it illustrated its materialistic partiality by saying this was not '*good news*'. However, the tangible bad news is assuredly the faith in evolution. i.e., everything made itself, which nullifies impartial science.

When liquids evaporate, they attract heat from their environments. This means that we have a valuable means of being cool: sweating. A vital phenomenon of this is water's high "latent heat" of vaporization. This means it consumes substantial amount of energy to evaporate water than most other liquids. Consequently, we require to perspire relatively little water to keep cool; if we sweat almost any other liquid, the volume we would need would be enormous.

## SUPER SOLVENT

Water is one of the closest things we have to a *"universal solvent"*. Numerous minerals and vitamins can be travelled all over the body after being dissolved. Dissolved sodium and potassium ions are vital for nerve impulses. Also, water dissolves gases, such as oxygen from the air, allowing water-living animals to utilize oxygen. Water, a main constituent in blood, also dissolves carbon dioxide, a waste product from energy production in all cells, and transports it to the lungs, where it can be breathed out.

Though, an actual universal solvent would be useless, as no container could store it! nevertheless, water is deterred by oily mixtures, so our cells have membranes made of these. Countless of our proteins contain partially oily regions, and they incline to fold together, deterred by the surrounding water. This is partially accountable for the numerous and varied shapes of proteins. These shapes are crucial for carrying out functions vivacious for life.

## UNDERSTANDING ICE

A vigorous and very rare property of water is that it swells as it freezes, dissimilar to most other constituents. That is the reason icebergs float. Actually, water contracts ordinarily as it is cooled, until it reaches 4°C (39.2°F), as it begins to expand again, Figure 6.47. This shows that icy-cold water is less dense, so tends to move upwards. This is very significant. Majority of liquids exposed to cold air would cool, and the cold liquid would sink, compelling more liquid to rise and be cooled by the air. In due course all the liquid would lose heat to the air and freeze, from the bottom up, till completely frozen. Nonetheless, with water, the cold regions, being less dense, stay on top, permitting the warmer regions to stay below and avoid losing heat to the air. This means that the surface may be frozen, but fish can still live in the water below. Nonetheless, if water were like other substances, huge bodies of water, such as North America's Great Lakes, would be frozen solid, with dire consequences on life on earth as a whole.

*Fig.6.47:* Understanding ICE – *Curtsy Wikipedia*

## Vital Knowledge

- The earth is 70% covered by water.
- Only 1% of the world's water is readily available for human consumption. Approximately 97% is too salty and 2% is ice.
- Australia is the world's driest inhabited continent having the least runoff and 70% desert.
- It takes about 150,000 liters of water to make a family car.
- Only 1% of household water consumption is for drinking. The rest goes on lawns, showers, etc.
- A household toilet flushes about 150 liters of water per day.
- A continuously dribbling tap wastes 600 liters of water per day. A dripping tap per day (1 drip per second) uses 30 liters.
- Garden mulching reduces evaporation by 75%.
- An average garden sprinkler uses 1000 liters per hour.

- Natural water has within itself small amounts of dissolved mineral salts, which provide it some taste. Pure water is tasteless.

*Fig.6.48*: National Oceanic - Curtsy - (NOAA), U.S. Dept Commerce

The earth' massive ice caps and glaciers hold a staggering 29 million km³ (7 million cubic miles) of water which is about 2% of the total on earth. The oceans hold 1,370 million km³ and the total amount of water falling as rain on land each year is roughly 110,300 km³, Figure 6.48.

## Exceptional Water

The slightest building block of water is the water molecule. This encompasses two hydrogen atoms clinging to an oxygen atom in a V-shape, with an angle of 104°. It is "polar," that is the oxygen atom has a negative electrical charge, though the two hydrogen atoms are positive, Figure 6.49. This is the reason water dissolves so many things, like salt, which also have electrically charged building blocks; though water would not dissolve oil which has uncharged molecules.

Also, water is attracted fairly strongly to other water molecules by "hydrogen bonds." These bonds are ten times weaker than typical chemical bonds, but sufficiently strong to make water liquid at room temperature, while a similar compound, hydrogen sulfide, lacking hydrogen bonds, is a gas. Hydrogen bonds are also accountable for water's high surface tension and specific and latent heats.

The profile of the molecule and hydrogen bonding indicates that ice has a very open hexagonal (six-sided) crystal structure, which is well exemplified by the endless diversity of snowflakes, Figure 6.49. This structure consumes up much of a space, but the structure breakdowns upon melting. Consequently, liquid water is denser. This is why ice floats. New research illustrates that water molecules produce clusters in the liquid, specifically a cage-like structure with six molecules. This is accountable for numerous of water's exclusive properties.

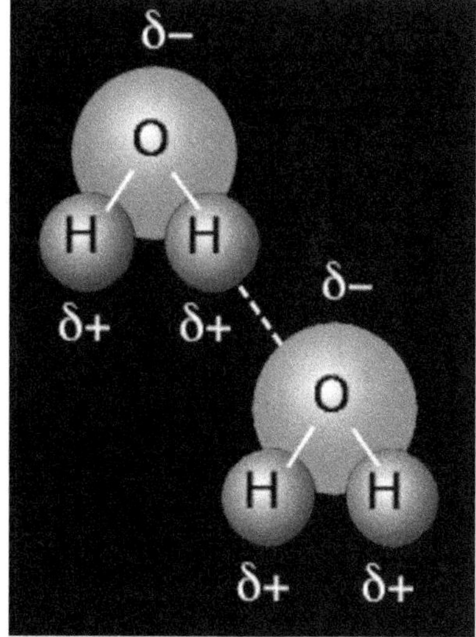


*Fig.6.49*: Hydrogen Bonding – 2 (H) Positive Charge – 1 (O) Negative Charge

Additional current research demonstrates that there are perhaps two kinds of hydrogen bond in water, one about twice as strong as the other. This could clarify the reason water is liquid over a fairly extensive range. Melting breaks only the weaker bonds, while boiling must break the stronger bonds too. This investigation also demonstrates that the alteration from strong to weak bonds necessitates specific temperatures, one of which is 37°C (98.6°F). This is our body temperature, signifying that this is one of the many elaborate design features we have.

# WATER THE BIBLICAL SCIENCE

There are at the minimum two biblical orientations about water which illustrate that the Bible predicted much about modern science. One is a orientation to the "water cycle"—evaporation, clouds, rain:

Job 36:26–28: '*Look, God is greater than we can understand. His years cannot be counted. He draws up the water vapor and then distills it into rain. The rain pours down from the clouds, and everyone benefits.*'

The other is the reference in *Psalm 8:8* of '*paths of the seas. "The birds in the sky, the fish in the sea, and everything that swims the ocean currents."*

The oceanography pioneer Matthew Fontaine Maury (1806–1873) was led by this verse to chart water currents. As "Maury" pointed out, '*The Bible is [the] authority for everything it touches*'—not just doctrine, but science and history as well. His work revolutionized shipping by drastically cutting travelling times.

"Maury" offered glory to God for his discoveries. And we should all offer God the glory for all the wonders of water, and be appreciative to Him for its many uses.

## Snow Flakes – Heavenly Organization

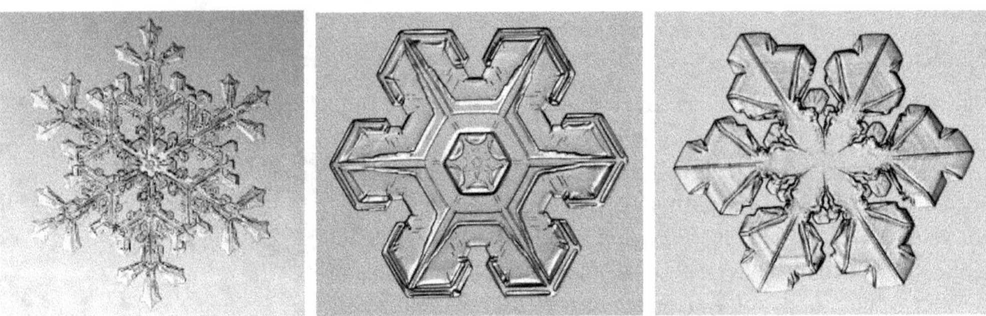

*Fig.6.50: Snow Crystals – Beautiful Shapes in Nature*

Snow crystals are nearly of the most beautiful configurations that nature has to offer, and no two flakes are alike, Figure 6.50. Many evolutionists have attempted to entitle the order of a crystal contracting contributed to atomic structures as proof for "*something coming out of nothing,*" due merely to natural laws. However, closer investigation of this argument demonstrates it does not yield to scientific scrutiny.

## Advanced Snowflake Research

Many scientists are attempting to cultivate their own crystals to comprehend and guide their development. Applications of this research influence way beyond climatology, with the goal of controlling the development of other crystals, such as silicon constructions, for the semiconductor industry.

Therefore, why do snow crystals configure this shape? Does it necessitate distinct design? No, snowflakes configurations is contributed to the properties of their construction blocks, the water molecules ($H_2O$). These are bent and polar, i.e., with positively and negatively charged ends. As Hydrogen and Oxygen combine together in solid form, they incline to shape the lowest-energy construction they can, which is crystals with hexagonal, i.e., six-fold symmetry. By dissimilarity, carbon dioxide ($CO_2$), a linear and more symmetrical mole, clue, forms *cubic* crystals in its solid form, i.e., "dry ice".

We recently recognize that not only temperature, but also humidity impacts crystal formation and configuration. The beautiful six-legged star-like crystals breed in air warmer than -3°C. Between -3°C and

-10°C, snow falls as little prisms. Between -10°C and -22°C, it is little stars again, and below that, prisms once more.

Nonetheless, scientists still cannot tell precisely the reason snow crystal configurations change depending on temperature. These configurations are contingent on how water vapor molecules are combined into the growing ice crystal, and the physical processes governing crystal growth are intricate and not well comprehended yet.

## Is Snowflakes a Proof of Evolution?

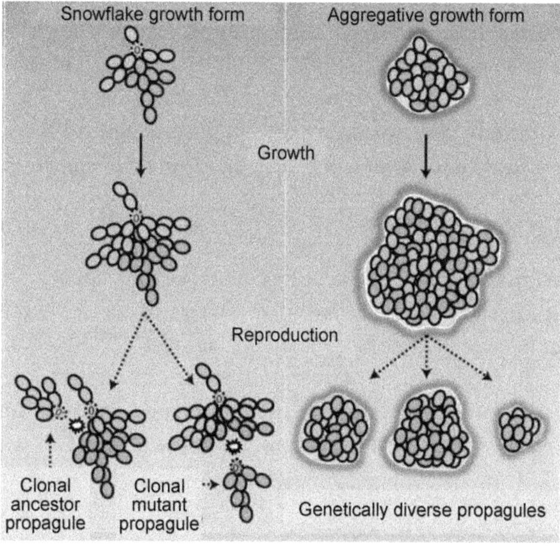


*Fig.6.51: Origins of Multicellular Evolvability in Snowflake Yeast*

Occasionally evolutionists claim that snowflakes demonstrate that "order" can ascend from "disorder," and more complex structures from simple ones, founded purely on the characteristic of the physical properties of matter, Figure 6.51. Consequently, the reasoning goes, life could have arisen from simple molecules that organize themselves in a manner that eventually guides to more complex structures, and ultimately the first living cell.

Nonetheless, crystals are nothing like a living cell. Fashioned by the "withdrawal" of heat from water, they are dead structures that encompass no more information than is in their component parts, the water molecules. Life forms, instead, brought to existence, evolutionists believe, through the "addition" of heat energy to some postulated primordial soup. Not only are these processes very dissimilar, but life necessitates the appearance of new information, i.e., a code, in order to take over the occupations of organization and reproduction of a cell. There is consequently, no analogy between snow crystals and the far, far superior complexity of living organisms.

More prominently, the organization in proteins and DNA is not caused by the properties of the constituent amino acids and nucleotides themselves, any more than forces between ink molecules make them join up into letters and words. "Michael Polanyi (1891–1976)," a former chairman of physical chemistry at the University of Manchester (UK) who turned to philosophy, confirmed this:

"As the arrangement of a printed page is extraneous to the chemistry of the printed page, so is the base sequence in a DNA molecule extraneous to the chemical forces at work in the DNA molecule. It is this physical indeterminacy of the sequence that produces the improbability of occurrence of any particular sequence and thereby enables it to have a meaning—a meaning that has a mathematically determinate information content …".

Crystals are not: direct" evidence for creation, either. Nonetheless, the philosophical argument can be made that a universe deprived of God cannot rationally be anticipated to create such order out of disorder. Consequently, when we observe "order and design" in the universe, as demonstrated by the six-cornered snowflake, doesn't this mandate a Creator who supplies this "order and design?"

Unquestionably, the physical characteristics of water are recognized to be essential preconditions for life to exist on Earth, which attests to a Creator who conceived the universe and its physical laws as conducive to life. For example, snow forms an insulating layer on the ground that guards plants and animals below it from the much severer temperatures above. But whereas this could have been accomplished with very simple forms, such as round or square disks, the extravagant beauty and diversity in snow crystals displays God's loving creativity in creating snow not only very advantageous, but also delightful to look at! As even evolutionists admit, *"One could almost convince oneself that snowflakes constitute a demonstration of supernatural power."*

## NO TWO SNOWFLAKES ALIKE

Essentially, lesser snowflakes that take the shape of hexagonal prisms appear pretty much identical. On the other hand, bigger, star-shaped crystals are all dissimilar. To comprehend the reason, consider of how many different ways 15 books can be arranged on a bookshelf. You have 15 choices for the first book, 14 for the second, 13 for the third, etc. The total number of potentials is thus $15 \times 14 \times 13 \ldots (15!)$, or over a trillion ways to arrange those books. Crystals can effortlessly have 100 or more features that can be recombined in different ways—leading to at least a staggering $10^{158}$ different possibilities. This is $10^{70}$ times the number of atoms in the entire universe![1]

### The Snowflake Man from Vermont

Astronomer "Johannes Kepler" appears to have been the first scientist to investigate snow crystals. He composed a booklet on the subject in 1611. Nevertheless, the real "Snowflake Man" was "Wilson Alwyn Bentley," born 1865 in Vermont, USA. Bentley was the first to photograph snowflakes. He printed more than 5,000 photographs, and wrote numerous articles on snow, rain, dew and other natural phenomena associated with water and precipitation.

"WA Bentley" was the first to photograph snowflakes. He devoted his life to investigating snow, dew and rain and though he was a farmer without formal scientific training, he was years in advance of his time with his meteorological hypotheses.

Bentley narrates that it was his mother who imparted the love of scientific investigation into him: he was home schooled until he was 14 years old, and in his pursuit for learning he even read an encyclopedia! *'It was my mother that made it possible for me, at fifteen, to begin the work to which I have devoted my life. She had a small microscope, which she had used in her school teaching. When the other boys of my age were playing with popguns and sling-shots, I was absorbed in studying things under this microscope: drops of water, tiny fragments of stone, a feather dropped from a bird's wing, a delicately veined petal from some flower. But always, from the very beginning, it was snowflakes that fascinated me most'.*

Bentley learnt nothing about photography and for the longest time could not manage to capture pictures of snowflakes. Nonetheless, through perseverance and learning by trial and error he learned how to work fast before the ice crystal altered its shape, how to use transmitted light by pointing the camera to the sky, and how to get sharpness of detail on the crystal by utilizing a large f-stop. Finally, during a January snowstorm in 1885, he captured the first photomicrographs ever taken of an ice crystal.

He reserved comprehensive meteorological records, and considered over the meaning of the shapes and sizes of the crystals and the reasons they often diverse from one storm to the next. Starting in 1898, he published his discoveries in scientific journals. "Bentley" significantly donated to what is today common knowledge, i.e., that temperature variations and movements in the storm clouds influence on the form and type of the crystals shaped. With his investigation, he was years in advance of the meteorological thinking of his time.

Bentley loved people, nonetheless, was misinterpreted by them, and the scientific world valued, i.e., or caught up with the value of his work only much later. When he assembled a meeting in his birthplace to present on his work, only six people attended.

One of his "National Geographic" (Jan. 1923) articles, 'The magic beauty of snow and dew', is attended by over 100 photomicrographs of ice crystals, frost patterns, and dew. In it, he denotes to the Bible and says the beauty of snow was known long ago, for the book of Job asks, '*Hast thou entered into the treasures of the Snow?*'

Although his photos were sold for jewelry and other drives, "Bentley" did not become wealthy through his work. Nonetheless, he said that he would not alter places with Ford or Rockefeller: he sensed he was serving the Great Designer; capturing the fleeting loveliness which, but for him, would be unappreciated— even unseen by most of his fellow men. And with that role he was content.

When he died of pneumonia in 1931, his obituary read, '*Truly, greatness blooms in quiet corners and flourishes under strange circumstances. For "Wilson Bentley" was a greater man than many a millionaire who lives in luxury of which the 'Snowflake Man' never dreamed'.*

## APPENDIX

Explanation of how pressure and temperature are related during a phase change.

This can be illustrated by applying the "Clausius–Clapeyron equation," which clarifies how pressure and temperature are related during a phase change:

$$\Delta p / \Delta T = L / T \Delta v$$

p = pressure,

T = temperature,

L = latent heat of fusion of water = $3.35 \times 10^5$ J/kg,

$\Delta v$ = specific volume change when ice melts = $-9.05 \times 10^{-5}$ m³/kg.

Putting those figures into the equation results in:

$$\Delta p / \Delta T = -13.5 \text{ MPa/K} = -133 \text{ atm/K}.$$

The negative sign indicates that a pressure of 13.5 MPa (133 atm = 1,960 psi) is needed per kelvin (Celsius degree) of freezing point depression. A typical hockey or ice-skating rink is around –4°C. Even a 100-kg man skating on one foot with a typical skating boot with a blade of 30 cm × 0.5 mm exerts a pressure of about 980 N / (0.3 m × 0.0005 m) = 6.53 MPa (64.5 atm = 948 psi), which would lower the melting point by only half a Celsius degree.

## REFERENCES


1. Irwin, J. and Emerson, W.A., To Rule the Night, A.J. Holman Company, Nashville, 1973 (first edition), p. 11.

2. Catchpoole, D., In pursuit of plant power, 25 September 2012; creation.com/plantpower.

3. Sarfati, J. The wonders of water, Creation 20(1):44–47, 1997; creation.com/water.

4. Cousteau, J., The Ocean World of Jacques Cousteau—Oasis in Space, Angus & Robertson (U.K.) Ltd. London, England, p.17, 1973.

5. Blanc, M., Kallenbach, R., and Erkaev, N.V., Solar system magnetospheres, Space Science Reviews 116(1–2):227–298, 2005.

6. Tascione, T., Introduction to the Space Environment, 2nd edn, Krieger Publishing, Malabar, FL, 2010.

7. Van Allen, J.A., Origins of Magnetospheric Physics: An Expanded Edition, University of Iowa Press, 2004.

8. Blanchett, A., Space radiation is risky business for the human body, nasa.gov, 16 Aug 2018.

9. Mishra, B. and Luderer, U., Reproductive hazards of space travel in women and men, Nature Reviews Endocrinology, 15:713–730, 2019.

10. Buis, A., Earth's magnetosphere: protecting our planet from harmful space energy, climate.nasa.gov/news, 3 Aug 2021.

11. Greshko, M., Interstellar space even weirder than expected, NASA probe reveals, nationalgeographic.com, 4 Nov 2019. See also creation.com/heliopause.

12. Baumgardner, J., 3-D finite element simulation of the global tectonic changes accompanying Noah's Flood, 2nd ICC, Creation Science Fellowship, Pittsburgh, pp. 35–45, 1990.

13. Vardiman, L., Climates before and after the Genesis Flood, ICR, California, pp. 7–21, 2001.

14. Baumgardner, J., Catastrophic Plate Tectonics: The physics behind the Genesis Flood, 5th ICC, Creation Science Fellowship, Pittsburgh, pp. 113–126, 2003.

15. Woodmorappe, J., Hypercanes: rainfall generators during the Flood? Journal of Creation 14(2):123–127, 2000.

16. Faulkner, D., A biblically-based cratering theory, Journal of Creation 13(1):100–104, 1999.

18. Ocean floor bathymetry, <www.waterencyclopedia.com/Oc-Po/Ocean-Floor-Bathymetry.html>, 24 May 2006.

19. Oard, M.J., The mountains rose, *Journal of Creation* 16(3):40–43, 2002.

20. Whitcomb, J.C., and Morris, H.M., The Genesis Flood, The Presbyterian and Reformed Publishing Company, pp. 2–3, 1961.

21. King, L.C., Wandering Continents and Spreading Sea Floors on an Expanding Earth, John Wiley and Sons, New York, NY, pp. 168, 71, 1983.

22. Macdonald, K.C., Fox, P.J., Alexander, R.T., Pockalny, R., and Gente, P., Volcanic growth faults and the origin of Pacific abyssal hills. Nature 380:125–129, 1996.

23. Gansser, A., Geology of the Himalayas, Inter-science Publishers, New York, NY, p. 164, 1964.

24. Coghlan, A., Massive 'ocean' discovered towards Earth's core, www.newscientist.com/article/dn25723-massive-ocean-discovered-towards-earths-core.html#.U6OltkDFDNn; 12 June 2014.

25. Schmandt, B., Jacobsen, S.D., Becker, T.W., Liu, Z., and Dueker, K.G., Dehydration melting at the top of the lower mantle, Science 344(6189):1265–1268, 13 June 2014; DOI: 10.1126/science.1253358; www.sciencemag.org/content/344/6189/1265.abstract

26. Bergeron, L., Deep Waters, New Scientist 155(2097):22–26, August 30, 1997.

27. Fellman, M., New Evidence for Oceans of Water Deep in the Earth: Water bound in mantle rock alters our view of the Earth's composition, www.northwestern.edu/newscenter/stories/2014/06/new-evidence-for-oceans-of-water-deep-in-the-earth.html 12 June 2014.

28. An associate professor of Earth and planetary sciences at Northwestern's Weinberg College of Arts and Sciences.

29. An assistant professor of geophysics at the University of New Mexico.

30. The 'image' produced from such an analysis would be a bit like the image from an x-ray, and would require skill and experience to interpret, and could be ambiguous.

31. Bui, Hoai-Tran, Water discovered deep beneath Earth's surface, www.usatoday.com/story/news/nation/2014/06/12/water-earth-reservoir-science-geology-magma-mantle/10368943/; 12 June 2014.

32. Oksin, B., Rare Diamond Confirms That Earth's Mantle Holds an Ocean's Worth of Water: The diamond contains ringwoodite, which is water-rich but only forms naturally under the extreme pressure found in Earth's mantle, scientificamerican.com/article/rare-diamond-confirms-that-earths-mantle-holds-an-oceans-worth-of-water/; 12 March 2014. Nature paper was published on the same day.

33. Whitcomb, J.C. and Morris, H., The Genesis Flood, Presbyterian and Reformed Publishing Company, Phillipsburg, 1961.

34. Whitcomb, J.C., The World that Perished, Revised Edition, Baker, 1996.

35. Dillow, J.C., The Waters Above: Earth's Pre-Flood Vapor Canopy, Revised Edition, Moody Press, Chicago, 1982.

36. Milton, R., The Facts of Life—Shattering the Myth of Darwinism, 4th. Estate, London, pp. 85–86, 1992.

37. Trewin, N.H., Mass mortalities of Devonian fish—the Achananas Fish Bed, Caithness, Geology Today, pp. 45–49, Mar/April, 1985.

38. Garner, P., Where is the Flood/post Flood boundary? Implications of dinosaur nests in the Mesozoic, Journal of Creation 10(1):101–106, 1996. Return to text.

39. Garton, M., The pattern of fossil tracks in the geological record, Journal of Creation 10(1):82–100, 1996.

40. Tyler, D.J., A post-Flood solution to the chalk problem, Journal of Creation 10(1):107–113, 1996.

41. Tyler, D.J., Flood models and trends in creationist thinking, Creation Matters 2(3), 1997.

42. Robinson, S.J., Was the Flood initiated by catastrophic plate tectonics? Origins 21:9–16, 1996.

43. Robinson, S.J., Can Flood geology explain the fossil record? Journal of Creation 10(1):32–69, 1996.

44. Other objections to the vapor canopy model concern the latent heat of such vaporization—see Vardiman, L., An analytical young-earth flow model of the ice sheet formation during the 'ice age'; in: Proceedings of the Fourth International Conference on Creationism, Pittsburgh Pennsylvania, pp. 561–568, 1994, and Vardiman, L., A conceptual transition model of the atmospheric global circulation following the Genesis Flood; in: Proceedings of the Fourth International Conference on Creationism, Pittsburgh Pennsylvania, pp. 569–580, 1994.

45. McIntosh, A.C., Edmondson, T. and Taylor, S.C., Genesis and catastrophe: the Flood as the major Biblical cataclysm, Journal of Creation 14(1):101–109, 2000.

46. Froede, C.R., Sequence stratigraphy and Creation geology, Creation Res. Soc. Quart. 31(3):138–147, 1994.

47. Froede, C.R., A proposal for a Creationist geological timescale, Creation Res. Soc. Quart., 32(2):90–94, 1995.

48. Woodmorappe, J., The geologic column: does it exist? Journal of Creation 13(2):77–82, 1999.

49. Berthault, G., Experiments in Stratification, Sarong Productions, Jersey, video, 1999.

50. Brown, W., In the Beginning: Compelling Evidence for the Creation and the Flood, 6th (special) edition, Centre for Scientific Creationism, Phoenix, Arizona, 1996.

51. Selbrede, M.G., Dr Walt Brown's Hydro-plate Theory, Chalcedon Report (Sept.), pp. 37–45, 1998.

52. McWhirter, N. (ed.), Guinness Book of Records, Guinness, London, p. 61, 1983. Return to text.

53. Whitcomb and Morris, Ref. 1, p. 264. Return to text.

54. Morris, J.D., The Young Earth, Master Books, p. 95, 1994.

55. Austin, S.A., Grand Canyon: Monument to Catastrophe, ICR, Santee, California, 1994.

56. Baumgardner, J.R., Runaway subduction as the driving mechanism for the Genesis Flood, in: Proceedings of the Third International Conference on Creationism, Creation Science Fellowship, Pittsburgh, Pennsylvania, pp. 63–76, 1994.

57. Austin, S.A., Baumgardner, J.R., Humphreys, D.R., Snelling, A.A., Vardiman, L. and Wise, K.P., Catastrophic plate tectonics: a global Flood model of Earth history; in: Proceedings of the Third International Conference on Creationism, Creation Science Fellowship, Pittsburgh, Pennsylvania, pp. 609–622, 1994.

58. Reed, J.K., Bennett, C.B., Froede, C.R., Oard, M.J. and Woodmorappe, J., An introduction to modern uniformitarian and catastrophic plate tectonic concepts, Creation Res. Soc. Quart. 33(4):202–210, 1996.

59. Humphreys, D.R., Has the earth's magnetic field flipped? Creation Res. Soc. Quart. 25(3):130–137, 1988.

60. Humphreys, D.R., Physical mechanisms for reversals of the earth's magnetic field during the Flood; in: Proceedings of the Second International Conference on Creationism, Creation Science Fellowship, Pittsburgh, Pennsylvania, pp. 129–142, 1990.

61. Brown, Ref. 18, pp. 82–92.

62. Berthault, G., Genesis and historical geology: a personal perspective, Journal of Creation 12(2):213–217, 1998.

63. Julien, P.Y., Lan, Y. and Berthault, G., Experiments on stratification of heterogeneous sand mixtures, Bulletin of the Society of Geology, France, 164(5):649–660, 1993.

64. Baumgardner, J.R. and Barnette, D.W., Patterns of ocean circulation over the continents during Noah's Flood; in: Proceedings of the Third International Conference on Creationism, Creation Science Fellowship, Pittsburgh, Pennsylvania, pp. 77–86, 1994.

65. Walker, T., A biblical geologic model; in: Proceedings of the Third International Conference on Creationism, Creation Science Fellowship, Pittsburgh, Pennsylvania, pp. 581–592, 1994.

66. Garner, P., Continental flood basalts indicate a pre-Mesozoic Flood/post-Flood boundary, Journal of Creation 10(1):114–127, 1996.

67. Hunter, M.J., Is the pre-Flood/Flood boundary in the earth's mantle? Journal of Creation 10(3):344–357, 1996.

68. Garton, M., The real lifestyle of the dinosaurs, Origins 24:14–22, 1998.

69. Robinson, S.J., Dinosaurs in the Oardic Flood, Journal of Creation 12(1):55–86, 63, 1998.

70. Taylor, J., Fossil Facts and Fantasies, Mt Blanco Publishing Co., ch. 4, pp. 36–47, 1999.

71. Snelling, A.A., Catastrophic sedimentation: giant submarine landslides, Journal of Creation 12(2):135–136, 1998.

72. Snelling, A.A., Waterborne gravity flows buried Mongolian dinosaurs, Journal of Creation 12(2):133–134, 1998.

73. Scheven, J., The Carboniferous floating forest—an extinct pre-Flood eco-system, Journal of Creation 10(1):70–81, 1996.

74. It has been argued that Psalm 104:6–9 is referring to the third day of the Creation Week. However, the context of Psalm 104:8–9 is of a perpetual decree that the waters 'may not pass over' (Hebrew abhar). The parallel passages Isaiah 54:9 and Jeremiah 5:22 use the same word (abhar) and refer to the Rainbow Covenant concerning the sea not being allowed to cross over the boundary of the shore. Thus there is a good case that although Psalm 104:1–5 is referring to Creation, Psalm 104:6–9 is speaking of the Flood. Whitcomb, Ref. 2, p. 40–41. See also Taylor, C.V., Did mountains really rise according to Psalm 104:8? Journal of Creation 12(3):312–313, 1998, and discussion in Journal of Creation 13(1):68–71, 1999.

75. Rush, D.E. and Vardiman, L., Pre–Flood vapor canopy radiative temperature profiles; in: Proceedings of the Second International Conference on Creationism, Volume 2, Creation Science Fellowship, Pittsburgh, Pennsylvania, pp. 231–245, 1990.

76. Brown, W.T., The fountains of the great deep; in: Proceedings of the First International Conference on Creationism, Volume 1, Creation Science Fellowship, Pittsburgh, Pennsylvania, pp. 23–38, 1986.

77. Baumgardner, J.R., Runaway subduction as the driving mechanism for the Genesis Flood; in: Proceedings of the Third International Conference on Creationism, Creation Science Fellowship, Pittsburgh, Pennsylvania, pp. 63–75, 1994.

78. Fitzpatrick, T., 3-D seismic model of vast water reservoir revealed Earth mantle 'ocean', news-info.wustl.edu/news/page/normal/8222.html

79. Johnson, R.A., Geotechnical classification of deep and ultra-deep Witwatersrand mining areas, South Africa, Mineralium Deposita 32:335–348, 1997; . www.csir.co.za/publications/schweitzer_1997.pdf

80. Analysis of log and seismic data from the world's deepest Kola Borehole, http://asuwlink.uwyo.edu/~seismic/kola.

81. Baumgardner, J., Catastrophic plate tectonics: the geophysical context of the Genesis Flood, Journal of Creation 16:58–63, 2002; creation.com/article/1596.

82. Rennie, G., The search for methane in Earth's mantle, S &TR, Lawrence Livermore National Laboratory, pp. 21–23, July/August 2005; www.llnl.gov/str/JulAug05/pdfs/07_05.3.pdf

83. Bergeron, L., Deep Waters, New Scientist 155(2097):22–26, August 30, 1997.

84. The earth's crust is, on average, about 25 km (16 miles) thick, while the core is about 2,900 km (1800 miles) below the surface. The mantle transition zone extends roughly from a depth of 400 km (250 miles) to 670 km (420 miles).

85. Kepler, J., A new year's gift or on the six-cornered snowflake, Frankfurt on Main, 1611; believed to be the first scientific reference to snowflakes.

86. Banchard, D., The snowflake man. Weatherwise 23(6): 260–269, 1970.

87. Bentley, W., The magic beauty of snow and dew, National Geographic Magazine 43(1):103–112, January 1923.

88. According to the Laws of Thermodynamics: by forming the lowest energy structure, the maximum amount of heat is released to the surroundings, increasing overall entropy.

89. Vardiman, L., Microscopic masterpieces: discovering design in snow crystals, Institute for Creation Research, 1 Dec 2007; icr.org/article/3555.

90. See snowcrystals.com.

91. See also Sarfati, J., By Design: Evidence for nature's Intelligent Designer—the God of the Bible, pp. 227–229, Creation Book Publishers, Australia, 2008.

92. E.g., Bailey, D., Evolution and Probability, Report of National Center for Science Education 20(4), 2001. Return to text.

93. Polanyi, M., Life's irreducible structure, Science 160(3834):1308–1312, 1968; p. 1309.

94. See Sarfati, J., Why does science work at all? Creation 31(3):12–14, 2009.

95. Sarfati, J., The wonders of water, Creation 20(1): 44–47, 1997; creation.com/water.